Patrick Moore's Prac[illegible]

For other titles published in the series, go to
http://www.springer.com/series/3192

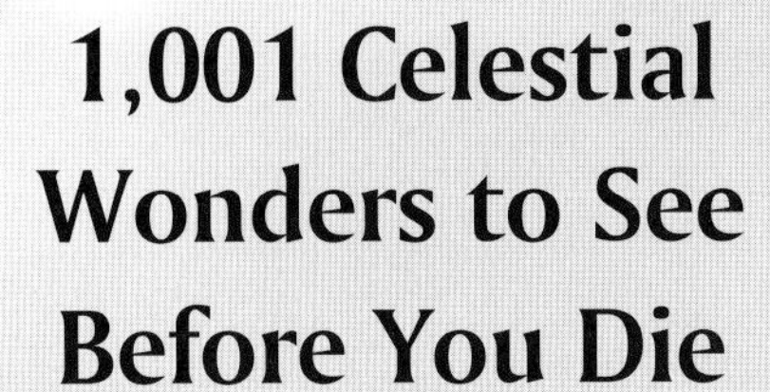

# 1,001 Celestial Wonders to See Before You Die

## The Best Sky Objects for Star Gazers

Michael E. Bakich

Michael E. Bakich
2001 W. Lapham St.
53204 Milwaukee Wisconsin
USA
mbakich@astronomy.com

ISSN 1431-9756
ISBN 978-1-4419-1776-8 e-ISBN 978-1-4419-1777-5
DOI 10.1007/978-1-4419-1777-5
Springer New York Dordrecht Heidelberg London

Library of Congress Control Number: 2010929261



Printed on acid-free paper

Springer is part of Springer Science+Business Media (www.springer.com)

# Foreword

There's nothing quite like deep-sky observing. Out under an inky black sky filled with stars, your telescope at your side, armed with lists and ideas about what to look at, the entire universe lies at your beck and call. And it's filled with exciting things to look at. The Milky Way Galaxy holds as many as 400 billion stars and the universe at least 125 billion galaxies. Does that mean amateur astronomers have 50,000 million billion targets for their telescopes? No — there's no need to get *greedy*. What can be seen with a medium-sized backyard telescope amounts to 10,000 nice targets, the objects that are closest to us in space and therefore the brightest. And in this impressive book you are holding, Michael Bakich presents more than 1,000 of these targets — star clusters, nebulae, and galaxies — 10% of the good stuff you can see with your scope, in a single book!

Before heading out under the stars, keep the basics in mind. Make sure you get as far away from city lights as possible, and observe close to the dark of the Moon. Choose areas or nights of good transparency, when little particle stuff is floating in the atmosphere, and nights of good seeing when the atmospheric turbulence is at a minimum. Don't be tempted to jump to high powers as a key to seeing things: most deep-sky objects are best viewed at relatively low magnifications. If you can, take along a set of light pollution reduction filters that will help with scattered artificial light. Make sure you use the basic techniques of backyard observers, such as averted vision — looking to the side of the eyepiece field, which engages your eyes' rods, its faint light receptors.

And make sure you enjoy the journey. I've owned many telescopes of various sizes but perhaps the most fun I ever had under the night sky came during my first year of observing in a country field in Ohio. Back then I had only binoculars and whatever it was — the Dumbbell Nebula, the Andromeda Galaxy, a star cluster like M7 — was a voyage of discovery. I had no idea what could be seen or what anything would look like, so the feeling of wonder and awe roughly equated to Galileo when he first gazed upon the heavens, unsure of what anything was at all. In some ways, that sense of wonder was greater than it is now observing very distant objects with a 30-inch telescope.

As you gain experience as an observer, you'll gain many friends in the sky — favorite objects you like to come back to again and again. One of my favorites is NGC 6888, known as the Crescent Nebula. This glowing cloud was cast off by a furiously hot central star, HD 192163, a Wolf-Rayet star with its hot, inner layer exposed. On a dark night, an 8-inch scope shows a weird, mottled texture along this object, floating in a rich field of faint stars. Another strange emission nebula, the Bubble Nebula, lies in the constellation Cassiopeia and is designated NGC 7635. Although this object has a somewhat low surface brightness, meaning its total light is spread out and made a little challenging to see, a 6-inch scope shows the Bubble on a dark night.

As you read Michael's book, you'll invariably draw up a list of your own favorites, either to go after before you've seen them once, or to return to on subsequent nights. After many nights under the stars you'll find this book a valuable reference and, I'm guessing, a constant companion under the stars.

David J. Eicher

---

David J. Eicher is editor-in-chief of *Astronomy* magazine and was the founder and editor of *Deep Sky* magazine. He has spent 30 years observing deep-sky objects from a variety of locations around the world.

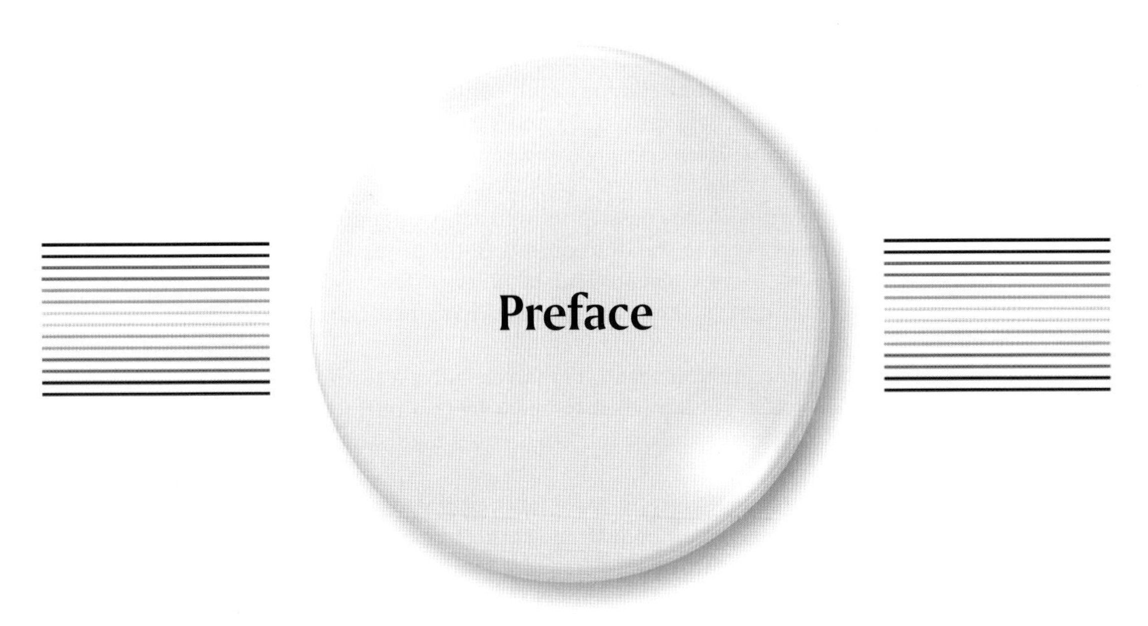

# Preface

Within the ranks of amateur astronomy, the phrase "What's up?" isn't a greeting. It's a call for help. Loosely, it translates as, "What should I look at?" or, "What can I see through my telescope?" And, it's not only beginning observers who pose these questions.

In September 2000, I attended the Great Plains Star Party near Kansas City, Missouri. One night near midnight, as I was walking across the observing field, a friend called down from his observing ladder, "Hey, Michael, I'm out of things to look at. Got any ideas?"

I was taken aback because my friend wasn't a newbie. He was a dedicated amateur astronomer, a long-time member of a large astronomy club, and a telescope owner. In fact, on this night he was using his new 24-inch Starmaster Dobsonian-mounted Newtonian reflector with full go-to capability. And yet, basically, he was asking, "What's up?"

I wrote this book to help answer that question.

I selected each of the 1,001 objects from past observations. Several go back more than 30 years. To choose these particular objects, I consulted personal observing logs, talked to friends about their favorites, and scanned my previous writings. Picking the first 218 targets was trivial — 109 came from Charles Messier's list and the other half originated in Sir Patrick Moore's Caldwell catalogue.

Two resources from *Astronomy* magazine also helped. One was Arizona amateur astronomer Tom Polakis' excellent series "Celestial Portraits," which ran from April 1998 to March 2004. If you missed it, the whole series is available for purchase (as PDF files) online at www.Astronomy.com. Also at *Astronomy*'s web site, you'll find the other resource I used: StarDome (or StarDome Plus for magazine subscribers). I incorporated almost every object from StarDome brighter than magnitude 11, and a few fainter ones as well.

A quick tally of the objects shows 357 galaxies, putting that class of celestial object at the top, number wise. Of those, I included 214 normal spirals, 61 barred spirals, 57 ellipticals, 19 irregulars, and six galaxy clusters. Star clusters took second place with 325 entries. Of those, 225 are open clusters, and 100 are globulars. You'll also find 140 nebulae, 114 double stars, and more. Because I envisioned beginning observers also using this book, I included nine constellations and 15 asterisms.

I made a serious effort to select objects from all over the sky, but I must admit to a Northern Hemisphere bias because most of my observing occurs north of the equator. Even so, I somehow managed to include at least one object in 86 of the 88 constellations that cover the sky. Only the star patterns of Phoenix and Pictor lack a target.

On the other end of the scale, 14 constellations each contain more than 20 of the 1,001 objects. Leading the way are Ursa Major (42), Sagittarius (38), and Virgo (33). Cassiopeia, Cygnus, Ophiuchus, and Puppis each contain 29 objects. Then come Leo and Scorpius (28 each), Cepheus and Monoceros (24 each), Coma Berenices (23), Orion (21), and Canes Venatici (20).

Not all deep-sky objects are equal. Some descriptions, like those of double stars, are short. To other descriptions, especially those of the true standout objects in the group, I dedicated more real estate. Along the way, I also defined some concepts and included a few historical notes. None of the descriptions, however, give all details about any object. That's because I intended this book as a beginning, not an end.

So, use this guide with a pencil in your hand. If you recognize a feature I describe (for example, "three times as long as it is wide"), put a small checkmark by that phrase. If you see a detail I didn't mention, by all means add it to that page. In so doing, you'll take what I started and make it your personal observing log.

If you find a mistake (I take full responsibility for any), or if you'd like to comment on an object, description, or omission, please feel free to e-mail me at m_bakich@yahoo.com.

I organized this book for easy use throughout the year. That's why the 12 chapters carry the names of the months. You'll find the objects in each chapter at their best during that month. You'll be able to observe every object, however, several months before or after the month in which it falls.

Finally, a word about object numbering. Although you could start in January and observe each object in the listed order (at least those in your hemisphere), you also can approach the list in a less formal way. Feel free to skip around in whatever way you choose. Perhaps for your next observing session, you'll select the objects in one constellation. A subsequent session might highlight spiral galaxies or planetary nebulae. It's all up to you.

# Acknowledgement

First, I want to thank my wife, Holley, for her indulgence and patience as I completed this manuscript. She understood when I excused myself from travel plans, missed art openings, or just headed upstairs to write. She's the best. Holley, I love you.

All the pictures of celestial objects reproduced in this book came from just two astroimagers: Adam Block from Tucson, Arizona, and Anthony Ayiomamitis, who lives in Athens, Greece.

Whether Adam's images originated from his years as Chief Observer at the Advanced Observer Program at Kitt Peak National Observatory or from his current position as Public Observing Programs Coordinator at the Mount Lemmon SkyCenter operated by the University of Arizona, the results have been terrific. Many have appeared in the pages of *Astronomy* magazine during the 7 years I have worked there. In fact, during that time, we've published more of Adam's images than those of any other astrophotographer. Although he has imaged every type of celestial object, Adam's specialty is spiral galaxies.

I've yet to meet Anthony, but we've corresponded via e-mail for years. His specialty involves intricate planning that lets him know when a sky event (like a sunrise or moonset) will occur behind one of Greece's many historical monuments. He then takes a single exposure to capture the scene. He also has produced more year-long analemma shots than any other photographer. And, thankfully, he loves to shoot star clusters. I have run many of his high-quality images in *Astronomy*. He's always been willing to shoot objects I needed for stories, and he often asks if I have other requests. I offer my sincere gratitude to both Adam and Anthony.

Thanks also to Tom and Steve Bisque of Software Bisque. They have generously provided me with copies of *TheSky* software for nearly two decades. I've used *TheSky* (now *TheSkyX*) at home and at work for all my astronomical projects starting with my first book, which appeared in 1995. All angular measurements for this book, as well as every star chart consultation, came from *TheSky*. If you use *TheSky*, be sure to download the new Sky Database File (.sdf) at www.bisque.com, which contains all 1,001 objects. That resource will make it a lot easier to find and identify the targets you choose. Thanks to Tom Bisque for helping me create that file.

A special thanks to Gene Turner, a land developer who started the Arizona Sky Village in Portal, Arizona, and the Rancho Hidalgo astronomy and equestrian community near Animas, New Mexico. Turner has hosted me for numerous multi-day observing sessions during which I used his 30-inch

Starmaster Dobsonian-mounted reflector to observe many of the objects in this book. In addition to Gene's hospitality, what made each trip special was the clear, ultra-dark sky — one of the finest anywhere — under which both developments sit.

I've saved an extra-special thank-you for *Astronomy* magazine Editor David J. Eicher, and not because he's my boss. Dave, the founder of *Deep-Sky* magazine more than 30 years ago, knows more about the 1,001 objects in this book than just about anyone. He provided sound counsel regarding this project (and others) whenever I asked. Observing with Dave, whether through a 30-inch telescope or a 3-inch one, always is fun and a learning experience.

I also extend heartfelt thanks to John Watson, Maury Solomon, and Megan Ernst at Springer. Their willingness to move this project along, to offer wise counsel, and to answer any and all of my questions made producing this manuscript much easier than I anticipated.

Finally, I dedicate this book to the memory of two dear friends, Jeff Medkeff and Vic Winter. Both were superb observers who understood and shared my excitement for the sky. I treasure the many hours I spent with both of them under dark skies observing both the familiar and the exotic. Rest in peace.

Contents

# About the Author

Michael E. Bakich lives in Milwaukee, Wisconsin, and is a Senior Editor at *Astronomy* magazine. He has previously published *The Cambridge Guide to the Constellations* (1995), *The Cambridge Planetary Handbook* (2000), and *The Cambridge Encyclopedia of Amateur Astronomy* (2003). In addition he has written and edited the following *Astronomy bookazines* of around 100 pgs., each: *Atlas of the Stars* (2006), *Hubble's Greatest Pictures* (2007), and *The 100 Most Spectacular Sky Wonders* (2008). He is also the author of many recent articles.

# January

M.E. Bakich, *1,001 Celestial Wonders to See Before You Die*, Patrick Moore's Practical Astronomy Series,
DOI 10.1007/978-1-4419-1777-5_1, © Springer Science+Business Media, LLC 2010

**Object #1** The Cone Nebula (NGC 2264) Michael Gariepy/Adam Block/NOAO/AURA/NSF

| OBJECT #1 | NGC 2264 |
|---|---|
| Constellation | Monoceros |
| Right ascension | 6h41m |
| Declination | 9°53′ |
| Magnitude | 3.9 |
| Size | 20′ |
| Type | Open cluster |
| Other name | The Christmas Tree Cluster |

Our first target, the Christmas Tree Cluster, is a fine deep-sky object in the faint constellation Monoceros the Unicorn. At magnitude 3.9, this grouping of stars is easily bright enough for you to spot with your naked eyes, albeit as an indistinct fuzz ball. To find it, look 3.2° south-southwest of magnitude 3.4 Xi ($\xi$) Geminorum.

It's easy to see why observers gave this deep-sky object its common name. At a magnification of 50×, you'll see a dozen or so stars to the east and west of the magnitude 4.7 star 15 Monocerotis. This line forms the half-degree-long base of the Christmas tree. Its top points to the south.

Although the southern stars form the tree's top, they don't belong to this cluster. That is, they're not moving through space with the main cluster. The stars only lie in the same direction. The true cluster lies approximately 2,500 light-years from Earth.

Through a 12-inch or larger telescope, you'll see a bright strip of nebulosity some 5′ long. It seems to radiate westward from the brightest star. This gas belongs to the emission nebula Sharpless 2–273, which stretches an additional 2° to the west.

At the top of the Christmas Tree Cluster lies the Cone Nebula, an obscuring cloud of dust visible only through the largest amateur telescopes, although you'll spot it easily on astroimages.

| OBJECT #2 | NGC 2266 |
|---|---|
| Constellation | Gemini |
| Right ascension | 6h43m |
| Declination | 26°58′ |
| Magnitude | 9.5 |
| Size | 5′ |
| Type | Open cluster |

This nice little cluster lies 1.8° north of magnitude 3.3 Propus (Eta [$\eta$] Geminorum) in a rich Milky Way star field. Through a 4-inch telescope at 100×, you'll spot two dozen stars, the brightest of which is magnitude 8.9 SAO 78670, which lies at NGC 2266's southwest end.

| OBJECT #3 | NGC 2280 |
|---|---|
| Constellation | Canis Major |
| Right ascension | 6h45m |
| Declination | –27°38′ |
| Magnitude | 10.5 |
| Size | 6.3′ by 2.8′ |
| Type | Spiral galaxy |

Our next target lies 3.3° west-northwest of magnitude 1.5 Adhara (Epsilon [$\varepsilon$] Canis Majoris). A small telescope reveals an oval twice as long as it is wide with barely noticeable dark markings paralleling its long axis. The central region appears broad and uniform. Through a 12-inch scope, you'll see two thin spiral arms to the east and west of the core. Although they're tightly wrapped around NGC 2280, magnifications above 300× will show the dark gap between them and the nucleus.

| OBJECT #4 | 12 Lyncis |
|---|---|
| Constellation | Lynx |
| Right ascension | 6h46m |
| Declination | 59°27′ |
| Magnitudes | 5.4/7.3 |
| Separation | 8.7″ |
| Type | Double star |

As a constellation, Lynx is difficult to find, so trying to locate a 5th-magnitude star in it can prove tough. The best star-hop starts at magnitude 3.7 Delta ($\delta$) Aurigae. The binary lies 8.2° to the northeast. Once you do find it, even a small telescope will show a bluish primary and a yellow companion. Use a magnification around 100×.

**Object #5** NGC 2281 Anthony Ayiomamitis

| OBJECT #5 | NGC 2281 |
|---|---|
| Constellation | Auriga |
| Right ascension | 6h49m |
| Declination | 41°04′ |
| Magnitude | 5.4 |
| Size | 14′ |
| Type | Open cluster |

You'll find our next object 0.8° south-southwest of magnitude 5.0 Psi$^7$ ($\psi^7$) Aurigae. Through a 4-inch scope at 100×, you'll spot two dozen stars. Four stars forming a parallelogram sit at the center of the cluster. They range in magnitudes from 8.8 to 10.1. Through a 12-inch scope, you'll count more than 50 member stars.

| OBJECT #6 | Alpha ($\alpha$) Canis Majoris |
|---|---|
| Constellation | Canis Major |
| Right ascension | 6h45m |
| Declination | −16°43′ |
| Magnitudes | −1.5/8.5 |

| (continued) | |
|---|---|
| Separation | 8.8″ |
| Type | Double star |
| Other name | Sirius |

American telescope-maker Alvin Graham Clark (1804–1887) discovered that Sirius was double in 1862. Clark was testing an 18-inch lens at the Dearborn Observatory in Illinois.

At times, Sirius is a nearly impossible binary star to split. At other times, separating the two components is not too difficult. Currently, the gap between the two stars is widening, and it won't start to close until 2025. By then, 11″ will lie between Sirius A and Sirius B, also known as the Pup. After that, the separation will close until 2043, when it will measure only 2.5″.

To spot Sirius B you'll need terrific seeing at your site. Let Sirius get to (or quite near) the meridian. You need every advantage, so look through the least amount of air. Be familiar with the separation of Sirius A and B. The separation between Rigel A and B (Object #939) is approximately 9″, the same as Sirius A and B in 2010 and 2011. Crank up the power. I use no less than 250× to attempt to separate the Pup from the brilliant primary.

The common name Sirius comes from the Greek *σειριοσ*, which means sparkling or scorching. Richard Hinckley Allen, writing in *Star Names and Their Meanings* (G. E. Stechert, 1899) attributes this star's name to the Greek poet Hesiod, who lived in the latter half of the 8th century B.C.

Admiral William Henry Smyth (1788–1865), who authored one of the great 19th-century observing guides, *A Cycle of Celestial Objects* (John W. Parker, 1844), offers an alternative explanation. He writes that, "Dr. Hutton [probably English mathematician Charles Hutton (1737–1823)] gravely informs us that the term is from *Siris*, which he says is the most ancient appellation of the Nile, for when this star rose heliacally, and became visible to the Egyptians and Ethiopians, their year commenced, and with it the inundation of their fecundating river. As that beneficial flood was attributed to the influence of the beautiful star, it was therefore worshipped as Sothis, Osiris, and Latrator Anubis; and was viewed as the abode of the soul of Isis".

| OBJECT #7 | NGC 2286 |
|---|---|
| Constellation | Monoceros |
| Right ascension | 6h48m |
| Declination | −3°09′ |
| Magnitude | 7.5 |
| Size | 15′ |
| Type | Open cluster |

You'll find our next object 6.1° northeast of magnitude 5.0 Beta (*β*) Monocerotis. A 4-inch telescope reveals 30 stars scattered about an area half the diameter of the Full Moon. A nice double star with components of magnitude 9.7 and 10.2 sits inside the southeastern edge of the cluster. The two stars lie 34″ apart and are NGC 2286's brightest stars.

| OBJECT #8 | M41 (NGC 2287) |
|---|---|
| Constellation | Canis Major |
| Right ascension | 6h47m |
| Declination | −20°44′ |
| Magnitude | 4.5 |
| Size | 38′ |
| Type | Open cluster |

Our next object is one of the easiest deep-sky targets because you can use the sky's brightest star as a guide to it. Open cluster M41 lies 4° south of Sirius (Alpha [α] Canis Majoris). From a dark site you'll easily spot M41 with your naked eyes.

It's possible that Greek philosopher Aristotle (384–322 B.C.) knew of this cluster around 325 B.C. The modern discovery goes to Italian astronomer Giovanni Batista Hodierna (1597–1660) who saw it before 1654. M41 is one of the brightest deep-sky objects without a common name. Why that's so is a mystery to me.

Through a 6-inch telescope, you'll see about 50 stars. With twice the aperture, you'll dramatically increase that number of stars. Star counts by astronomers indicate M41 contains slightly more than 100 member stars strewn over an area slightly larger than that covered by the Full Moon. M41's stars range in brightness from 7th to 13th magnitude.

The brightest star in the field of view, 12 Canis Majoris, does not belong to the cluster. At a distance of 1,100 light-years, 12 CMa lies less than half as far away as M41. This magnitude 6.1 star sits 21′ southeast of M41's center.

At first glance, this cluster appears roughly circular. Closer inspection reveals several chains of stars running north-south. Through binoculars or a low-power telescope/eyepiece combination, look for these curving arrangements.

| OBJECT #9 | NGC 2298 |
|---|---|
| Constellation | Puppis |
| Right ascension | 6h49m |
| Declination | −36°00′ |
| Magnitude | 9.2 |
| Size | 6.8′ |
| Type | Globular cluster |

You'll find our next target 5.8° west of magnitude 2.7 Pi (π) Puppis. Because it lies 40,000 light-years away, this globular does not appear dazzling. Telescopes smaller than 8′ in diameter show a relatively smooth object with a broad, concentrated halo. At magnifications above 250×, you might see a graininess that suggests the presence of stars. Through a 14-inch scope, the cluster looks much larger. At 300×, about 30 stars pop into view around a still-unresolved core.

| OBJECT #10 | NGC 2301 |
|---|---|
| Constellation | Monoceros |
| Right ascension | 6h52m |
| Declination | 0°28′ |
| Magnitude | 6.0 |
| Size | 15′ |
| Type | Open cluster |
| Other name | Hagrid's Dragon |

Our next target, which sharp-eyed observers can spot without optical aid from a dark site, sits 5.1° west of magnitude 4.2 Delta (δ) Monocerotis. It's a great object through any size telescope, and wide-angle views will show a rich surrounding star field. A 6-inch scope reveals some 50 stars. Crank the magnification past 200×, and look for a double star dead-center in the cluster. The two components have magnitudes of 8.0 and 8.8.

The common name, Hagrid's Dragon, is a recent one. *Astronomy* magazine Contributing Editor Stephen James O'Meara sees a dragon in flight when he looks at this cluster. He named it Hagrid's Dragon after a creature in the *Harry Potter* series of novels by J. K. Rowling. Why not call NGC 2301 Norbert, the name Rowling gave to the fictional dragon? Probably, and I certainly mean no disrespect to anyone named Norbert, because "Hagrid's Dragon" sounds more impressive.

| OBJECT #11 | NGC 2302 |
| --- | --- |
| Constellation | Monoceros |
| Right ascension | 6h52m |
| Declination | −7°04′ |
| Magnitude | 9.0 |
| Size | 2.5′ |
| Type | Open cluster |

This object lies 5.7° east of magnitude 5.0 Beta ($\beta$) Monocerotis. I could only coax 10 member stars out of this tiny cluster through an 8-inch telescope. Larger scopes may show more stars, but not all that many more. The magnitude 6.6 star SAO 133781 sits 9′ northwest of the cluster's center.

| OBJECT #12 | Epsilon ($\varepsilon$) Canis Majoris |
| --- | --- |
| Constellation | Canis Major |
| Right ascension | 6h59m |
| Declination | −28°58′ |
| Magnitudes | 1.5/7.4 |
| Separation | 7.5″ |
| Type | Double star |
| Other name | Adhara |

Usually, binary stars are interesting because of their color contrast. In this case, however, it's the brightness contrast. The blue primary outshines the white secondary by some 230 times. Be patient when you observe this pair. Crank the magnification past 150× to separate the components enough so the primary doesn't overwhelm its companion.

The common name Adhara (sometimes spelled Adara), comes from the Arabic "al Adhara," meaning "the virgins." That name refers to a now-extinct constellation that included several other nearby bright stars.

Adhara is the second-brightest star in Canis Major, but its Greek-letter label, Epsilon, indicates it should be fifth-brightest because Epsilon is the fifth letter in the Greek alphabet. The practice of labeling bright stars in constellations with Greek letters began with German mapmaker Johannes Bayer (1572–1625) when he published *Uranometria* in 1603. As Adhara demonstrates, Bayer was not infallible. He should have designated it Beta.

| OBJECT #13 | NGC 2311 |
| --- | --- |
| Constellation | Monoceros |
| Right ascension | 6h58m |
| Declination | −4°35′ |
| Magnitude | 9.6 |
| Size | 7′ |
| Type | Open cluster |

The brightest 15 stars in this cluster form a swath that stretches from the southeast to the northwest. Through a 4-inch telescope, you'll count these stars and perhaps a few more. An 8-inch scope bumps the star count to 30. You'll find this cluster 1.3° west-southwest of the magnitude 5.0 star 19 Monocerotis.

| OBJECT #14 | NGC 2316 |
| --- | --- |
| Constellation | Monoceros |
| Right ascension | 7h00m |

| (continued) | |
|---|---|
| Declination | −7°46′ |
| Size | 4′ by 3′ |
| Type | Emission nebula |

Our next target is a small, comet-shaped nebula that lies 1° northwest of open cluster M50 (Object #15). Through a 10-inch telescope at 200×, NGC 2316 has a uniform brightness across its face. A nebula filter will help you pull out what details are here.

**Object #15** M50 Anthony Ayiomamitis

| OBJECT #15 | M50: (NGC 2323) |
|---|---|
| Constellation | Monoceros |
| Right ascension | 7h03m |
| Declination | −8°20′ |
| Magnitude | 5.9 |
| Size | 16′ |
| Type | Open cluster |
| Other name | The Heart-Shaped Cluster |

You know, Messier's list is full of spectacular objects—the Andromeda Galaxy is M31; the Orion Nebula is M42; and M51 is the Whirlpool Galaxy. In such illustrious company, M50 might feel a bit intimidated.

Well, it shouldn't. At magnitude 5.9, sharp-eyed observers under a dark sky can spot this cluster with their naked eyes. Through a small telescope at 100×, you'll spot 50 stars in an area 12′ across. The brightest glows at 8th magnitude, and many more 8th- to 10th-magnitude stars form curving chains within the cluster.

Admiral Smyth in *Cycle of Celestial Objects* talks about the bright star, its companion, and the rest of the cluster: "A delicate and close double star in a cluster of the Via Lactea [Milky Way], on the Unicorn's right shoulder. A 8 and B 13, both pale white. This is an irregularly round and very rich mass, occupying with its numerous outliers more than the field, and composed of stars from the 8th to the 16th magnitudes; and there are certain spots of splendour which indicate minute masses beyond the power of my telescope."

You may not have heard much about M50, but don't overlook it. It's a hidden gem that will reward you for observing it. Its common name, the Heart-Shaped Cluster, refers to how the oval-shaped central region combines with two trails of stars that move outward.

| OBJECT #16 | NGC 2324 |
|---|---|
| Constellation | Monoceros |
| Right ascension | 7h04m |
| Declination | 1°03′ |
| Magnitude | 8.4 |
| Size | 7′ |
| Type | Open cluster |

Our next target lies 2.5° northwest of magnitude 4.2 Delta ($\delta$) Monocerotis. This cluster contains stars of mostly 12th and 13th magnitude. All Through a 4-inch telescope, you'll easily discern the cluster's slightly oval shape. You'll count more than 30 stars at 100×. An 8-inch scope will reveal more than 50 stars irregularly scattered across the field of view. Although NGC 2324 sits in a rich star field, the cluster shows a distinct edge.

| OBJECT #17 | NGC 2331 |
|---|---|
| Constellation | Gemini |
| Right ascension | 7h07m |
| Declination | 27°21′ |
| Magnitude | 8.5 |
| Size | 19′ |
| Type | Open cluster |

You'll find our next target 3.1° south-southwest of magnitude 4.4 Tau ($\tau$) Geminorum. This scattered cluster measures more than half the diameter of the Full Moon, but it is sparse. A 6-inch telescope will reveal only about 20 members.

| OBJECT #18 | NGC 2335 |
|---|---|
| Constellation | Monoceros |
| Right ascension | 7h07m |
| Declination | −10°02′ |
| Magnitude | 7.2 |
| Size | 7′ |
| Type | Open cluster |

The easiest way to locate our next object is to look 3.7° east-northeast of magnitude 4.4 Theta ($\theta$) Canis Majoris. One of the first things you'll notice is magnitude 7.0 SAO 134220, which lies 8′ east-northeast of NGC 2335's core.

Through a 4-inch telescope, two dozen stars immediately pop into view. You'll see lots of curves and geometric shapes formed by the brightest members. A 12-inch scope at 150× will increase your star count to 50.

**Object #19** NGC 2336 Adam Block/NOAO/AURA/NSF

| OBJECT #19 | NGC 2336 |
|---|---|
| Constellation | Camelopardalis |
| Right ascension | 7h27m |
| Declination | 80°11′ |
| Magnitude | 10.4 |
| Size | 6.4′ by 3.3′ |
| Type | Barred spiral galaxy |

Our next target sits in a star-poor region. Find it by heading 15° north-northeast of magnitude 4.6 Gamma ($\gamma$) Camelopardalis. A small telescope shows a bright central region surrounded by an easily seen halo. Through a 10-inch telescope at 200×, you'll see this galaxy's bar. When you've finished with NGC 2336, head to magnitude 12.3 IC 467, which sits 20′ to the south-southeast.

| OBJECT #20 | NGC 2343 |
|---|---|
| Constellation | Monoceros |
| Right ascension | 7h08m |
| Declination | −10°37′ |
| Magnitude | 6.7 |
| Size | 6′ |
| Type | Open cluster |

Our next object lies within and on the northeastern end of the huge IC 2177 complex (Object #23). It's easily seen through binoculars or a finder scope, and sharp-eyed observers have at least an even chance of spotting it from a true-dark site without optical aid. Through a 4-inch telescope, you'll count 15 stars in a small region. A 12-inch scope will add another 10 stars to the cluster.

To find NGC 2343, look 3.7° east-northeast of magnitude 4.1 Theta ($\theta$) Canis Majoris.

| OBJECT #21 | NGC 2345 |
|---|---|
| Constellation | Canis Major |
| Right ascension | 7h08m |
| Declination | −13°10′ |
| Magnitude | 7.7 |
| Size | 12′ |
| Type | Open cluster |

To find this object, look 3° east-northeast of magnitude 5.3 Mu ($\mu$) Canis Majoris. A 6-inch telescope at 150× will reveal 30 stars splayed unevenly across the field of view. The eastern side of the cluster appears denser. Within it, an arc of the cluster's brightest stars moves from the south to the north. Increase to a 12-inch scope, and your star count will top 50. Despite the rich surrounding field, the cluster stands out well.

| OBJECT #22 | NGC 2348 |
|---|---|
| Constellation | Volans |
| Right ascension | 7h03m |
| Declination | −67°24′ |
| Size | 11′ |
| Type | Open cluster |

Is our next target a true open cluster or not? Astronomers are still unsure, so have a look and hazard your own guess. Star counts here are just enough to warrant open cluster status, but the real question is, "Do the stars move through space together?" Look for a couple dozen stars ranging from 10th to 14th magnitude over an area 10′ in diameter 1.4° west-northwest of magnitude 4.0 Delta ($\delta$) Volantis.

**Object #23** Cederblad 90 Tad Denton/Adam Block/NOAO/AURA/NSF

| OBJECT #23 | IC 2177 |
|---|---|
| Constellation | Monoceros |
| Right ascension | 7h05m |
| Declination | −10°38′ |
| Size | 120′ by 40′ |

| (continued) | |
|---|---|
| Type | Emission nebula |
| Other name | The Seagull Nebula |

Our next target is huge, so it looks best through a telescope/eyepiece combination that gives a wide field of view. A nebula filter will help a lot. Star chains extend along the length of the nebula. Disengage an 8-inch telescope's drive, and slowly sweep across the field at low power. You'll spot the nebula easily.

The detached part of the Seagull Nebula to the south represents the bird's head. Astronomers designated that object NGC 2327. It measures 19′ by 17′. Near the nebula's center lies an 8th-magnitude star that lies at one end of a thin channel of dark nebulosity. Could that dark division signify the seagull's mouth?

Lying 1° farther to the south is the reflection nebula Cederblad 90, which glows brightly as it reflects the light of the 8th-magnitude star at its center. Ced 90 measures 3′ across and appears round with a hazy edge. Because it's a reflection nebula don't use a nebula filter when you view it.

Through a 12-inch or larger scope, the boundary between IC 2177's eastern edge and the dark sky appears complex. Also look for several embedded open clusters, such as NGC 2335 (Object #18) and NGC 2343 (Object #20).

| OBJECT #24 | Gamma ($\gamma$) Volantis |
|---|---|
| Constellation | Volans |
| Right ascension | 7h09m |
| Declination | –70°30′ |
| Magnitudes | 3.8/5.7 |
| Separation | 13.6″ |
| Type | Double star |

To view our next target, you'll have to travel to a latitude south of 20° north latitude. If you do, you'll see a nice yellow primary with a white companion star. Gamma Volantis lies about 9° east-southeast of the Large Magellanic Cloud.

| OBJECT #25 | Sharpless 2-301 |
|---|---|
| Constellation | Canis Major |
| Right ascension | 7h10m |
| Declination | –18°29′ |
| Size | 8′ by 7′ |
| Type | Emission nebula |

Our next target lies 3.2° south-southeast of magnitude 4.1 Gamma ($\gamma$) Canis Majoris. This nebula lies 42,000 light-years from the Milky Way's center. An unfiltered 8-inch telescope shows a hazy object with a bright center lying in a rich star field. Insert a nebula filter, and your view will improve dramatically. Larger apertures reveal dark regions scattered about the nebula's face.

| OBJECT #26 | NGC 2353 |
|---|---|
| Constellation | Monoceros |
| Right ascension | 7h15m |
| Declination | –10°18′ |
| Magnitude | 7.1 |
| Size | 20′ |
| Type | Open cluster |
| Other name | Avery's Island |

The first thing you'll notice about our next target is the magnitude 6.0 star SAO 152598, which sits just south of the cluster's center. Even a 4-inch scope will reveal three dozen stars. Step up to a 12-inch instrument, and you'll count more than 100 stars. Most shine between magnitudes 9 and 11. Even without the 6th-magnitude luminary, the southern half outshines the northern part.

To find NGC 2353, look 6.6° west of magnitude 3.9 Alpha (α) Monocerotis.

*Astronomy* magazine Contributing Editor Stephen James O'Meara named this treasure trove of stars after Captain Avery, a native of Devonshire, England. In 1695, Avery had captured a ship belonging to the Great Mogul of India. After looting the treasure aboard, Avery retired to an island a rich man.

| OBJECT #27 | NGC 2354 |
|---|---|
| Constellation | Canis Major |
| Right ascension | 7h14m |
| Declination | −25°44′ |
| Magnitude | 6.5 |
| Size | 20′ |
| Type | Open cluster |

Here's an easy target to find. Just locate magnitude 1.8 Delta (δ) Canis Majoris, and then move 1.5° to the east-northeast. A 4-inch telescope reveals three dozen stars, but through a 10-inch instrument, you'll count 100. Look for an oval dark void at the cluster's center elongated north-south.

| OBJECT #28 | ADS 5951 |
|---|---|
| Constellation | Canis Major |
| Right ascension | 7h17m |
| Declination | −23°19′ |
| Magnitudes | 4.8/6.8 |
| Separation | 26.8″ |
| Type | Double star |

Find this binary, and you'll see a nice color contrast. The brighter component shines yellow or golden, and the secondary is blue. You'll find this star 3° east of magnitude 3.0 Omicron$^2$ ($o^2$) Canis Majoris.

This star's designation (ADS) makes it apparent to astronomers that it is a double star. The label comes from American astronomer Robert Aitken's (1864–1951) massive two volume *New General Catalogue of Double Stars within 120 degrees of the North Pole*, published by the Carnegie Institution in 1932. Aitken's catalog contained measurements of 17,180 double stars north of declination −30°.

| OBJECT #29 | NGC 2355 |
|---|---|
| Constellation | Gemini |
| Right ascension | 7h17m |
| Declination | 13°47′ |
| Magnitude | 9.7 |
| Size | 8′ |
| Type | Open cluster |

Our next target lies in southern Gemini only 1.5° from the Canis Major border. To find it, look 2.8° south of magnitude 3.6 Lambda (λ) Geminorum. This cluster really benefits from increased telescope aperture. Some 20 stars are visible through a 4-inch telescope. An 8-inch scope, on the other hand, shows nearly 50.

**Object #30** Thor's Helmet (NGC 2359) Christine and David Smith/Steve Mandel/Adam Block/NOAO/AURA/NSF

| OBJECT #30 | NGC 2359 |
|---|---|
| Constellation | Canis Major |
| Right ascension | 7h19m |
| Declination | –13°12′ |

| (continued) | |
|---|---|
| Size | 9′ by 6′ |
| Type | Emission nebula |
| Common names | Thor's Helmet, the Duck Nebula |

Thor's Helmet—perhaps the most impressive name for any deep-sky object—is a cosmic bubble sculpted by radiation from a type of luminous, massive star called a Wolf-Rayet star. These short-lived supergiant stars are rare; astronomers have discovered less than 250 of them within the Milky Way.

As the star enters the Wolf-Rayet stage, a powerful stellar wind of up to 6 million mph (10 million km/h) ejects the star's outer envelope. This gas slams into the surrounding interstellar medium and sculpts a beautiful shocked shell of ionized gas.

NGC 2359 lies 8.8° east-northeast of Canis Major's brightest star, Sirius. Alternatively, you'll find it 4.3° northeast of magnitude 4.1 Muliphain (Gamma [$\gamma$] Canis Majoris). The nebula responds well to narrowband and nebula filters. Through a 12-inch telescope, you'll see the circular central area and the helmet's two "wings." The brightest part measures 1′ wide and extends to the south approximately 4′.

Recently, I observed Thor's Helmet through a 30-inch telescope. The intricate details I thought visible only in photographs amazed me. My view differed from a CCD image only because it's tough to see the nebula's striking colors with your eyes. If you have the opportunity to observe NGC 2359 through a large telescope, take it. You won't be disappointed.

Although the common name "Thor's Helmet" seems to indicate a mythological tie, it's more likely this moniker has a more recent history. Why? Because Thor, the Norse god of thunder, did not wear a helmet, let alone a winged one. Only since August 1962 has any publication pictured Thor wearing such a cap. His first appearance so adorned occurred in Marvel Comics' *Journey into Mystery* #83 (a copy of which I happen to own).

| OBJECT #31 | NGC 2360 |
|---|---|
| Constellation | Canis Major |
| Right ascension | 7h18m |
| Declination | −15°37′ |
| Magnitude | 7.2 |
| Size | 12′ |
| Type | Open cluster |
| Other name | Caldwell 58 |

Our next object lies 3.3° east of magnitude 4.1 Gamma ($\gamma$) Canis Majoris in a rich star field. Through a 4-inch telescope at 100×, you'll spot what looks like a stellar bar stretching east to west through the cluster's center. NGC 2360's brightest member is magnitude 8.9 SAO 152691, which sits on the eastern end.

An 8-inch telescope under a dark sky reveals more than 50 stars. The background is so dense here that you may have trouble defining where the cluster's eastern edge ends and the field stars begin.

| OBJECT #32 | NGC 2362 |
|---|---|
| Constellation | Canis Major |
| Right ascension | 7h19m |
| Declination | −24°57′ |
| Magnitude | 4.1 |
| Size | 8′ |

| (continued) | |
|---|---|
| Type | Open cluster |
| Other names | The Tau Canis Majoris Cluster, the Mexican Jumping Star, Caldwell 64 |

Our next object ranks as the sky's 9th-brightest open cluster. Named for the brightest star it contains, the Tau Canis Majoris Cluster contains magnitude 4.4 Tau ($\tau$) CMa and lots of fainter stars. In fact, no cluster star shines within 3 magnitudes of Tau.

Tau is one of the most luminous supergiants known. Its absolute magnitude (the brightness it would have if its distance were 32.6 light-years) is –7. That makes it roughly 50,000 times brighter than the Sun.

One of Tau CMa's common names, the Mexican Jumping Star, probably arose when Northern Hemisphere amateur astronomers observed it twinkling wildly. Because the star sits so low in the sky (it stands a maximum of only 25° high as viewed from 40° north latitude), its light passes through a lot of atmosphere before it reaches northern telescopes.

To locate NGC 2362, find magnitude 1.8 Wezen (Delta [$\delta$] Canis Majoris), and move 2.7° to the east-northeast. You'll spot the cluster easily with your naked eyes. Use binoculars, and you'll see that many stars pack this area of the winter Milky Way.

Through a telescope, Tau dominates the view, partially obscuring many of the 10th-magnitude stars surrounding it. To get an accurate count of the stars in NGC 2362, place Tau just outside the field of view, first to the north, then to each of the other three cardinal directions. Then, simply count the suddenly visible fainter stars and add the four numbers together.

| OBJECT #33 | Delta ($\delta$) Geminorum |
|---|---|
| Constellation | Gemini |
| Right ascension | 7h20m |
| Declination | 21°59′ |
| Magnitudes | 3.5/8.2 |
| Separation | 6.8″ |
| Type | Double star |
| Other name | Wasat |

This easy naked-eye star sits in the middle of Gemini. The primary is white and its companion glows much fainter and orange. Be sure to use a magnification of 100× or more to split it.

The name Wasat comes from the Arabic "Al Wasat," which means "middle." This may refer to the center of the constellation (more specifically, the twin Pollux), or to its position (less than 0.2°) from the ecliptic.

| OBJECT #34 | NGC 2366 |
|---|---|
| Constellation | Camelopardalis |
| Right ascension | 7h29m |
| Declination | 69°13′ |
| Magnitude | 10.8 |
| Size | 8.2′ by 3.3′ |
| Type | Irregular galaxy |

Our next target is a faint, rather large galaxy. Because of its size, NGC 2366's surface brightness is low. You'll need at least a 12-inch telescope under a dark sky to see anything more than a faint dull glow.

What you can see through smaller scopes is a cloud of ionized hydrogen, called an HII region, that lies off the galaxy's southwestern edge. You may see this feature cataloged as NGC 2363. That's an error.

Unlike the galaxy, the nebula has a high surface brightness. Two star clusters supply the energy to make it glow, the more energetic one of which lies embedded in the cloud. Crank the magnification up to 200×, and insert an Oxygen-III filter. The galaxy will disappear, but the HII region remains bright. Look for the nebula's bright center.

| OBJECT #35 | NGC 2367 |
|---|---|
| Constellation | Canis Major |
| Right ascension | 7h20m |
| Declination | −21°53′ |
| Magnitude | 7.9 |
| Size | 5′ |
| Type | Open cluster |

This nice object sits in a rich starfield 4.4° east-northeast of magnitude 3.0 Omicron$^2$ ($o^2$) Canis Majoris. Most of the cluster's bright members lie on its eastern side. The luminary is a double star barely east of center with a separation of 5″ and magnitudes of 9.4 and 9.7.

| OBJECT #36 | 19 Lyncis |
|---|---|
| Constellation | Lynx |
| Right ascension | 7h23m |
| Declination | 55°17′ |
| Magnitudes | 5.6/6.5 |
| Separation | 14.8″ |
| Type | Double star |

You'll split this double through any telescope. The primary shines sunflower yellow, while the secondary appears medium-blue. Find this star 10.5° west-southwest of magnitude 3.4 Muscida (Omicron [$o$] Ursae Majoris). Deep-sky observers might want to note that the rich galaxy cluster Abell 576 lies 0.5° to the north-northwest.

| OBJECT #37 | NGC 2371–2 |
|---|---|
| Constellation | Gemini |
| Right ascension | 7h26m |
| Declination | 29°29′ |
| Magnitude | 11.3 |
| Size | 54″ by 35″ |
| Type | Planetary nebula |
| Other name | The Double Bubble Nebula |

The Double Bubble Nebula, also known as NGC 2371 and NGC 2372, is a twin-lobed planetary nebula that glows at 11th magnitude.

This object's common name comes from its unusual appearance: Two rounded puffs of gas lie side by side, with each lobe getting brighter toward the middle.

You'll find the Double Bubble 1.7° north of magnitude 3.8 Iota ($\iota$) Geminorum. And although you can spot it through an 8-inch telescope, a 12-inch or larger instrument will help you see more of its details.

When you're first hunting NGC 2371–2, use low power. This object measures 54″ by 35″, making it nearly as big as the Ring Nebula (M57) in Lyra. An Oxygen-III eyepiece filter definitely helps.

If the seeing is good, crank the magnification past 200×, and look for a brightness difference between the two lobes. Then try to spot the central star. If you can observe it, you'll see the object whose radiation fuels the glow of the Double Bubble Nebula.

| OBJECT #38 | NGC 2374 |
|---|---|
| Constellation | Canis Major |
| Right ascension | 7h24m |
| Declination | –13°16′ |
| Magnitude | 8.0 |
| Size | 12′ |
| Type | Open cluster |

To find this object, look 5.6° southwest of magnitude 3.9 Alpha ($\alpha$) Monocerotms. Through a 4-inch telescope at 100×, you'll spot 20 stars of equal brightness. An 8-inch scope shows twice that number of stars and also reveals some dark lanes snaking between lines of stars.

| OBJECT #39 | NGC 2383 |
|---|---|
| Constellation | Canis Major |
| Right ascension | 7h25m |
| Declination | –20°57′ |
| Magnitude | 8.4 |
| Size | 5′ |
| Type | Open cluster |

You'll find this target 5.8° east-northeast of magnitude 3.0 Omicron$^2$ ($o^2$) Canis Majoris. It's a small cluster in which most telescopes will reveal 20 stars. Two magnitude 9.7 stars sit at the east and west ends. Our next object, NGC 2384, lies 8′ to the east-southeast.

| OBJECT #40 | NGC 2384 |
|---|---|
| Constellation | Canis Major |
| Right ascension | 7h25m |
| Declination | –21°01′ |
| Magnitude | 7.4 |
| Size | 5′ |
| Type | Open cluster |

Our next target lies 8′ west-northwest of our previous object, NGC 2383. You'll first spot the cluster's two brightest stars, magnitude 8.6 SAO 173685 and magnitude 8.9 HD 58465. Then look for an irregular flow of stars that has an east-west orientation.

| OBJECT #41 | NGC 2395 |
|---|---|
| Constellation | Gemini |
| Right ascension | 7h27m |
| Declination | 13°37′ |
| Magnitude | 8.0 |
| Size | 15′ |
| Type | Open cluster |

You'll find our next target 3.7° southeast of magnitude 3.6 Lambda ($\lambda$) Geminorum. Through a 6-inch telescope, you'll see a gentle wash of stars starting in the southeast and progressing toward the northwest. Three 10th-magnitude stars, one near the center and two at the southeastern edge, gently dominate the cluster. The rest, about 50 stars brighter than magnitude 15, form a nice background.

**Object #42** The Medusa Nebula (Abell 21) Al and Andy Ferayomi/Adam Block/NOAO/AURA/NSF

| OBJECT #42 | Abell 21 |
|---|---|
| Constellation | Gemini |
| Right ascension | 7h29m |
| Declination | 13°15′ |
| Magnitude | 10.3 |
| Size | 615″ |
| Type | Planetary nebula |
| Other name | The Medusa Nebula |

Our next object lies 0.5° southeast of our previous treat, NGC 2395. Although the Medusa Nebula (also known as Sharpless 2–274) has a moderate listed brightness, it can be tough to spot through an 8-inch telescope unless your sky conditions are ideal. Expect to see a fat, discontinuous arc of nebulous material with numerous dark gaps. A wedge on the northern end and a round region due south are the nebula's brightest areas. An Oxygen-III filter really will help with this object. Step up to a 16-inch telescope to get a really good look.

This object's common name comes from the braided filaments of glowing hydrogen that, on long-exposure images, resemble the Gorgon Medusa's dreadful locks.

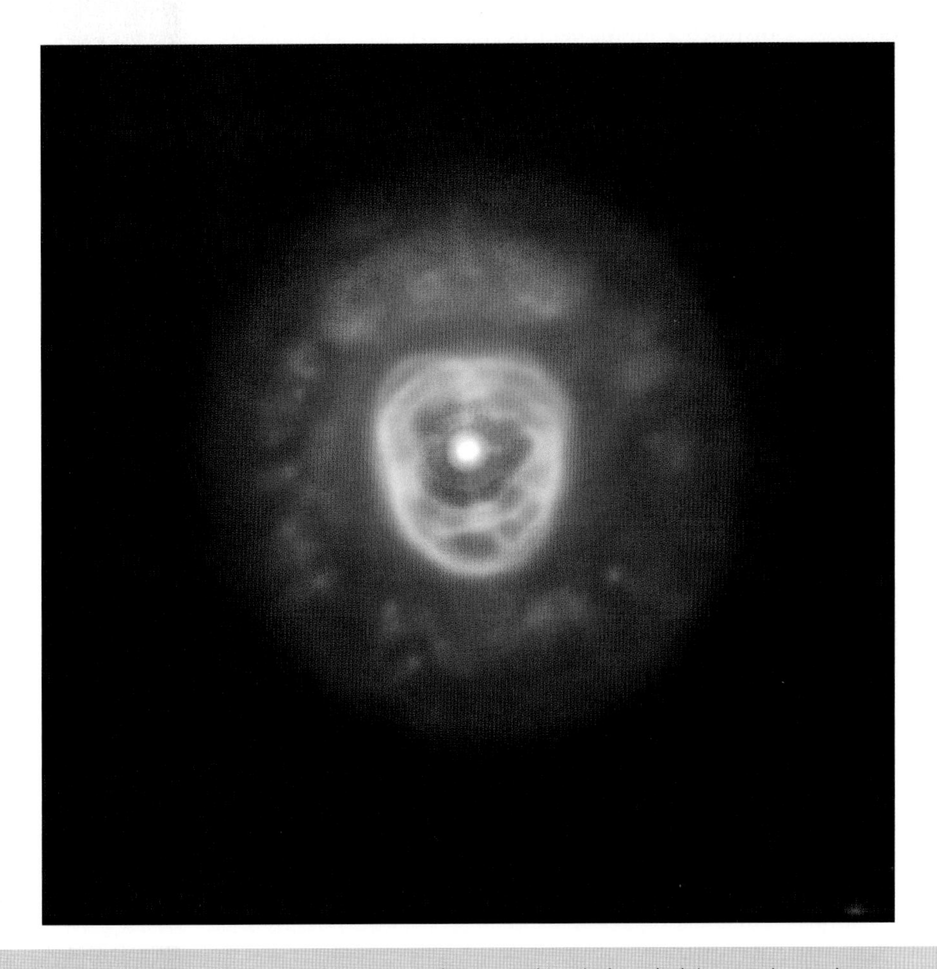

**Object #43** The Eskimo Nebula (NGC 2392) Peter and Suzie Erickson/Adam Block/NOAO/AURA/NSF

| OBJECT #43 | NGC 2392 |
|---|---|
| Constellation | Gemini |
| Right ascension | 7h29m |
| Declination | 20°55′ |
| Magnitude | 9.2 |
| Size | 15″ |
| Type | Planetary nebula |
| Other names | The Eskimo Nebula, the Clown Face Nebula, Caldwell 39 |

This deep-sky object is a planetary nebula called the Eskimo. Through a medium-size telescope, this object resembles a face surrounded by a fur parka.

To find the Eskimo Nebula, point your telescope 2.4° east-southeast of magnitude 3.5 Delta ($\delta$) Geminorum. The planetary glows at magnitude 9.1, which means you can spot it through just about any scope.

Seeing detail in this object, especially through telescopes with apertures of 10 inches or larger, is easy. Use as high a magnification as sky conditions warrant. You'll easily spot the planetary's 10th-magnitude central star.

NGC 2392 has a double-shell appearance. The inner shell appears bright, with a mottled texture. The outer shell looks fainter than the inner one, and a dark ring separates the two. The outer shell also dims with increasing distance from the central star.

| OBJECT #44 | Alpha ($\alpha$) Geminorum |
|---|---|
| Constellation | Gemini |
| Right ascension | 7h35m |
| Declination | 31°53′ |
| Magnitudes | 1.9/2.9 |
| Separation | 3.9″ |
| Type | Double star |
| Other name | Castor |

Castor is one of the brightest double stars in the sky, but the separation between the two stars is quite close, so crank the magnification to 150× or above. To most observers, the primary is white, and the secondary appears pink or orangish.

"Castor" is a proper name. He was the mortal of the Twins. His brother, Polydeuces (Pollux), was immortal. According to legend, both had the same mother, Leda, but Pollux's father was Zeus, while Castor's was Tyndareus, a king of Sparta.

**Object #45** NGC 2403 Adam Block/Mount Lemmon SkyCenter/University of Arizona

| OBJECT #45 | NGC 2403 |
|---|---|
| Constellation | Camelopardalis |
| Right ascension | 7h37m |
| Declination | 65°36′ |
| Magnitude | 8.5 |
| Size | 25.5′ by 13.0′ |
| Type | Barred spiral galaxy |
| Other name | Caldwell 7 |

NGC 2403 is one of the sky's brightest galaxies, shining with a magnitude of 8.5. It's large, however, so that brightness spreads over an area defined by the galaxy's 25.5′ by 13′ dimensions. Its size, by the way, makes NGC 2403's area 47% as large as that of the Full Moon.

Small telescopes show this object as an indistinct haze roughly twice as long as wide, with a bright central region. Through a 12-inch scope, you'll begin to see the galaxy's spiral arms, but you'll need an even larger instrument to trace them all the way back to the nucleus.

Look for stellar associations in NGC 2403's spiral arms. Associations are a type of star cluster resembling open clusters but larger and holding only up to about 100 stars. The presence of these clusters indicates star formation is ongoing within this galaxy.

Plan to spend a lot of time observing NGC 2403. It's one of the sky's most spectacular wonders. You'll find it 7.7° northwest of magnitude 3.4 Muscida (Omicron [*o*] Ursae Majoris).

| OBJECT #46 | NGC 2414 |
|---|---|
| Constellation | Puppis |
| Right ascension | 7h33m |
| Declination | −15°27′ |
| Magnitude | 7.9 |
| Size | 6′ |
| Type | Open cluster |

Our next object is a small cluster with a bright star centered. The star is magnitude 8.2 SAO 153056. A 4-inch telescope at 150× will reveal about 15 stars around it, while an 8-inch scope will double the count to 30.

**Object #47** The Intergalactic Wanderer (NGC 2419) Anthony Ayiomamitis

| OBJECT #47 | NGC 2419 |
|---|---|
| Constellation | Lynx |
| Right ascension | 7h38m |
| Declination | 38°53′ |
| Magnitude | 10.3 |
| Size | 4.1′ |
| Type | Globular cluster |
| Other names | The Intergalactic Wanderer (or Tramp), Caldwell 25 |

The Intergalactic Wanderer lies in a region of southwestern Lynx devoid of bright stars. To find it, use magnitude 1.6 Castor (Alpha [α] Geminorum). From that brilliant luminary, move 7° due north.

This object isn't famous for its brightness or beauty through a telescope, but rather because it's one of the Milky Way's most remote globular clusters. It lies some 300,000 light-years from our galaxy's core and only 25,000 light-years closer to us. That places NGC 2419 more than 100,000 light-years beyond the Milky Way's most famous satellite galaxy, the Large Magellanic Cloud and even farther away than the Small Magellanic Cloud

Astronomers christened it the Intergalactic Tramp, then the Intergalactic Wanderer (Was the change because "tramp" is now a politically incorrect word?), because they thought its position placed it in the space between galaxies. We now know that gravity binds NGC 2419 to the Milky Way, although it takes some 3 billion years to complete one orbit.

A 4-inch telescope reveals scant detail in NGC 2419, but you will see it. Through an 8-inch or larger scope at 200× or above, look for an ever-so-slightly brighter center ringed by an irregularly lit halo. Despite the lack of details, seeing the most distant globular visible through most amateur instruments makes the Intergalactic Wanderer a worthy catch.

| OBJECT #48 | NGC 2420 |
|---|---|
| Constellation | Gemini |
| Right ascension | 7h39m |
| Declination | 21°34′ |
| Magnitude | 8.3 |
| Size | 10′ |
| Type | Open cluster |

Our next target lies 4.3° east of magnitude 3.5 Wasat (Delta [δ] Geminorum). Through an 8-inch telescope, you'll spot about two dozen stars here. Most shine at magnitude 12 or 13. The magnitude 9.4 star on the cluster's edge is not part of NGC 2420. That star is GSC 1373:1207.

| OBJECT #49 | NGC 2421 |
|---|---|
| Constellation | Puppis |
| Right ascension | 7h36m |
| Declination | –20°36′ |
| Magnitude | 8.3 |
| Size | 8′ |
| Type | Open cluster |

You'll find our next treat 5.2° northwest of magnitude 3.3 Xi (ξ) Puppis. It's a rich open cluster that looks good through any size telescope, especially if your location allows you to see it more than halfway up in the sky. A 4-inch scope at 150× will show the cluster's 30 brightest stars. Step up to a 10-inch instrument at 200×, and a new tier of stellar brightness appears. In addition to the bright 30, another, slightly fainter 30 stars tremble into view.

**Object #50** M47 Anthony Ayiomamitis

| OBJECT #50 | M47 (NGC 2422) |
|---|---|
| Constellation | Puppis |
| Right ascension | 7h37m |
| Declination | −14°30′ |
| Magnitude | 4.4 |
| Size | 29′ |
| Type | Open cluster |

Our next target is one you'll see easily from a dark site without optical aid. Still, some observers like landmarks, so M47 lies 5° south-southwest of magnitude 3.9 Alpha (α) Monocerotis.

This cluster ranks as the sky's 14th-brightest open cluster. Most of that brightness comes from just six stars. They range in brightness from magnitude 5.7 to magnitude 8.0.

Although it's terrific through binoculars or a finder scope, M47 is somewhat disappointing through a telescope, especially at magnifications above about 75×. That's probably because the stars spread out over an area equal to that covered by the Full Moon. The cluster's six luminaries lie in a field of about another 75 stars.

| OBJECT #51 | NGC 2423 |
|---|---|
| Constellation | Puppis |
| Right ascension | 7h37m |
| Declination | −13°52′ |
| Magnitude | 6.7 |
| Size | 12′ |
| Type | Open cluster |
| Gemini | |

This object lies just 0.6° north of our previous target, M47. It's a moderately rich cluster that sharp-eyed observers can just detect from a dark site with their naked eyes. A 4-inch telescope will show you 30 stars, and the star count goes up from there as you increase your aperture. Through a 12-inch scope at 200×, you'll count 100 stars. The brightest is magnitude 9.0 HD 61098, which sits at NGC 2423's center.

**Object #52** M46 Anthony Ayiomamitis

| OBJECT #52 | M46 (NGC 2437) |
|---|---|
| Constellation | Puppis |
| Right ascension | 7h42m |
| Declination | −14°49′ |
| Magnitude | 6.1 |
| Size | 27′ |
| Type | Open cluster |

Several open clusters within Puppis are more prominent than M46. M47, for example, outshines M46 by more than a magnitude and a half, and NGC 2451 is more than 3 magnitudes brighter. But I think M46 is one of the top 100 celestial objects in the sky. Point a 4-inch or larger telescope its way, and you'll see why.

The star cluster, discovered by French comet-hunter Charles Messier (1730–1817) in 1771, contains several hundred stars. You'll see 100 of them through an 8-inch scope. The stars appear evenly distributed throughout a circle slightly less than half a degree across. Look closely, however, and you'll see a slightly denser concentration of stars at M46's southern edge. A dark lane separates this region from the rest of the cluster.

To continue your enjoyment of M46, go immediately to our next object.

**Object #53** NGC 2438 Nicole Bies and Esidro Hernandez/Adam Block/NOAO/AURA/NSF

| OBJECT #53 | NGC 2438 |
|---|---|
| Constellation | Puppis |
| Right ascension | 7h42m |
| Declination | –14°44′ |
| Magnitude | 11.0 |
| Size | 66″ |
| Type | Planetary nebula |

Within the boundaries of M46 (Object #52) resides planetary nebula NGC 2438. It sits 7′ north of the cluster's center and measures about 1′ across. Use a 10-inch telescope and high magnification, and you may detect the planetary's donut-like appearance. Several stars lie within the donut's boundary, but none is the object's central star, which glows dimly at magnitude 17.7.

Most astronomers agree that NGC 2438 is a foreground object, somewhat fortunately superimposed upon M46. Distance calculations place NGC 2438 several thousand light-years closer than M46.

| OBJECT #54 | NGC 2439 |
|---|---|
| Constellation | Puppis |
| Right ascension | 7h41m |
| Declination | −31°39′ |
| Magnitude | 6.9 |
| Size | 10′ |
| Type | Open cluster |

The easiest way to find our next target is to draw a line from magnitude 1.8 Wezen (Delta [$\delta$] Canis Majoris) through magnitude 2.5 Aludra (Eta [$\eta$] Canis Majoris). Extend it about an equal distance in the same direction, and you'll land on NGC 2439.

Through a 4-inch telescope, you'll see 15 stars forming a well-defined ring. R Puppis, a variable star that hovers around magnitude 6.6, lies at the northeastern edge of the ring. An additional 20 fainter stars surround the ring.

On the best nights with at least a 12-inch scope, look for the faint cluster Ruprecht 30. It appears as a light stellar dusting half as large as NGC 2439 and lying 23′ to the brighter cluster's northeast.

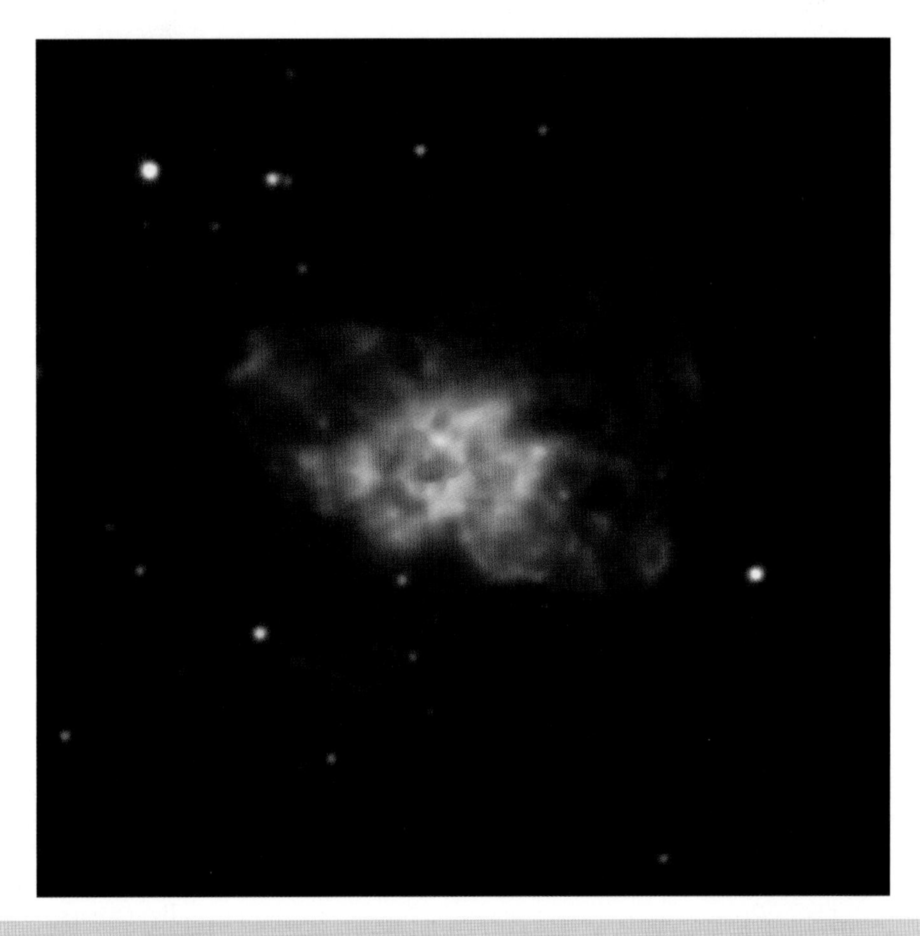

**Object #55** The Albino Butterfly Nebula (NGC 2440) Jeff Cremer/Adam Block/NOAO/AURA/NSF

| OBJECT #55 | NGC 2440 |
|---|---|
| Constellation | Puppis |
| Right ascension | 7h42m |
| Declination | −18°13′ |
| Magnitude | 9.4 |
| Size | 14″ |
| Type | Planetary nebula |
| Other names | The Albino Butterfly Nebula, the Kiss Nebula |

Our next target lies 8.5° northwest of magnitude 2.8 Rho ($\rho$) Puppis, or a little more than 3° south of M46. This planetary nebula lies some 3,500 light-years from Earth.

A 4-inch telescope shows NGC 2440 as an oval disk oriented northwest to southeast with a high surface brightness. Because it is so bright, you can use high magnifications on it. A 12-inch scope at 300× will help you begin to resolve two lobes, with the northeastern one appearing brighter. A faint haze surrounds the bright inner disk.

Don't confuse this planetary with Minkowski's Butterfly, which has the designation M2–9. The Albino Butterfly Nebula got its name quite naturally. Many planetary nebulae take a butterfly, or hourglass, shape as they form. This one's white, so that solves the mystery of its name.

| OBJECT #56 | NGC 2442 |
|---|---|
| Constellation | Volans |
| Right ascension | 7h36m |
| Declination | −69°32′ |
| Magnitude | 10.4 |
| Size | 5.4′ by 2.6′ |
| Type | Barred spiral galaxy |
| Other name | The Meat Hook Galaxy |

One look at our next treat, and you'll understand how it got its common name. The Meat Hook Galaxy lies 2.3° southeast of magnitude 4.0 Delta ($\delta$) Volantis. Through a 10-inch telescope, NGC 2442 shows symmetrical hooks curving from a 4′-long faint, thick bar. Except for its bright core, the body of the galaxy has a uniform brightness. Its distorted form hints at past interaction with other galaxies. It's a joy to observe this object through a 16-inch or larger scope.

If you can use such an instrument, look 10′ east-northeast of NGC 2442 for the magnitude 13.4 spiral galaxy PGC 21457. At magnifications above 150×, this galaxy has an unusual rectangular shape.

| OBJECT #57 | Melotte 71 |
|---|---|
| Constellation | Puppis |
| Right ascension | 7h38m |
| Declination | −12°04′ |
| Magnitude | 7.1 |
| Size | 9′ |
| Type | Open cluster |

Despite its obscure designation, our next treat ranks as one of the showpiece clusters in Puppis. Look for it 2.7° south-southwest of magnitude 3.9 Alpha ($\alpha$) Monocerotis, just 1° south of Puppis' border with Monoceros.

Through a 10-inch telescope, Melotte 71 shows up as a well-defined group of 30 stars magnitude 10 and fainter. The star count doubles through a 16-inch scope.

| OBJECT #58 | Kappa (κ) Puppis |
|---|---|
| Constellation | Puppis |
| Right ascension | 7h39m |
| Declination | −26°48′ |
| Magnitudes | 4.5/4.7 |
| Separation | 9.9″ |
| Type | Double star |

This nicely spaced binary combines two white stars of nearly identical magnitudes. I've often imagined a pair of car headlights as I've observed this pair.

| OBJECT #59 | Canis Minor |
|---|---|
| Right ascension (approx.) | 7h36m |
| Declination (approx.) | 6°30′ |
| Size (approx.) | 183.37 square degrees |
| Type | Constellation |

The tiny constellation Canis Minor the Small Dog is small — its size ranks 71st out of the 88 star patterns that cover the sky. Canis Minor occupies only 183 square degrees, or just 0.4% of the sky.

Its big brother, Canis Major, is a well-formed constellation that contains seven of the 200 brightest stars. The Small Dog, on the other hand, contains only two.

Luckily, one of those is Procyon (Alpha [α] Canis Minoris, the sky's eighth-brightest star. It shines at magnitude 0.34, and lies 11.4 light-years away. The name Procyon is Greek and means "before the Dog." This refers to the fact that, in mid-northern latitudes, the star rises slightly before brilliant Sirius, known for millennia as the Dog Star. Procyon, therefore, is Sirius' herald.

The other notable star in this constellation in Gomeisa (Beta [β] Canis Minoris). This luminary glows at magnitude 2.9, making it the sky's 149th-brightest star.

When you go out at night, these two stars comprise the constellation. Procyon marks the dog's head and Gomeisa its tail. And that's all there is of the Small Dog.

The best date to spot Canis Minor is January 14. Around that date, it sits opposite the Sun as seen from Earth. Conversely, you'll have no luck spotting the Small Dog around July 16, which is when the Sun is in its area.

| OBJECT #60 | Alpha (α) Canis Minoris |
|---|---|
| Constellation | Canis Minor |
| Right ascension | 7h39m |
| Declination | 5°14′ |
| Magnitudes | 0.4/10.0 |
| Separation | 4.8″ |
| Type | Double star |
| Other name | Procyon |

"Procyon" comes from the Greek, and it means "before the dog." This refers to the fact that Procyon's rising occurs slightly before that of Sirius, the Dog Star.

Astronomers measuring Procyon's motion in the mid-nineteenth century noticed it didn't move through space in a straight line. Instead, the star seemed to wobble back and forth as if acted upon by an outside force. It turned out that the outside force was a companion star. In 1896, American astronomer James M. Schaeberle discovered Procyon B at Lick Observatory's 36-inch refractor. Like Sirius B (Object #6), Procyon B is a white dwarf.

Visually, splitting Procyon is tough, but you can try my most recent method. At the 2001 Texas Star Party, I observed it during mid-evening twilight. The seeing was excellent, and I made my sighting through a superb 11-inch Starmaster Dobsonian-mounted Newtonian reflector using an eyepiece that magnified about 300×. I confirmed the sighting by asking three observing friends to view the stars and provide a position for the fainter one. Each verified Procyon B in the position I had seen it.

| OBJECT #61 | M93 (NGC 2447) |
|---|---|
| Constellation | Puppis |
| Right ascension | 7h45m |
| Declination | –23°52′ |
| Magnitude | 6.2 |
| Size | 22′ |
| Type | Open cluster |

Our next treat comes from Charles Messier's catalog. It lies 1.5° northwest of magnitude 3.3 Xi (ξ) Puppis. Odds are that you may be able to glimpse M93 without optical aid from a dark site.

Through a 4-inch telescope at 100×, you'll see an arrowhead shape formed by the cluster's 30 brightest stars pointing toward the southwest. An 8-inch scope adds about two dozen more stars around the arrowhead.

| OBJECT #62 | NGC 2451 |
|---|---|
| Constellation | Puppis |
| Right ascension | 7h45m |
| Declination | –37°58′ |
| Magnitude | 2.8 |
| Size | 45′ |
| Type | Open cluster |
| Other name | The Stinging Scorpion |

To find our next target, look 4.1° west-northwest of magnitude 2.2 Naos (Zeta [ζ] Puppis). There you'll find one of the sky's brightest apparent open clusters. I say "apparent" because astronomers have plotted the motions of the individual stars here, and they are not traveling through space together. Still, it's a nice target that's so bright you can even spot it without optical aid from locations with low levels of light pollution.

Binoculars with a magnification of 15× really bring out the best in this object. When you switch to your telescope, use a low-power eyepiece. You'll see 15 bright stars around orange, magnitude 3.6 SAO 198398, which sits dead-center.

*Astronomy* magazine Contributing Editor Stephen James O'Meara bestowed the common, although little-used, name on this object. He said it appears to him like a scorpion approaching head-on, with two outstretched claws and an upright tail ready to sting.

| OBJECT #63 | 2 Puppis |
|---|---|
| Constellation | Puppis |
| Right ascension | 7h46m |
| Declination | –14°41′ |
| Magnitudes | 6.1/6.8 |
| Separation | 17″ |
| Type | Double star |

The easiest way to find 2 Puppis is to look 0.9° east of open cluster M46. This is a moderately wide pair, and both components appear white.

| OBJECT #64 | NGC 2452 |
|---|---|
| Constellation | Puppis |
| Right ascension | 7h47m |

| (continued) | |
|---|---|
| Declination | −27°20′ |
| Magnitude | 12.0 |
| Size | 19″ |
| Type | Planetary nebula |

To find our next target, look 2.5° south of magnitude 2.2 Naos (Zeta [ζ] Puppis). Through an 8-inch telescope at 200×, you'll easily see the bright rectangular shape, 50% longer than it is wide, stretched in a north-south orientation.

A 14-inch scope at 300× reveals the rectangle to be two fuzzy lobes. A bit of mottling (small dark areas visible on the bright surface) and an irregular edge are the only visible details.

| OBJECT #65 | NGC 2453 |
|---|---|
| Constellation | Puppis |
| Right ascension | 7h48m |
| Declination | −27°12′ |
| Magnitude | 8.3 |
| Size | 4′ |
| Type | Open cluster |

You'll find this object in the same telescopic field at NGC 2452 (Object #64), just 8′ to the planetary nebula's north-northeast. About a dozen stars huddle together in the center of the cluster. The brightest star, magnitude 9.4 SAO174539, lies in the northwest corner. At a distance of some 19,000 light-years, the cluster lies about twice as far away as the planetary.

| OBJECT #66 | NGC 2467 |
|---|---|
| Constellation | Puppis |
| Right ascension | 7h53m |
| Declination | −26°23′ |
| Magnitude | 7.1 |
| Size | 14′ |
| Type | Open cluster |

Our next treat lies 1.7° south-southeast of magnitude 3.3 Xi (ξ) Puppis. It's a combination object that pairs an open cluster with emission nebula Sharpless 2–311.

The cluster appears as little more than a spray of randomly distributed stars enveloped by gas. Just to the northwest of the nebula's main mass is magnitude 9.4 Haffner 19, a tiny, separate cluster. But stars are not why you're pointing your telescope here.

Through an 8-inch scope, use a magnification of 150×, and insert an Oxygen-III filter. That will effectively eliminate the stars, making Sh 2–311 really stand out. Its high surface brightness will let you really crank up the magnification. The bright central clump, which surrounds an 8th-magnitude star, appears mottled with a hollow center and a bright southern rim. A 14-inch scope at 350× will reveal brightness differences along the face of the nebula, as well as several gaps and disassociated nebular regions.

| OBJECT #67 | NGC 2477 |
|---|---|
| Constellation | Puppis |
| Right ascension | 7h52m |
| Declination | −38°33′ |
| Magnitude | 5.8 |
| Size | 27′ |
| Type | Open cluster |
| Other name | Caldwell 71 |

You'll find this object 2.6° west-northwest of magnitude 2.2 Naos (Zeta [ζ] Puppis). From a dark observing location, most of you will spot this cluster with your naked eyes. It lies in a rich star field, however, so using averted vision really helps make it pop.

This is a spectacular target through any size telescope. Even a 4-inch instrument will reveal 60 or more stars. Most lie tightly packed near the center, and all have roughly equal brightnesses. Through a 12-inch scope, you'll have to segment the cluster to count all its stars.

Segmenting involves counting the stars that lie within a pie-shaped region. You can divide the cluster into thirds, fourths, fifths, etc. If you choose a quadrant, count the stars that lie in the section of NGC 2477 that covers a clock's face from noon to 3 o'clock. Then multiply by 4. What's your tally? 150? 200? More?

This cluster is huge. It covers almost as much sky as the Full Moon. Oh, and the magnitude 4.5 foreground star 20′ to the south-southeast is SAO 198545.

| OBJECT #68 | NGC 2482 |
|---|---|
| Constellation | Puppis |
| Right ascension | 7h55m |
| Declination | −24°15′ |
| Magnitude | 7.3 |
| Size | 10′ |
| Type | Open cluster |

Our next target lies 1.5° east-northeast of magnitude 3.3 Xi (ξ) Puppis. Through an 8-inch or smaller telescope, use a magnification of 75× or lower, and try to pick out this cluster's distinctive "Y" asterism. The Y opens toward the northwest, and its base points southeastward. It's definitely there, so try different eyepieces until you spot it.

The cluster appears as a rich but irregular grouping of 50 stars. Step up to a 12-inch scope, and your star count will double.

| OBJECT #69 | NGC 2489 |
|---|---|
| Constellation | Puppis |
| Right ascension | 7h56m |
| Declination | −30°04′ |
| Magnitude | 7.9 |
| Size | 5′ |
| Type | Open cluster |

Look 5.4° south-southeast of magnitude 3.3 Xi (ξ) Puppis, and you'll sweep up our next target. Small telescopes show only 15 or 20 stars scattered about the field of view. An 8-inch scope at 200× will let you count 50 members.

Three nice field stars lie generally south and nearby. Magnitude 6.3 SAO 198609, which glows with a deep-orange hue, appears brightest. Another 7′ south from this star lies another open cluster, magnitude 11.0 Haffner 20. Your best approach to viewing this dim object is to place SAO 198609 just out of the field of view to the north.

| OBJECT #70 | NGC 2506 |
|---|---|
| Constellation | Monoceros |
| Right ascension | 8h00m |
| Declination | −10°46′ |
| Magnitude | 7.6 |
| Size | 12′ |
| Type | Open cluster |
| Other name | Caldwell 54 |

Our next target lies in the far southeastern corner of Monoceros 0.5° from that constellation's border with Puppis and less than 3° from its Hydra border. From magnitude 3.9 Alpha (α) Monocerotis, move 4.8° east-southeast, and you'll land on NGC 2506.

Through a 4-inch telescope, this cluster isn't all that impressive. Its stars all appear to be about the same brightness, but they have a wildly uneven distribution. Spread them out with a magnification of 150×, and you'll see a clumpy center and lots and lots of patterns: streamers, spiral "arms," letters, and more.

Move up to a 12-inch scope at 200×, and the scene gets interesting. The same 30 or 40 stars you saw through the smaller aperture now hang before a background glow that glistens like a diamond-encrusted black velvet sheet. The gaps between twisting lines of stars also appear wider and darker.

| OBJECT #71 | NGC 2516 |
|---|---|
| Constellation | Carina |
| Right ascension | 7h58m |
| Declination | −60°52′ |
| Magnitude | 3.8 |
| Size | 30′ |
| Type | Open cluster |
| Other name | Caldwell 96 |

You'll find this spectacular cluster 3.3° west-southwest of magnitude 1.9 Avior (Epsilon [ε] Carinae), the westernmost and southernmost star of the False Cross asterism. You'll have no trouble spotting this object with your naked eyes — it's one of the sky's 10 brightest open cluster.

Through a 6-inch telescope, you'll count 75 stars, but it won't be easy. Here, the stars divide into two brightness ranges. The "upper class" ranges from magnitude 5.8 SAO 250055, the cluster's brightest star, through magnitude 8. Unless you use high magnification — and I mean above 250× — all those bright stars will mask the many faint stars this cluster contains.

| OBJECT #72 | Gamma (γ) Velorum |
|---|---|
| Constellation | Vela |
| Right ascension | 8h10m |
| Declination | −47°20′ |
| Magnitudes | 1.9/4.2 |
| Separation | 41.2″ |
| Type | Double star |
| Other name | Al Suhail (al Muhlif), Regor |

Our next target is one you'll have no trouble finding if you're far enough south. The components are bright, and both shine with a blue light.

The Arabic name for this star means the "Plain of the Oath," and is one of many Suhails in the region (the most notable being Canopus (Alpha [α] Carinae). This star's other name, Regor, is "Roger" spelled backward. It honors American astronaut Roger Chaffee, who died in the Apollo 1 fire at the Kennedy Space Center.

| OBJECT #73 | NGC 2525 |
|---|---|
| Constellation | Puppis |
| Right ascension | 8h06m |
| Declination | −11°26′ |
| Magnitude | 11.6 |
| Size | 3′ by 2′ |
| Type | Spiral galaxy |

You'll find our next target 6.3° east-southeast of magnitude 3.9 Alpha (α) Monocerotis. Through an 8-inch telescope at 100×, you'll see an evenly illuminated object 50% longer than it is wide, oriented east-west. A 16-inch scope shows a slight darkening between the central region and the thick southern spiral arm, which curves toward the west.

| OBJECT #74 | NGC 2527 |
|---|---|
| Constellation | Puppis |
| Right ascension | 8h05m |
| Declination | −28°09′ |
| Magnitude | 6.5 |
| Size | 10′ |
| Type | Open cluster |

This sparse cluster lies 4.8° southeast of magnitude 3.3 Xi (ξ) Puppis. A 4-inch telescope at 150× will reveal two dozen stars, haphazardly strewn about the field of view. Larger apertures add a few fainter stars, but not many. The most interesting section of NGC 2527 is its eastern side, where a U-shaped chain of closely spaced stars reside.

| OBJECT #75 | NGC 2533 |
|---|---|
| Constellation | Puppis |
| Right ascension | 8h07m |
| Declination | −29°52′ |
| Magnitude | 7.6 |
| Size | 6′ |
| Type | Open cluster |

Our next object is a loose cluster that features the magnitude 9.0 star SAO 175203. The magnitude 10.8 star HIP 39707 sits 1′ to its southwest. About a dozen much fainter points surround this pair, mainly to the north. NGC 2533 isn't close to any bright star. You'll find it 6.4° southeast of magnitude 3.3 Xi (ξ) Puppis.

| OBJECT #76 | NGC 2539 |
|---|---|
| Constellation | Puppis |
| Right ascension | 8h11m |
| Declination | −12°50′ |
| Magnitude | 6.5 |
| Size | 21′ |
| Type | Open cluster |
| Other name | The Dish Cluster |

Our next target sits nearly 8° east-southeast of magnitude 3.9 Alpha (α) Monocerotis. When you move into the area, look for the yellow-white magnitude 4.7 star 19 Puppis. NGC 2539 sits less than 12′ to the west-northwest.

The common name the Dish Cluster comes from a description of NGC 2539 by *Astronomy* magazine Contributing Editor Stephen James O'Meara. He sees an oval shape in the brightest stars of this cluster when he observes it at a magnification of 23×.

A 4-inch telescope shows 75 stars between 9th and 13th magnitude, while a 10-inch scope reveals more than 100 stars. The brightest stars appear clumped in an oval a bit south of center that stretches east-west.

| OBJECT #77 | NGC 2546 |
|---|---|
| Constellation | Puppis |
| Right ascension | 8h12m |
| Declination | −37°37′ |
| Magnitude | 6.3 |
| Size | 70′ |
| Type | Open cluster |
| Other names | The Heart and Dagger Cluster, the Wounded Heart Cluster |

Here's a target that really benefits from increased aperture. Through a 4-inch telescope, you'll see only about 15 stars generally strewn in a southeast to northwest direction. An 8-inch scope doubles the star count, and through a 14-inch instrument at 150×, you'll count 75 stellar points. Two magnitude 6.4 stars will attract your attention immediately. One, SAO 198942, lies halfway from the cluster's center to its southern border. The other, SAO 198848, sits off NGC 2546's western edge.

You'll find this object not quite 3° northeast of magnitude 2.3 Naos (Zeta [ζ] Puppis).

Both of the common names for this object originate with *Astronomy* magazine Contributing Editor Stephen James O'Meara. I've tried to see the complex figure of a heart pierced by a dagger he describes. If, like me, you can't imagine that, try seeing a Greek letter Phi ($\phi$).

| OBJECT #78 | NGC 2547 |
|---|---|
| Constellation | Vela |
| Right ascension | 8h11m |
| Declination | −49°16′ |
| Magnitude | 4.7 |
| Size | 74′ |
| Type | Open cluster |
| Other name | The Golden Earring |

Oh, my! Here's a gorgeous object that, unfortunately, many Northern Hemisphere observers have never seen. NGC 2547 is easily visible to the naked eye from a dark site. It lies in a dense star field 1.9° south of magnitude 1.7 Regor (Gamma [γ] Velorum).

Through any size telescope, you'll see a dozen stars that shine at magnitude 9 or brighter. The brightest is magnitude 6.5 SAO 219538, which sits just to the east of center. Note that the given size exceeds 1°. Because of the richness of the surrounding star field, you'll have trouble following the cluster's boundary to anywhere near that size. Still, what you detect will cover more area than the Full Moon.

*Astronomy* magazine Contributing Editor Stephen James O'Meara coined the common name (among others) for this cluster from a drawing by Australian astronomer James Dunlop (1793–1848). Dunlop drew the cluster as an oval of stars dangling from two rows of parallel stars. O'Meara likened the image to a golden earring dangling from a pirate's ear.

| OBJECT #79 | Zeta (ζ) Cancri |
|---|---|
| Constellation | Cancer |
| Right ascension | 8h12m |
| Declination | 17°39′ |
| Magnitudes | 5.6/6.0 |
| Separation | 5.9″ |
| Type | Double star |
| Other name | Tegmeni |

This nice binary sits 7° west-southwest of the Beehive Cluster (M44). Both of its components shine with an attractive yellow light.

The star's name, sometimes spelled Tegmen, means "in the covering," and may refer to its position on the rear edge of the Crab's shell.

**Object #80** M48 Anthony Ayiomamitis

| OBJECT #80 | M48 (NGC 2548) |
|---|---|
| Constellation | Hydra |
| Right ascension | 8h14m |
| Declination | –5°48′ |
| Magnitude | 5.8 |
| Size | 54′ |
| Type | Open cluster |

Our next target is visible to sharp-eyed observers at a dark site without optical aid. Large binoculars resolve a couple dozen of its brightest stars scattered over 1° of sky.

A 6-inch telescope reveals about 75 stars sprinkled across the entire field of view. Increasing the magnification increases the star count, although the appearance that this object is a cluster becomes lost. Look for a zigzag chain of 9th- and 10th-magnitude stars running from the south-southwest to the north-northeast through the cluster's center.

You'll find the 48th entry in Charles Messier's catalog 3° south-southeast of magnitude 4.4 Zeta (ζ) Monocerotis only 0.6° from Hydra's border with that constellation.

| OBJECT #81 | NGC 2559 |
|---|---|
| Constellation | Puppis |
| Right ascension | 8h17m |
| Declination | –27°28′ |
| Magnitude | 10.9 |

| (continued) | |
|---|---|
| Size | 4.2′ by 2.3′ |
| Type | Barred spiral galaxy |

This object appears twice as long as it is wide, with the long axis oriented north-south. Small telescopes show a nearly rectangular shape. Through an 8-inch scope, you'll see a faint, broad central concentration. Both NGC 2559 and our next target, NGC 2566, belong to a cluster of galaxies known as the Puppis concentration. Less than 1′ east-southeast of NGC 2559's core, you'll spot magnitude 9.4 SAO 175514.

Look for this galaxy 3.8° southeast of magnitude 2.8 Rho (ρ) Puppis.

| OBJECT #82 | NGC 2566 |
|---|---|
| Constellation | Puppis |
| Right ascension | 8h19m |
| Declination | −25°29′ |
| Magnitude | 11.0 |
| Size | 4.1′ by 2.0′ |
| Type | Barred spiral galaxy |

Our next target lies 2.8° east-southeast of magnitude 2.8 Rho (ρ) Puppis. Through a 12-inch telescope it appears small and bright with little central concentration save for a faint stellar core. The elliptical galaxy IC 2311, located 7′ to the northeast, looks a little brighter than NGC 2566 and shows a strong brightening toward its center. It glows at magnitude 11.5.

| OBJECT #83 | NGC 2567 |
|---|---|
| Constellation | Puppis |
| Right ascension | 8h19m |
| Declination | −30°38′ |
| Magnitude | 7.4 |
| Size | 11′ |
| Type | Open cluster |

This attractive cluster lies 6.8° south-southeast of magnitude 2.8 Rho (ρ) Puppis. Through a 6-inch telescope, you'll see 30 stars. Half of them form two figures. The first is an ever-so-slightly curved line of stars that runs north-south and terminates near the cluster's center. The second is a U-shaped group that sits in the southwest quadrant.

The brightest star in your field of view, magnitude 8.9 SAO 199057, is not part of NGC 2567. It lies 7′ to the south-southwest.

| OBJECT #84 | NGC 2571 |
|---|---|
| Constellation | Puppis |
| Right ascension | 8h19m |
| Declination | −29°45′ |
| Magnitude | 7.0 |
| Size | 7′ |
| Type | Open cluster |

You'll find our next target 6.3° west-northwest of magnitude 3.7 Alpha (α) Pyxidis. Train any telescope on this cluster, and you'll first see its two brightest stars at dead-center. They appear evenly bright, but the easternmost, magnitude 8.8 SAO 175580, shines slightly brighter than its companion, magnitude 8.9 SAO 175577. These stars lie 1′ apart.

| OBJECT #85 | NGC 2610 |
|---|---|
| Constellation | Hydra |
| Right ascension | 8h33m |
| Declination | −16°09′ |
| Magnitude | 12.8 |
| Size | 37″ |
| Type | Planetary nebula |

You'll find this bright planetary nebula in Hydra's southwestern corner. Find it by sweeping 2° west of the magnitude 4.9 star 9 Hydrae. Through anything less than a 16-inch telescope, this object appears uniformly bright, circular, and featureless. Really big scopes show it as a thick ring with an elusive central star.

| OBJECT #86 | NGC 2613 |
|---|---|
| Constellation | Pyxis |
| Right ascension | 8h33m |
| Declination | −22°58′ |
| Magnitude | 10.5 |
| Size | 7.6′ by 1.9′ |
| Type | Spiral galaxy |

Our next target lies 6.1° east of magnitude 2.8 Rho ($\rho$) Puppis. Through a 10-inch telescope at 200×, this relatively bright galaxy appears hazy with a slightly brighter center. It stretches four times as long as it is wide. Like NGC 2559 (Object #81), NGC 2613 belongs to a galaxy cluster called the Puppis Concentration. Although this cluster has a density comparable to that of the Virgo Cluster, our intervening Milky Way blocks the light from most of its members.

| OBJECT #87 | Vela SNR |
|---|---|
| Constellation | Vela |
| Right ascension | 8h34m |
| Declination | −45°45′ |
| Size | 5° |
| Type | Supernova remnant |
| Other name | The Vela supernova remnant |

The Vela supernova remnant (SNR) formed as the result of a massive star exploding approximately 11,000 years ago. The progenitor star was about 800 light-years away. The most observed part of the Vela SNR is the Pencil Nebula (Object #101). The Vela SNR is the sky's largest supernova remnant and one of the closest, covering an incredible 5°. The Pencil Nebula forms the east-southeast section.

As you observe the Vela SNR, you'll see a pretty double star. That's DUN 70. Its blue and white components shine at magnitudes 5.2 and 6.8, respectively, and they lie 4.5″ apart.

The designation "DUN" refers to the double star catalog *Approximate Places of Double Stars in the Southern Hemisphere, observed at Paramatta in New South Wales* by Australian astronomer James Dunlop (1793–1848). His catalog contains 256 binary stars below declination −30°.

To find the Vela SNR, look roughly 2° south of magnitude 2.2 Lambda ($\lambda$) Velorum.

| OBJECT #88 | NGC 2627 |
|---|---|
| Constellation | Pyxis |
| Right ascension | 8h37m |
| Declination | −29°57′ |

| (continued) | |
|---|---|
| Magnitude | 8.4 |
| Size | 11′ |
| Type | Open cluster |

Our next object lies only 0.7° southwest of magnitude 4.9 Zeta (ζ) Pyxis. Through an 8-inch telescope, you'll see three dozen stars. Most fill a wide swath that stretches east-west across the cluster. A faint background glow hints at the presence of additional stars.

| OBJECT #89 | IC 2391 |
|---|---|
| Constellation | Vela |
| Right ascension | 8h40m |
| Declination | −53°04′ |
| Magnitude | 2.5 |
| Size | 50′ |
| Type | Open cluster |
| Other names | The Omicron Velorum Cluster, Caldwell 85 |

Here's an object you won't have any trouble locating. Being one of the sky's half-dozen brightest open clusters, it will jump out at your naked eyes from a dark site. As its common name implies, this star group centers on magnitude 3.6 Omicron (ο) Velorum. If that's not a bright enough signpost, then locate magnitude 1.9 Delta (δ) Velorum, and move not quite 2° to the north-northwest.

My preferred view of this cluster is through 15× binoculars or through a 3-inch telescope at a magnification around 30×. The brilliant stars dominate the field of view, but there are many fainter stars hovering about as well.

Two stars at magnitudes 4.8 and 5.5 on IC 2391's eastern side form a terrific pair separated by 75″. Surrounding them is a group of four stars, magnitudes 7.4–8.7, that resemble the constellation Corvus the Crow.

**Object #90** The Beehive Cluster (M44) Tom Bash and John Fox/Adam Block/NOAO/AURA/NSF

| OBJECT #90 | M44 (NGC 2632) |
|---|---|
| Constellation | Cancer |
| Right ascension | 8h40m |
| Declination | 19°40′ |
| Magnitude | 3.1 |
| Size | 95′ |
| Type | Open cluster |
| Common names | The Beehive Cluster; The Praesepe, the Manger |

The richness of the Beehive Cluster — a celestial object known since antiquity — makes up for the faintness of the constellation that contains it. M44 sits midway between Castor and Pollux, the two brightest stars of Gemini, and Regulus, Leo the Lion's luminary.

The usual common name for this object, the Beehive Cluster, is self explanatory. The meaning of "Praesepe," however, may be less obvious. That term is the Latin word for "manger," and it (and the English equivalent) refers to the birthplace of Jesus of Nazareth.

If your sky isn't all that dark, use binoculars to locate the Beehive. With an apparent size of some three Full Moons side by side, M44 looks best to some observers through binoculars with magnifications between 10× and 16×.

To the unaided eye, the Beehive appears nebulous, but the telescope's invention revealed its true nature. Galileo wrote in *Sidereus Nuncius* that he counted more than 40 stars in M44 as early as 1610. Astronomers today list upwards of 350 stars belonging to the cluster.

The Beehive's brightest star is Epsilon (ε) Cancri, which shines at magnitude 6.3. Some 80 of the cluster's stars are brighter than 10th magnitude.

| OBJECT #91 | NGC 2655 |
|---|---|
| Constellation | Camelopardalis |
| Right ascension | 8h56m |
| Declination | 78°13′ |
| Magnitude | 10.1 |
| Size | 6.0′ by 5.3′ |
| Type | Spiral galaxy |

Through small telescopes you won't see much of this bright galaxy except an evenly illuminated face and an oval shape. It's 50% longer than it is wide, stretched in an east-west orientation. A 12-inch scope shows a bit more: NGC 2655's central region is brighter, and a thin outer halo surrounds it.

NGC 2655 effectively sits in the middle of nowhere. You'll find it 13.5° northwest of magnitude 3.8 Lambda (λ) Draconis.

| OBJECT #92 | Eta Chamaeleontis Cluster |
|---|---|
| Constellation | Chamaeleon |
| Right ascension | 8h41m |
| Declination | −78°58′ |
| Size | 30′ |
| Type | Open cluster |

Our next object has just 13 stars lying within a 0.5° circle. It took astronomers until 1999 to recognize this group as an open cluster. The Eta Chamaeleontis Cluster is quite close, only 315 light-years away. Through a small telescope, three stars dominate. Eta (η) Chamaeleontis shines at magnitude 5.5. Only 5′ to the south-southeast sits magnitude 7.4 SAO 256544. Brighter SAO 256549, which glows at magnitude 6.1, lies 8′ southeast of Eta.

This cluster sits 2.4° south-southeast of magnitude 4.1 Alpha (α) Chamaeleontis.

| OBJECT #93 | The Head of Hydra |
|---|---|
| Right ascension | 8h42m |
| Declination | 4°39′ |
| Type | Asterism |

Our next object is a good one for beginning stargazers. It's the Head of Hydra. This asterism marks the westernmost part of the sky's largest constellation.

The Head of Hydra lies 2° due south of the midpoint of a line that joins Procyon (Alpha [$\alpha$] Canis Minoris) and Regulus (Alpha Leonis). Unless you live under the worst light pollution, you'll see the Head with your naked eyes.

Six stars form the asterism. The brightest is magnitude 3.1 Zeta ($\zeta$) Hydrae. From there, move west to Epsilon ($\varepsilon$) and Delta ($\delta$) Hydrae. Then swing back east to Rho ($\rho$) Hydrae. Drop 3.5° southwest to Sigma ($\sigma$) Hydrae. At magnitude 4.4, this is the faintest star in the asterism. Finally, head east again to the sixth and final star, Eta ($\eta$) Hydrae.

| OBJECT #94 | Iota ($\iota$) Cancri |
|---|---|
| Constellation | Cancer |
| Right ascension | 8h47m |
| Declination | 28°46′ |
| Magnitudes | 4.2/6.6 |
| Separation | 30″ |
| Type | Double star |

Iota Cancri is the northernmost star in this constellation's inverted Y shape. It's a great target for any size scope. The primary shines yellow and its companion is blue.

| OBJECT #95 | Epsilon ($\varepsilon$) Hydrae |
|---|---|
| Constellation | Hydra |
| Right ascension | 8h47m |
| Declination | 6°25′ |
| Magnitudes | 3.4/6.8 |
| Separation | 2.7″ |
| Type | Double star |

This nice binary has a lemon-yellow primary and a grayish-blue secondary. The separation is close, so crank the magnification up to 150× or beyond.

| OBJECT #96 | NGC 2659 |
|---|---|
| Constellation | Vela |
| Right ascension | 8h43m |
| Declination | −45°00′ |
| Magnitude | 8.6 |
| Size | 15′ |
| Type | Open cluster |

You'll find our next target between two really bright stars. It lies slightly more than halfway from magnitude 1.8 Regor (Gamma [$\gamma$] Velorum) to magnitude 2.2 Suhail (Lambda [$\lambda$] Velorum).

Through a 4-inch telescope at 100×, you'll see 30 stars of relatively equivalent brightness scattered across the field of view. The brightest, magnitude 9.7 GSC 8151:259, sits at the cluster's southeast corner.

| OBJECT #97 | NGC 2681 |
|---|---|
| Constellation | Ursa Major |
| Right ascension | 8h54m |
| Declination | 51°19′ |
| Magnitude | 10.2 |
| Size | 3.6′ by 3.3′ |
| Type | Spiral galaxy |

Our next treat is a small, round galaxy, but it's quite bright. The central region takes up most of the diameter, with a faint haze surrounding it. You'll need magnifications above 400× and at least a 16-inch scope to identify the diaphanous spiral arms. They certainly do hug the galaxy's core tightly.

**Object #98** M67 Anthony Ayiomamitis

| OBJECT #98 | M67 (NGC 2682) |
|---|---|
| Constellation | Cancer |
| Right ascension | 8h51m |
| Declination | 11°50′ |
| Magnitude | 6.9 |
| Size | 29′ |
| Type | Open cluster |

Our next target is the "other" open cluster in Cancer, M67. (It's the one that's not the Beehive [M44]). You'll find M67 easily through binoculars or a small telescope 1.7° due west of magnitude 4.3 Alpha (α) Cancri.

Through a 4-inch telescope, you'll resolve roughly two dozen stars in M67 across an area two-thirds the width of the Full Moon. Increase the aperture to 6′, and 50 stars will shine forth.

A dozen of M67's stars shine brighter than 11th magnitude. When you view the Beehive through a telescope, you'll note the yellow star on its northeastern edge. Identified as SAO 98178, this star shines at magnitude 7.8 but is not a member of the cluster.

**Object #99** The UFO Galaxy (NGC 2683) Doug Matthews/Adam Block/NOAO/AURA/NSF

| OBJECT #99 | NGC 2683 |
|---|---|
| Constellation | Lynx |
| Right ascension | 8h53m |
| Declination | 33°25′ |
| Magnitude | 9.8 |
| Size | 8.4′ by 2.4′ |
| Type | Spiral galaxy |
| Other name | The UFO Galaxy |

This spectacular deep-sky object lies in a constellation even more difficult to find than some deep-sky objects — Lynx. The obscure star group lies due north of Cancer and stretches to the northwest from there.

NGC 2683 is a spiral galaxy, and a relatively bright one at that. You can spot it through a 3-inch telescope from a dark observing site. To pull out its details, however, you'll need a bigger scope.

This galaxy is a classic edge-on spiral that orients exactly northeast to southwest. Its common name derives from the resemblance of its shape to descriptions of unidentified flying objects from the 1950s. It appears more than three times as long as it is wide with an extended, bright central region.

The faint spiral arms begin to show alternate dark and bright patches called mottling through a 12-inch telescope. Through even larger scopes, you'll notice that the northeastern arm extends a bit farther than the southwestern one.

| OBJECT #100 | NGC 2685 |
|---|---|
| Constellation | Ursa Major |
| Right ascension | 8h56m |
| Declination | 58°44′ |
| Magnitude | 11.1 |
| Size | 4.9′ by 2.4′ |
| Type | Spiral galaxy |
| Other names | The Helix Galaxy, the Pancake Galaxy |

Astronomers classify this unusual lenticular galaxy as a polar ring galaxy. Several filamentary strands, made up of knots of luminous star-forming regions, form a helical band perpendicular to the galaxy's main disk and centered on its nucleus.

These structures suggest that NGC 2685 once had a companion, perhaps like one of the Milky Way's neighboring Magellanic Clouds. The main galaxy captured the satellite into a polar orbit, and its stars eventually merged with those of the larger system. What was left? Only the gas and dust of the smaller galaxy. New stars formed from this material to produce the luminous ring. It is possible that if the Magellanic Clouds had been closer to the Milky Way, they too would have created a polar ring around our galaxy.

To find NGC 2685, look 3.8° east-southeast of magnitude 3.4 Muscida (Omicron [*o*] Ursae Majoris). Although interesting, the galaxy isn't bright. At low magnifications, you'll see a disk-shaped, evenly illuminated glow three times as long as it is wide.

Point a 14-inch or larger telescope at this object, and go after the ephemeral ring that gives the Helix Galaxy its common name. Use a magnification around 200× to start, and increase the power if sky conditions are good enough. I've had some luck viewing the ring by moving NGC 2685's main mass just out of the eyepiece's field of view.

M.E. Bakich, *1,001 Celestial Wonders to See Before You Die*, Patrick Moore's Practical Astronomy Series, DOI 10.1007/978-1-4419-1777-5_2, 

| OBJECT #101 | NGC 2736 |
|---|---|
| Constellation | Vela |
| Right ascension | 9h00m |
| Declination | −45°57′ |
| Size | 20′ |
| Type | Supernova remnant |
| Other names | The Pencil Nebula, Herschel's Ray |

The Pencil Nebula forms a small part of the Vela supernova remnant (SNR), which lies in the southern constellation Vela the Sails. Amateur astronomers dubbed this object the Pencil because, through a telescope, it appears long, straight, and one end looks "sharpened."

British astronomer Sir John Herschel (1792–1871) discovered the Pencil Nebula in 1835 while he was staying in South Africa. He described it as "an extraordinary long narrow ray of excessively feeble light."

The Pencil Nebula measures about 0.75 light-year across, while the Vela supernova remnant spans 114 light-years. The remnant lies about 815 light-years away.

The best way to observe the Pencil Nebula is to use a 12-inch or larger telescope with a low-power eyepiece equipped with a nebula filter, such as an OIII. Either disengage your telescope's drive motor or set its slewing speed at "medium," and scan the area.

| OBJECT #102 | NGC 2768 |
|---|---|
| Constellation | Ursa Major |
| Right ascension | 9h12m |
| Declination | 60°02′ |
| Magnitude | 9.9 |
| Size | 6.4′ by 3.0′ |
| Type | Elliptical galaxy |

Astronomers classify this galaxy as elliptical, but it's definitely lenticular. It's easily three times as long as it is wide. The large central region takes up two-thirds of the galaxy's width. NGC 2768's core has even illumination, as does the object's outer halo. Although this galaxy shone brightly when I observed it through a 30-inch telescope, I didn't see any details.

**Object #103** NGC 2775 Jeff Newton/Adam Block/NOAO/AURA/NSF

| OBJECT #103 | NGC 2775 |
|---|---|
| Constellation | Cancer |
| Right ascension | 9h10m |
| Declination | 7°02′ |
| Magnitude | 10.1 |
| Size | 4.6′ by 3.7′ |
| Type | Spiral galaxy |
| Other name | Caldwell 48 |

You'll find our next target 3.8° east-northeast of magnitude 3.1 Zeta (ζ) Hydrae. Seen through an 8-inch telescope, this galaxy appears oval oriented north-northwest to south-southeast. A 12-inch scope at 250× reveals the evanescent outer halo.

Through even larger instruments, look for two other faint galaxies, magnitude 13.6 NGC 2773, which sits 12′ to the northwest, and magnitude 13.1 NGC 2777, which lies 12′ to the north-northeast. NGC 2777 travels through space with NGC 2775, but NGC 2773 lies four times as far away and isn't associated with the others.

| OBJECT #104 | NGC 2784 |
|---|---|
| Constellation | Hydra |
| Right ascension | 9h12m |
| Declination | −24°10′ |
| Magnitude | 10.0 |
| Size | 5.5′ by 2.4′ |
| Type | Spiral galaxy |

Our next target lies 1.9° north-northeast of magnitude 4.6 Kappa (κ) Pyxidis. I imagine a disk within a disk when I view this galaxy. Through an 8-inch telescope, you'll see the elongated central region surrounded by a similarly shaped halo. The halo appears quite thick. Both the core and the halo tilt east-northeast to south-southwest. Less than 1′ from the galaxy's northeastern tip lies the magnitude 12.9 star GSC 6586:357.

| OBJECT #105 | NGC 2787 |
|---|---|
| Constellation | Ursa Major |
| Right ascension | 9h19m |
| Declination | 69°12′ |
| Magnitude | 10.9 |
| Size | 3.1′ by 1.8′ |
| Type | Spiral galaxy |

Now here's a bit of an odd duck. Although NGC 2787 appears bright, it's better seen through large apertures. When I first viewed this galaxy through a 30-inch telescope at the Rancho Hidalgo astronomy and equestrian village in Animas, New Mexico, I thought it might be a barred spiral. Astronomers, however, put it in a different class — barred lenticular. Well, I saw the bar. It puzzled me at first, because the bar's long axis tilts a bit to the overall long axis of NGC 2787. The central region appeared much brighter than the outer areas, but I saw no other details.

| OBJECT #106 | NGC 2805 |
|---|---|
| Constellation | Ursa Major |
| Right ascension | 9h20m |
| Declination | 64°06′ |
| Magnitude | 10.9 |
| Size | 6.3′ by 4.8′ |
| Type | Spiral galaxy |

You'll find our next object 1.2° east-northeast of magnitude 4.7 Tau (τ) Ursae Majoris. Although its magnitude is relatively bright, it spreads out quite a bit. Through a 12-inch telescope at 200×, you'll see a nearly stellar central region surrounded by irregularly bright haze. The overall shape is oval in an east-west orientation.

| OBJECT #107 | NGC 2808 |
|---|---|
| Constellation | Carina |
| Right ascension | 9h12m |
| Declination | −64°52′ |
| Magnitude | 6.3 |
| Size | 13.8′ |
| Type | Globular cluster |

This magnificent object — the sky's 10th-brightest globular cluster — is visible to sharp-eyed observers under a dark sky without optical aid. It lies in an incredible star field 3.7° west of magnitude 3.1 Upsilon (υ) Carinae.

Although brilliant, you'll have difficulty resolving this cluster's stars through any telescope smaller than a 14-inch. An 8-inch scope shows a blazing core surrounded by an unevenly illuminated halo.

| OBJECT #108 | NGC 2811 |
|---|---|
| Constellation | Hydra |
| Right ascension | 9h16m |
| Declination | −16°19′ |
| Magnitude | 11.4 |
| Size | 2.5′ by 0.9′ |
| Type | Spiral galaxy |

Our next object lies 6.1° west-southwest of magnitude 5.1 Kappa ($\kappa$) Hydrae. Through telescopes smaller than 10′ in aperture, this object appears evenly illuminated, tilted in a north-northeast to south-southwest orientation. Magnifications below 200× show a sliver four times as long as it is wide. Only through larger scopes will you begin to fatten out the central region.

| OBJECT #109 | NGC 2818 |
|---|---|
| Constellation | Pyxis |
| Right ascension | 9h16m |
| Declination | −36°37′ |
| Magnitude | 8.2 |
| Size | 9′ |
| Type | Open cluster |

Our next target lies 7.4° east-southeast of magnitude 4.0 Beta ($\beta$) Pyxidis. It's an open star cluster with a planetary nebula inside. Through most telescopes, the open cluster (NGC 2818) appears as a loose group of two dozen stars. Most of its members glow below 12th magnitude, but the cluster stands out thanks to the sparse surrounding star field.

The planetary nebula (NGC 2818A) appears as a small, moderately bright object with a dumbbell shape. An 8-inch scope shows both lobes, but you'll need a 20-inch or larger instrument to see any other details.

| OBJECT #110 | 38 Lyncis |
|---|---|
| Constellation | Lynx |
| Right ascension | 9h19m |
| Declination | 36°48′ |
| Magnitudes | 3.9/6.6 |
| Separation | 2.7″ |
| Type | Double star |

Each of the stars in this pair appears white. The separation is close, so crank the magnification past 150×. The star 38 Lyn sits in a no-man's land of faint stars. Look for it not quite 2.5° north of magnitude 3.1 Alpha ($\alpha$) Lyncis.

| OBJECT #111 | NGC 2832 |
|---|---|
| Constellation | Lynx |
| Right ascension | 9h20m |
| Declination | 33°44′ |
| Magnitude | 11.9 |
| Size | 3.0′ by 2.1′ |
| Type | Elliptical galaxy |
| Notes | in Abell 779 |

Our next object resides in galaxy cluster Abell 779, which lies less than 0.7° south-southwest of magnitude 3.1 Alpha ($\alpha$) Lyncis. NGC 2832 is the brightest member of the cluster. It has an oval shape and measures 50% longer than it is wide oriented roughly northwest to southeast.

If you have access to a 16-inch or larger telescope, study the area around NGC 2832, and see how many other faint galaxies you can spot. The magnitude 13.4 elliptical galaxy NGC 2831 lies only 24″ to the southwest. Even fainter, the magnitude 13.9 lens-shaped spiral NGC 2830 lies a bit more than 1′ to the west-southwest.

Want more? Just 5′ west of NGC 2832 lies magnitude 14.4 NGC 2825, and magnitude 14.5 NGC 2834 lies 4′ southeast of our starting point, NGC 2832.

| OBJECT #112 | NGC 2835 |
|---|---|
| Constellation | Hydra |
| Right ascension | 9h18m |
| Declination | −22°21′ |
| Magnitude | 10.3 |
| Size | 6.6′ by 4.4′ |
| Type | Spiral galaxy |

To find our next target, look 3.7° north-northwest of magnitude 4.7 Theta ($\theta$) Pyxidis. Through an 8-inch telescope at 150×, this object appears round, faint, and stretched a bit on a north-south line. A 16-inch scope at 300× starts to reveal the galaxy's spiral structure. Its arms appear thin and broken. The magnitude 12.1 star GSC 6040:550 lies on NGC 2835's eastern edge.

| OBJECT #113 | NGC 2841 |
|---|---|
| Constellation | Ursa Major |
| Right ascension | 9h22m |
| Declination | 50°59′ |
| Magnitude | 9.3 |
| Size | 8.1′ by 3.5′ |
| Type | Spiral galaxy |

What a gorgeous object! This galaxy tilts southeast to northwest and displays a classic disk appearance. Its nucleus is wide and bright. Through an 8-inch telescope, you'll see several dark regions within the tightly wound spiral arms, but the arms themselves are tough to see even at high powers. You'll find this treat 1.8° west-southwest of magnitude 3.2 Theta ($\theta$) Ursae Majoris.

| OBJECT #114 | NGC 2859 |
|---|---|
| Constellation | Leo Minor |
| Right ascension | 9h24m |
| Declination | 34°31′ |
| Magnitude | 10.9 |
| Size | 4.6′ by 4.1′ |
| Type | Spiral galaxy |

To find this object, first locate magnitude 3.1 Alpha ($\alpha$) Lyncis. Then scan just 0.7° east. Through a 10-inch telescope, this galaxy tips north-northwest to east-southeast. Its thick outer halo takes up 25% of the galaxy's overall diameter. The featureless central region appears evenly illuminated.

| OBJECT #115 | NGC 2867 |
|---|---|
| Constellation | Carina |
| Right ascension | 9h21m |
| Declination | −58°19′ |
| Magnitude | 9.7 |
| Size | 11″ |
| Type | Planetary nebula |
| Other name | Caldwell 90 |

You'll find our next object 1.1° north-northeast of magnitude 2.2 Aspidiske (Iota [$\iota$] Carinae). This bright planetary takes all the magnification you can throw at it. Through 4-inch and larger telescopes, it appears robin's-egg blue to most observers. A 12-inch scope at 300× reveals a bright edge with an ever-so-slightly darker central region. A magnitude 10.2 star lies a bit more than 2″ east of the nebula.

| OBJECT #116 | NGC 2899 |
|---|---|
| Constellation | Vela |
| Right ascension | 9h27m |
| Declination | −56°06′ |
| Magnitude | 11.8 |
| Size | 120″ |
| Type | Planetary nebula |

Our next object lies 1.3° south-southeast of magnitude 2.5 Kappa ($\kappa$) Velorum. Astronomers classify this object as a bipolar planetary nebula. What you'll see through a 12-inch telescope looks more like an evenly illuminated rectangle, 50% longer than it is wide. Its long axis stretches in an east-west direction.

When you're done observing NGC 2899, take a look at the large, magnitude 7.4 open cluster IC 2488. This object, dubbed the Hoopskirt Cluster or the String of Pearls Cluster, lies 0.9° to the south.

| OBJECT #117 | NGC 2903 |
|---|---|
| Constellation | Leo |
| Right ascension | 9h32m |
| Declination | 21°30′ |
| Magnitude | 9.0 |
| Size | 12.0′ by 5.6′ |
| Type | Spiral galaxy |

Leo the Lion contains many fine galaxies, five of which (M65, M66, M95, M96, and M105) made Messier's list. Don't overlook NGC 2903, however, which shines brighter than any Messier galaxy except M66. German-born English astronomer William Herschel (1738–1822) discovered NGC 2903 in 1784. I have no idea why this bright, easily seen celestial wonder lacks a common name.

Astronomers classify NGC 2903 as a "hotspot" galaxy, a term coined in the 1950s that describes a ring of infrared-luminous knots near a galaxy's core. The knots are hot star clusters only 6–9 million years old.

Although NGC 2903 is a bright galaxy, you won't see much detail through a telescope with an aperture smaller than 10′. Through a 10-inch scope, look for a halo 4′ by 2′ that surrounds a bright core. Close examination at high magnification reveals the galaxy's central bar and the spiral arms, which aren't much brighter than its halo. Through larger scopes, look for dust lanes and emission nebulae spread throughout NGC 2903's spiral arms.

A second deep-sky object — NGC 2905 — appears as a bright knot within NGC 2903. NGC 2905 is a star-forming region, which Herschel assigned a second designation. It lies slightly more than 1′ north-northeast of NGC 2903's core.

| OBJECT #118 | NGC 2964 |
|---|---|
| Constellation | Leo |
| Right ascension | 9h43m |
| Declination | 31°51′ |
| Magnitude | 11.2 |
| Size | 3′ by 1.7′ |
| Type | Spiral galaxy |

Our next target lies in far-northern Leo. Point your telescope 6.2° north-northwest of magnitude 3.9 Mu ($\mu$) Leonis, or 1.9° north of the magnitude 5.6 star 15 Leonis. An 8-inch scope reveals an evenly illuminated oval. A 14-inch at 350× lets you see the ultra-thin outer halo, which is our clue to this object's spiral structure.

You won't see the tightly wound spiral arms through anything less than a 24-inch instrument, and even through such an instrument they will appear only as stubs. The magnitude 11.9 spiral galaxy NGC 2968 lies 6′ to the northeast. For a real challenge, magnitude 14.6 NGC 2970 lies 11′ northeast of NGC 2964.

| OBJECT #119 | NGC 2974 |
|---|---|
| Constellation | Sextans |
| Right ascension | 9h43m |
| Declination | –3°42′ |
| Magnitude | 10.9 |
| Size | 3.4′ by 2.1′ |
| Type | Elliptical galaxy |

This object lies 2.6° south-southeast of magnitude 3.9 Iota ($\iota$) Hydrae. Probably the first thing you'll notice about NGC 2974 is the magnitude 9.4 star that sits less than 1′ southwest of the galaxy's core. It's a bit distracting. The galaxy has an oval shape and stretches 50% longer than it is wide in a northeast-to-southwest orientation. Through a 10-inch telescope at 250× or more, you can see the faint outer halo.

| OBJECT #120 | NGC 2976 |
|---|---|
| Constellation | Ursa Major |
| Right ascension | 9h47m |
| Declination | 67°55′ |
| Magnitude | 10.2 |
| Size | 5.0′ by 2.8′ |
| Type | Spiral galaxy |

Our next target is a member of the M81 galaxy group, which lies approximately 12 million light-years away. NGC 2976 lies 1.4° south-southwest of M81, and 2.2° south-southeast of the magnitude 4.5 star 24 Ursae Majoris. Through an 8-inch telescope, the galaxy's oval shape stretches twice as long as it is wide in a northwest-to-southeast orientation. Larger instruments show a mottled appearance with many tiny dark regions within the galaxy's disk.

| OBJECT #121 | NGC 2985 |
|---|---|
| Constellation | Ursa Major |
| Right ascension | 9h50m |
| Declination | 72°17′ |
| Magnitude | 10.4 |
| Size | 4.6′ by 3.4′ |
| Type | Spiral galaxy |

Look toward the northern part of the Great Bear for our next target, which lies 0.6° east of the magnitude 5.2 star 27 Ursae Majoris. The slightly oval shape is easy to see at magnifications above 150×. Through a 10-inch or larger telescope, you'll see the wide, faint outer halo. Look for the magnitude 11.5 spiral galaxy NGC 3027, which lies 0.4° east of NGC 2985.

| OBJECT #122 | NGC 2986 |
|---|---|
| Constellation | Hydra |
| Right ascension | 9h44m |
| Declination | –21°17′ |
| Magnitude | 10.7 |
| Size | 3.2′ by 2.6′ |
| Type | Elliptical galaxy |

This target lies 6.7° south-southwest of magnitude 4.1 Upsilon$^1$ ($\upsilon^1$) Hydrae. No amateur telescope reveals vast detail in this object, but what you will see is an evenly illuminated central region that spans more than three-quarters of the galaxy's total diameter. The thin outer halo is visible at high magnification through 10-inch and larger scopes. When you're through with NGC 2986, crank up the magnification and observe the magnitude 14.4 spiral PGC 27873, which lies 2′ west-southwest of NGC 2986.

| OBJECT #123 | NGC 2997 |
|---|---|
| Constellation | Antlia |
| Right ascension | 9h46m |
| Declination | −31°11′ |
| Magnitude | 9.3 |
| Size | 10.0′ by 6.3′ |
| Type | Spiral galaxy |

You'll find this target 3° east-northeast of magnitude 5.9 Zeta$^2$ ($\zeta^2$) Antliae. A 10-inch telescope shows an oval haze elongated east-west with a bright core. Several broken, dark spaces that mark the regions between spiral arms surround the core. The faint halo looks like a slightly brighter ring around the central region.

| OBJECT #124 | Upsilon ($\upsilon$) Carinae |
|---|---|
| Constellation | Carina |
| Right ascension | 9h47m |
| Declination | −65°04′ |
| Magnitudes | 3.1/6.1 |
| Separation | 5″ |
| Type | Double star |

You'll have no trouble splitting this binary through even a 2.4-inch scope. Both components appear white.

**Object #125** Bode's Galaxy (M81) Stefan Seip/Adam Block/NOAO/AURA/NSF

| OBJECT #125 | M81 (NGC 3031) |
|---|---|
| Constellation | Ursa Major |
| Right ascension | 9h56m |
| Declination | 69°04′ |
| Magnitude | 6.9 |
| Size | 24.0′ by 13.0′ |
| Type | Spiral galaxy |
| Other name | Bode's Galaxy |

In the northwest section of Ursa Major the Great Bear sits M81, one of the sky's brightest galaxies. You'll find it 2° east-southeast of the magnitude 4.5 star 24 Ursae Majoris.

German astronomer and celestial cartographer Johann Elert Bode (1747–1826) discovered this object, and nearby irregular galaxy M82, December 31, 1774. French astronomer Pierre Francois André Méchain (1744–1804) independently discovered both galaxies in August 1779 and reported them to Messier, who added them to his list.

Bode's Galaxy glows brightly enough to show up through binoculars, but the larger the telescope you can point at it, the better. Through an 8-inch scope, you'll see a large, bright central region surrounding the much brighter core. Through a 12-inch instrument, you'll detect how the spiral arms wind tightly around the core. The easternmost appears brighter. Unfortunately, you won't detect any dust lanes or star-forming regions through amateur scopes of any size.

M81 is the brightest member of the M81 Group, one of the closest galactic groups to our own Local Group (see Object #245). The M81 Group contains about a dozen galaxies and lies 12 million light-years away. Other members of this group include M82, NGC 2403, NGC 2366, and NGC 3077.

**Object #126** The Cigar Galaxy (M82) Adam Block/NOAO/AURA/NSF

| OBJECT #126 | M82 (NGC 3034) |
|---|---|
| Constellation | Ursa Major |
| Right ascension | 9h56m |
| Declination | 69°41′ |
| Magnitude | 8.4 |

| (continued) | |
|---|---|
| Size | 12.0′ by 5.6′ |
| Type | Starburst galaxy |
| Other name | The Cigar Galaxy |

Point your telescope 37′ due south of Bode's Galaxy (M81), and you might think you're seeing a galaxy explode. Not exactly. What you've found is M82, the classic example of a starburst galaxy. Its core is a complex of star-forming regions that dwarfs our Milky Way's Orion Nebula.

M82's appearance is due to a close interaction with M81 between 500 million and 600 million years ago. Radio telescope maps show a great deal of gas surrounding both objects.

Amateur astronomers call M82 the Cigar Galaxy because of its appearance through a small telescope. It appears 4 times as long as wide, and its long axis orients east-southeast to west-northwest. The galaxy's brightest part lies east of center. Farther east, a dark lane cuts diagonally across M82's minor axis (the short dimension).

M82 has a greater surface brightness than most galaxies. To illustrate what this means, compare M82 to M81 in an eyepiece that just frames them both. Although M82 shines a magnitude and a half fainter than M81, it appears about as bright. Use high magnification to reveal its details, even on less-than-perfect nights. This technique doesn't work on low-surface-brightness galaxies like M101.

| OBJECT #127 | Sextans B |
|---|---|
| Constellation | Sextans |
| Right ascension | 10h00m |
| Declination | 5°20′ |
| Magnitude | 11.3 |
| Size | 5.5′ by 3.7′ |
| Type | Irregular galaxy |

Sextans B is a member of our Local Group. At 4.5 million light-years, it sits right at the outer edge of the small collection of galaxies that contains our Milky Way. Through a 12-inch telescope, you'll see a rectangular smudge of light dotted here and there by faint foreground stars. The central region, about one-third the galaxy's diameter, is ever-so-slightly brighter than the outer regions.

To find this galaxy, look 6° north-northwest of magnitude 4.5 Alpha (α) Sextantis.

By the way, the name "Sextans B" is an example of a radioastronomical name. Such a label combines a constellation's name and an uppercase letter. Letters start at A and descend in order of brightness in radio wavelengths for such galaxies within a constellation. Astronomers John G. Bolton and Gordon J. Stanley introduced this system around 1950.

| OBJECT #128 | NGC 3077 |
|---|---|
| Constellation | Ursa Major |
| Right ascension | 10h03m |
| Declination | 68°44′ |
| Magnitude | 9.8 |
| Size | 5.5′ by 4.1′ |
| Type | Irregular galaxy |

This galaxy lies 0.8° east-southeast of Bode's Galaxy (M81) and 1.1° southeast of the Cigar Galaxy. If, therefore, your telescope/eyepiece combination gives a bit more than a 1° field of view, you'll see all three galaxies at once, albeit at low power.

This galaxy has a wide, bright nucleus. It's an oval with faint outer layers that I had no trouble seeing. I was expecting the often-described ray to be more of an extension to the northeast. What I have observed is short, thick, and quite faint.

NGC 3077 may be a satellite galaxy of M81. The two objects lie only 200,000 light-years away from each other.

Two 8th-magnitude stars lie near NGC 3077. One, less than 4′ away, lies in the direction of M82. The other, 10′ distant, lies just off a line that connects NGC 3077 with M81.

**Object #129** NGC 3079 Jeff Hapeman/Adam Block/NOAO/AURA/NSF

| OBJECT #129 | NGC 3079 |
|---|---|
| Constellation | Ursa Major |
| Right ascension | 10h02m |
| Declination | 55°41′ |
| Magnitude | 10.9 |
| Size | 8.0′ by 1.5′ |
| Type | Barred spiral galaxy |

You'll find this object 2.2° northeast of magnitude 4.6 Phi ($\phi$) Ursae Majoris. Yet another of the sky's galactic "splinters," NGC 3079 appears five times as long as it is wide. Through a 12-inch telescope at 300×, you'll see that the bright central region stretches nearly two-thirds of the galaxy's total length. If your sky is ultra-steady, you may spot the thin extensions that mark the spiral arms.

A nice triangle of relatively bright stars lies near the galaxy's southern end. The brightest, magnitude 7.9 SAO 27476, lies 6′ south-southwest, magnitude 9.6 SAO 27482 sits less than 4′ southeast, and magnitude 9.5 SAO 27486 lies 7′ south-southeast. Finally, you'll find the magnitude 13.0 elliptical galaxy NGC 3073 only 10′ to the west-southwest of NGC 3079.

| OBJECT #130 | NGC 3109 |
|---|---|
| Constellation | Hydra |
| Right ascension | 10h03m |
| Declination | −26°09′ |
| Magnitude | 9.8 |
| Size | 16.0′ by 2.9′ |
| Type | Barred spiral galaxy |

This target lies 7.2° northwest of magnitude 4.3 Alpha ($\alpha$) Antliae. This nearby barred spiral galaxy belongs to the Local Group. Through a 6-inch telescope, NGC 3109 appears as a faint mist measuring three to four times longer than it is wide, but you'll need to be under a really dark sky to see it through that aperture. A 16-inch scope, on the other hand, reveals a surface punctuated with knots of unresolved stars and clouds of ionized hydrogen.

| OBJECT #131 | NGC 3114 |
|---|---|
| Constellation | Carina |
| Right ascension | 10h03m |
| Declination | −60°07′ |
| Magnitude | 4.2 |
| Size | 35′ |
| Type | Open cluster |

Get ready for a glorious view. This terrific cluster lies within a spectacular star field. Specifically, you'll find it 5.8° east-southeast of magnitude 2.2 Aspidiske (Iota [$\iota$] Carinae). Although NGC 3114 is visible without optical aid, scan the area with binoculars for the best view.

Through a 4-inch telescope, you'll first spot two bright stars in the cluster's area, magnitude 6.2 SAO 237640 and magnitude 7.3 SAO 237655. Surrounding this pair are dozens of similarly bright stars you'll form into a variety of patterns. Larger scopes add another layer of faint stars to the background.

While you're in the area, be sure to take a look at the magnitude 8.8 open cluster Trumpler 12. It lies 0.5° east-southeast of NGC 3114.

| OBJECT #132 | NGC 3115 |
|---|---|
| Constellation | Sextans |
| Right ascension | 10h05m |
| Declination | −7°43′ |

| (continued) | |
|---|---|
| Magnitude | 8.9 |
| Size | 8.1′ by 2.8′ |
| Type | Lenticular galaxy |
| Common names | The Spindle Galaxy, Caldwell 53 |

Not only is the Spindle Galaxy the showpiece of Sextans, it's one of the sky's brightest galaxies. In fact, observers can hardly believe Charles Messier failed to include it in his catalog. It is the prototype S0 galaxy, a class that American astronomer Edwin Hubble used to bridge the gap between the flattest ellipticals and spirals. Such objects have a large central bulge and long extensions (like the type of spindle that holds thread or yarn) but no spiral arms.

NGC 3115 is so bright that you can spot it through binoculars or a finder scope. Through a 4-inch telescope, you'll see an object four times as long as it is wide with a bright center. Through a 12-inch scope at 300×, the central bulge appears more distinct, an oval bulge surrounds the center, and a faint oval glow surrounds the entire object.

You'll find the Spindle Galaxy 3.2° east of magnitude 5.1 Gamma (γ) Sextantis.

| OBJECT #133 | Alpha (α) Leonis |
|---|---|
| Constellation | Leo |
| Right ascension | 10h08m |
| Declination | 11°58′ |
| Magnitudes | 1.3/8.1/13.5 |
| Separation | 177″ and 4.2″ |
| Type | Double star |
| Other name | Regulus |

This triple star features magnitude 1.3 Regulus A and the much fainter pair B–C. B shines at magnitude 8.1, and C glows faintly at magnitude 13.5. Stars B and C lie 4.2″ apart, and the A component lies 177″ away from them. I've seen all three components through a 10-inch telescope from a dark site.

| OBJECT #134 | NGC 3132 |
|---|---|
| Constellation | Vela |
| Right ascension | 10h08m |
| Declination | −40°26′ |
| Magnitude | 9.7 |
| Size | 30″ |
| Type | Planetary nebula |
| Other names | The Eight Burst Nebula, the Southern Ring Nebula, Caldwell 74 |

Ten thousand years ago, a Sun-like star in the southern constellation Vela the Sails reached the end of its life and expelled its outer layers into space. What remained of the star's core shrank to the size of Earth, its surface heated up to 180,000°F (100,000°C), and it began emitting ultraviolet radiation. Scientists call such stellar corpses white dwarfs. The expanding shell of gas illuminated by the white dwarf's radiation formed a planetary nebula.

When you observe NGC 3132, you'll find an object about the same size and dimensions as the Ring Nebula (Object #567), and from this likeness comes one of the object's common names. NGC 3132's structure, however, is more complex. Several ovals appear superimposed and tilted at different angles. The nebula's outer boundary also looks more irregular, and the region around the central star contains more material than in the Ring Nebula's case.

NGC 3132 has a high surface brightness, so it responds well to high magnifications (in excess of 250×). A diffuse, irregularly bright inner shell surrounds the imposter central star. The brightest gaseous region forms the nebula's border.

Finally, the other common name for NGC 3132, the Eight-Burst Nebula (or Eight-Burst Planetary), comes from its complex structure as seen on photographs. That assessment comes from American astronomer Robert Burnham, Jr. (1931–1993) in *Burnham's Celestial Handbook: Volume Three, Pavo through Vulpecula* (Dover, 1978).

**Object #135** NGC 3147 Alex and Mike Beck/Adam Block/NOAO/AURA/NSF

| OBJECT #135 | NGC 3147 |
|---|---|
| Constellation | Draco |
| Right ascension | 10h17m |
| Declination | 73°24′ |
| Magnitude | 10.6 |
| Size | 4.3′ by 3.7′ |
| Type | Spiral galaxy |

Our next target is an attractive face-on spiral galaxy. NGC 3147 has a bright, small core surrounded by a circular haze. Although you won't see spiral structure directly, the aspect of the core and halo are telltale signs that this is a spiral galaxy. You'll find NGC 3147 slightly more than 7° northwest of magnitude 3.8 Lambda (λ) Draconis.

| OBJECT #136 | Leo I |
|---|---|
| Constellation | Leo |
| Right ascension | 10h09m |
| Declination | 12°18′ |
| Magnitude | 10.2 |
| Size | 12.0′ by 9.3′ |
| Type | Dwarf spheroidal galaxy |

Our next target is a personal favorite of mine. It demonstrates that a celestial object can be easy to find, but really difficult to observe details in. What makes it easy to find is its location: It lies only 20′ due north of magnitude 1.3 Regulus (Alpha [α] Leonis). But that brilliant star's glare through the eyepiece also makes Leo I difficult to see.

At a dark site, an 8-inch telescope at 150× reveals a faint mist that appears uniformly bright. Whatever you do, keep Regulus out of the field of view.

American astronomers Robert G. Harrington and Albert George Wilson discovered Leo I in 1950. They were searching photographic plates taken during the National Geographic Society Palomar Sky Survey using the 48-inch Samuel Oschin Schmidt Telescope.

**Object #137** NGC 3169 (upper right) and NGC 3166 Adam Block/NOAO/AURA/NSF

| OBJECT #137 | NGC 3169 |
|---|---|
| Constellation | Sextans |
| Right ascension | 10h14m |
| Declination | 3°28′ |
| Magnitude | 10.2 |
| Size | 5.0′ by 2.8′ |
| Type | Spiral galaxy |

Our next target is a two-in-one. It combines NGC 3169 with the magnitude 10.5 spiral NGC 3166, which lies only 8′ to the west-southwest. The pair forms a non-interacting quartet with two more spirals, magnitude 13.9 NGC 3165 and magnitude 12.1 NGC 3156.

Through an 8-inch telescope, NGC 3169 appears about twice as long as it is wide oriented northeast to southwest. The central region, also elongated, is much brighter than the halo.

Magnitude 10.5 NGC 3166 appears ever-so-slightly oval (4.8′ by 2.3′), stretched in an east-west direction. Its central region is wide, and its halo is thin.

| OBJECT #138 | NGC 3172 |
|---|---|
| Constellation | Ursa Minor |
| Right ascension | 11h47m |
| Declination | 89°06′ |
| Magnitude | 13.8 |
| Size | 1.1′ by 0.8′ |
| Type | Barred spiral galaxy |
| Other name | Polarissima Borealis |

This next deep-sky object is a tough one: spiral galaxy NGC 3172 in Ursa Minor. Because it glows dimly at magnitude 13.6 (and some estimates put it fainter than magnitude 14) you'll need at least a 10-inch scope and a really dark sky to spot it.

But if you live north of the equator, you do have an advantage. You can search for NGC 3172 any night of the year. That's because it lies only 1.5° from Polaris (Alpha [α] Ursae Minoris) — the North Star. In fact, it's the closest NGC object to the North Celestial Pole — only 0.9° from that heavenly marker. Because of its extreme northern position that keeps it above the horizon for just about the whole Northern Hemisphere, astronomers have dubbed NGC 3172 Polarissima Borealis.

Visually, this isn't the Whirlpool Galaxy (M51). In fact, it's pretty uninteresting apart from its location. If you spot it, you'll see a faint oval haze about 0.5′ across with a slightly brighter center.

| OBJECT #139 | NGC 3175 |
|---|---|
| Constellation | Antlia |
| Right ascension | 10h15m |
| Declination | –28°52′ |
| Magnitude | 11.3 |
| Size | 5.0′ by 1.3′ |
| Type | Spiral galaxy |

This object lies 3.5° northwest of magnitude 4.3 Alpha (α) Antliae. The disk-shaped object appears three times as long as it is wide. Through telescopes smaller than those with a 14-inch aperture, all you'll see is a uniformly bright splinter. Above that size, look for an irregularly bright halo, especially at the galaxy's ends.

| OBJECT #140 | NGC 3183 |
|---|---|
| Constellation | Draco |
| Right ascension | 10h22m |
| Declination | 74°11′ |
| Magnitude | 11.8 |
| Size | 2.3′ by 1.4′ |
| Type | Spiral galaxy |

You'll find our next target 7.2° northwest of magnitude 3.8 Lambda (λ) Draconis. Through a 10-inch telescope, it appears roughly twice as long as it is wide with an even illumination. Bigger scopes and higher powers don't reveal all that much more. Interestingly, a tight group of four 14th-magnitude stars lie within about 1′ of the northern edge.

| OBJECT #141 | NGC 3184 |
|---|---|
| Constellation | Ursa Major |
| Right ascension | 10h18m |

| (continued) | |
|---|---|
| Declination | 41°25′ |
| Magnitude | 9.8 |
| Size | 7.8′ by 7.2′ |
| Type | Spiral galaxy |
| Other name | The Little Pinwheel Galaxy |

Oh my! What a gorgeous galaxy through a large telescope. This face-on spiral reminds me of another, similar object in Ursa Major — M101 (Object #351). And, as its common name implies, many observers see similarities between it and the Pinwheel Galaxy (Object #799). NGC 3184's spiral arms are wide, so you'll have to use high power — past 400× — to spot the dark regions that divide them from the galaxy's nucleus. The magnitude 11.6 foreground star (GSC 3004:998) looks like a supernova blowing its top at the north end of NGC 3184.

NGC 3184 sits at the Great Bear's border with Leo Minor. If you're star-hopping, start at magnitude 3.0 Mu ($\mu$) Ursae Majoris, and move 0.8° west. Through a more modest scope, say one with a 6-inch aperture, you'll see a circular haze with a slightly brighter center.

| OBJECT #142 | NGC 3190 |
|---|---|
| Constellation | Leo |
| Right ascension | 10h18m |
| Declination | 21°50′ |
| Magnitude | 11.2 |
| Size | 4.1′ by 1.6′ |
| Type | Spiral galaxy |
| Other name | The Gamma Leonis Group |

You'll find this nice galaxy, and three others, 2° north-northwest of magnitude 2.0 Algieba (Gamma [$\gamma$] Leonis). Known as Hickson 44, the group is the brightest in Canadian astronomer Paul Hickson's catalog of 100 compact galaxy groups. The most famous entry in his catalog is number 92, Stephan's Quintet (Object #710).

NGC 3190, the largest galaxy in Hickson 44, appears oval, three times as long as it is wide. Its central region is long and bright. Through a 12-inch telescope at 250×, you'll see the dust lane south of the nucleus. The lane is least apparent near the nucleus. It broadens in both directions as you look away from the center.

The brightest galaxy in Hickson 44 is our next object, NGC 3193. The magnitude 12.0 barred spiral NGC 3185 resides 11′ southwest of NGC 3190. Finally, the magnitude 12.9 spiral NGC 3187 sits 5′ to the west-northwest of NGC 3190. Even through a large telescope, you won't pull much detail out of either of these two objects.

| OBJECT #143 | NGC 3193 |
|---|---|
| Constellation | Leo |
| Right ascension | 10h18m |
| Declination | 21°54′ |
| Magnitude | 10.8 |
| Size | 2′ by 2′ |
| Type | Elliptical galaxy |

The compact galaxy group Hickson 44, which I talked about with NGC 3190 (Object #142), also contains NGC 3193, which lies 6′ to the northeast of NGC 3190. Through an 8-inch telescope, you'll see that NGC 3193 contains a broad, evenly illuminated central region and a thin halo.

| OBJECT #144 | NGC 3195 |
|---|---|
| Constellation | Chamaeleon |
| Right ascension | 10h09m |
| Declination | −80°52′ |
| Magnitude | 11.6 |
| Size | 38″ |
| Type | Planetary nebula |
| Other name | Caldwell 109 |

This high surface-brightness object lies 1.5° west-southwest of magnitude 4.5 Delta$^2$ ($\delta^2$) Chamaeleontis. A 4-inch telescope at 100× will reveal this object as a slightly fat "star." Through a 10-inch scope, crank the power beyond 200×, and you'll have no problem seeing the planetary nature of NGC 3195. At this magnification, the planetary appears slightly stretched in a north-northeast to south-southwest orientation.

| OBJECT #145 | Sextans A |
|---|---|
| Constellation | Sextans |
| Right ascension | 10h11m |
| Declination | −4°42′ |
| Magnitude | 11.5 |
| Size | 5.9′ by 4.9′ |
| Type | Irregular galaxy |

Our next target is a member of our Local Group of galaxies. At 4.3 million light-years away, it sits on the outskirts of this assembly. You'll find it 5° west-southwest of magnitude 5.2 Delta ($\delta$) Sextantis.

Through an 8-inch or larger telescope this odd-looking galaxy appears square. If you use a 14-inch telescope to observe this object, use a magnification of 300× and a nebula filter, and try to spot the brightest feature, an HII region at the southeastern edge.

For a discussion of the name "Sextans A," see Sextans B (Object #127).

| OBJECT #146 | Sextans Dwarf |
|---|---|
| Constellation | Sextans |
| Right ascension | 10h13m |
| Declination | −1°36′ |
| Magnitude | 12.0 |
| Size | 26.9′ by 5.9′ |
| Type | Galaxy |

You'll find this faint target 1.8° southeast of magnitude 4.5 Alpha ($\alpha$) Sextantis. Astronomers discovered it in 1990 during a photographic survey. It lies only 320,000 light-years away, but it's so large that is has a surface brightness barely above that of the night sky. Through at least a 16-inch telescope, use your lowest-power eyepiece and slowly scan the area looking for a tiny increase in the background glow.

**Object #147** NGC 3198 John Vickery and Jim Matthes/Adam Block/NOAO/AURA/NSF

| OBJECT #147 | NGC 3198 |
|---|---|
| Constellation | Ursa Major |
| Right ascension | 10h20m |
| Declination | 45°33′ |
| Magnitude | 10.2 |
| Size | 8.5′ by 3.3′ |
| Type | Spiral galaxy |

Our next object lies 2.7° north of magnitude 3.5 Tania Borealis (Lambda [$\lambda$] Ursae Majoris). Through a 6-inch telescope, you'll see an irregularly illuminated oval that appears more than twice as long as it is wide oriented northeast to southwest. A 14-inch scope at 300× shows lots of detail. A small, bright central region lies within an irregular halo that looks like truncated spiral arms. Only 2′ north of the northeastern tip lies the magnitude 11.2 star GSC 3435:470.

| OBJECT #148 | NGC 3199 |
|---|---|
| Constellation | Carina |
| Right ascension | 10h17m |
| Declination | –57°55′ |
| Size | 20′ by 15′ |
| Type | Emission nebula |

Look for our next target 4.4° southeast of magnitude 3.5 Phi ($\phi$) Velorum. Through most telescopes, this nebula looks like a large, thick crescent that opens toward the east. Use a nebula filter for the best view. Only the largest amateur instruments show some of the additional nebulosity that fills in a roughly circular shape. Near the crescent's center lies a Wolf-Rayet star (HD 89358), a massive, hot star that generates lots of ultraviolet radiation and an intense stellar wind. In this case, the Wolf-Rayet star is sculpting out the gas surrounding it.

| OBJECT #149 | NGC 3201 |
|---|---|
| Constellation | Vela |
| Right ascension | 10h18m |
| Declination | −46°25′ |
| Magnitude | 6.8 |
| Size | 18.2′ |
| Type | Globular cluster |
| Other name | Caldwell 79 |

Our next target is one of the southern sky's great globular clusters. It lies 5.7° west-northwest of magnitude 2.7 Mu ($\mu$) Velorum. From a true-dark site, a sharp-eyed observer has an even chance to spot this object without optical aid. And although you'll see it through binoculars or a finder scope, set up your telescope to marvel at this swarm of suns. You won't be disappointed.

Through a 4-inch telescope at 100×, many unresolvable stars fill the brilliant core. Crank up the power to view a few outlier stars. Through an 11-inch scope, you'll count more than 100 stars — a lot more. Also note a moderately dark but shallow V-shaped indentation that encroaches on the southern side.

| OBJECT #150 | Gamma ($\gamma$) Leonis |
|---|---|
| Constellation | Leo |
| Right ascension | 10h20m |
| Declination | 19°50′ |
| Magnitudes | 2.3/3.5 |
| Separation | 4.4″ |
| Type | Double star |
| Other name | Algieba |

Algieba makes a nice sight through small telescopes. Both stars are yellow giants, so the difference between the two only depends on their brightnesses. The primary outshines the secondary by a factor of 3.

This star's name may come from the Arabic *al Jabbah*, meaning "the forehead." Al Jabbah was the 8th lunar mansion in ancient Arabia. Another possibility is that it originally comes from the Latin word *juba*, which means "lion's mane." "Algieba," therefore, would be an Arabicized version.

| OBJECT #151 | NGC 3227 |
|---|---|
| Constellation | Leo |
| Right ascension | 10h24m |
| Declination | 19°52′ |
| Magnitude | 10.3 |
| Size | 6.9′ by 5.4′ |
| Type | Spiral galaxy |

You'll find this object, and its companion, the magnitude 11.4 elliptical galaxy NGC 3226, 50′ east of magnitude 2.0 Gamma ($\gamma$) Leonis. The elliptical appears to be attached to the northern end of the brighter galaxy. Through 6-inch and larger telescopes, NGC 3227 appears oval shaped with a long, concentrated central region. Its outer halo has an abrupt edge. NGC 3226 appears circular with a broad central region punctuated by a tiny, dim core.

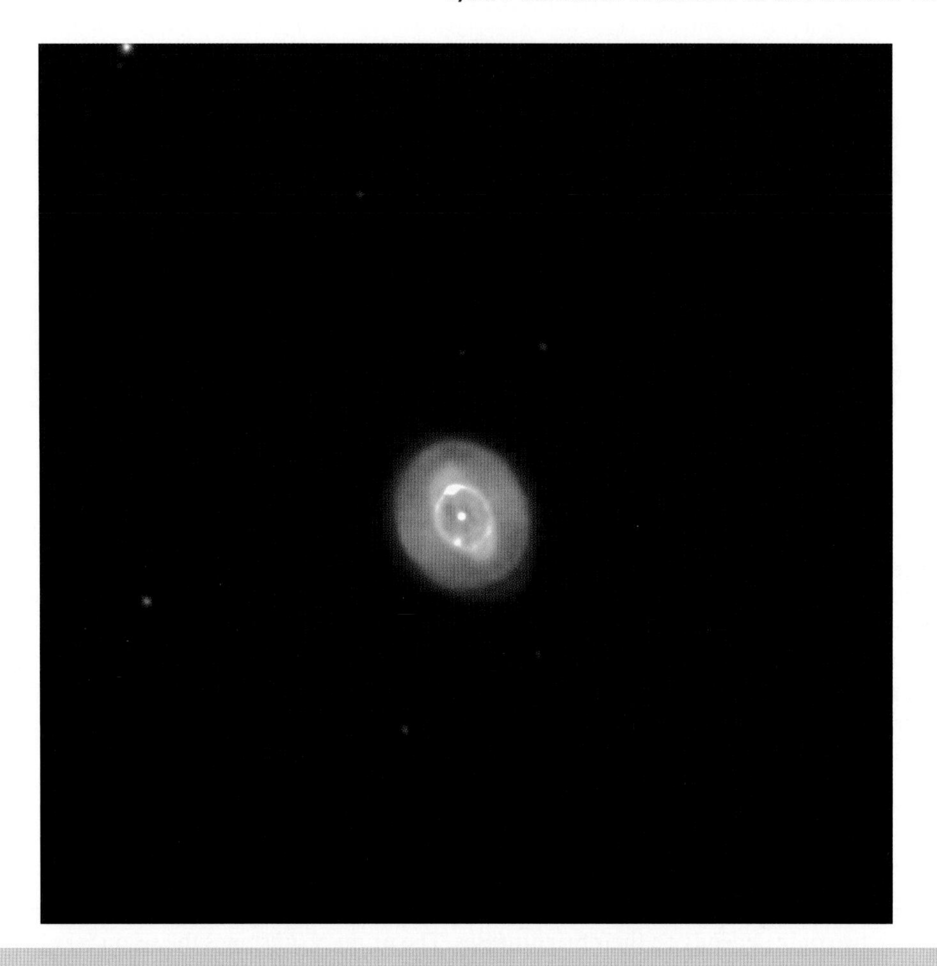

**Object #152** The Ghost of Jupiter (NGC 3242) Adam Block/NOAO/AURA/NSF

| OBJECT #152 | NGC 3242 |
|---|---|
| Constellation | Hydra |
| Right ascension | 10h25m |
| Declination | −18°38′ |
| Magnitude | 7.8 |
| Size | 16″ |
| Type | Planetary nebula |
| Other names | The Ghost of Jupiter, the CBS Eye, Caldwell 59 |

In Hydra's east-central section, you'll find NGC 3242, the spring sky's showpiece planetary nebula. This object received its common name, "Ghost of Jupiter," from its planet-like appearance through small telescopes, although based on its color and brightness, it mostly resembles Uranus or Neptune. Sir William Herschel discovered NGC 3242 February 7, 1785.

At low magnification through a 6-inch telescope, you'll see the Ghost's softly glowing, pale, blue-green disk. Through larger scopes, and at powers in excess of 200×, the interior appears oval, like an eye or a football. The inner 10″ appears hollow, except for the dim central star. A faint spherical shell 40″ across encloses the "eye." Your best bet for observing the outer shell is to use a 12-inch or larger scope, an eyepiece that provides about 100×, and a nebula filter.

Much of the Ghost's blue-green color comes from oxygen atoms absorbing ultraviolet radiation from the central star and reradiating it as visible light. Because this emission is strong, an OIII filter works best for observing NGC 3242. "OIII" stands for "doubly ionized oxygen." Chemists use "III" to represent double ionization because they label neutral oxygen OI. Nebula filters such as OIII screw directly into the threaded barrels of 1 1/4″ or 2″ eyepieces.

| OBJECT #153 | IC 2574 |
|---|---|
| Constellation | Ursa Major |
| Right ascension | 10h28m |
| Declination | 68°25′ |
| Magnitude | 10.4 |
| Size | 13.5′ by 8.3′ |
| Type | Spiral galaxy |
| Other name | Coddington's Nebula |

Although this object's popular name includes the word "nebula," it is, indeed, a galaxy. American astronomer Edwin Foster Coddington (1870–1950) discovered it April 17, 1898 while at Lick Observatory.

IC 2574 lies 5.7° west of magnitude 3.8 Lambda (λ) Draconis. Through an 8-inch telescope at 75×, it appears twice as long as it is wide oriented northeast to southwest. Because it's so huge, most of this galaxy has a low surface brightness. The central region glows a bit brighter, but seems offset to the southwest. In fact, most of the galaxy lies northeast of the wide core. Through a 12-inch or larger telescope, insert an eyepiece that magnifies 250× and a nebula filter, and view the star-forming region at the galaxy's northeastern tip.

| OBJECT #154 | NGC 3245 |
|---|---|
| Constellation | Leo Minor |
| Right ascension | 10h27m |
| Declination | 28°31′ |
| Magnitude | 10.7 |
| Size | 3.2′ by 1.8′ |
| Type | Spiral galaxy |

Our next target lies 5.6° north-northeast of magnitude 3.4 Zeta (ζ) Leonis. Through an 8-inch telescope, this lens-shaped galaxy appears nearly twice as long as it is wide, with a north-south orientation. The broad, bright central region takes up most of the galaxy's area. At high powers, and through larger apertures, you might just spot the thin halo.

| OBJECT #155 | NGC 3310 |
|---|---|
| Constellation | Ursa Major |
| Right ascension | 10h39m |
| Declination | 53°30′ |
| Magnitude | 10.6 |
| Size | 3.1′ by 2.4′ |
| Type | Spiral galaxy |

This object sits 4.4° southwest of magnitude 2.3 Merak (Beta [β] Ursae Majoris). Through an 8-inch telescope at 200×, this object appears slightly elongated in a north-northwest to south-southeast orientation. The galaxy's face has even illumination, except at the northern and southern ends, where you can detect the beginnings of wide spiral arms. Orange magnitude 5.6 SAO 27724 sits 10′ to the north-northeast. You'll want to keep it out of the field of view when you observe NGC 3310.

| OBJECT #156 | NGC 3311 |
|---|---|
| Constellation | Hydra |
| Right ascension | 10h37m |
| Declination | –27°32′ |
| Magnitude | 10.9 |
| Size | 4.0′ by 3.6′ |
| Type | Spiral galaxy |

Our next object, NGC 3311, is part of the Hydra I Galaxy Cluster. Look for it 4.1° north-northeast of magnitude 4.3 Alpha ($\alpha$) Antliae. The Hydra I Galaxy Cluster lies three times as far away as the Virgo Cluster, which it resembles.

Through at least a 12-inch telescope, this small region of sky teems with faint galaxies. But before you notice those, you'll spot two stars, magnitude 4.9 SAO 179041 and magnitude 7.3 GSC 6641:1410. Around these stars, you'll spot NGC 3307, NGC 3308, NGC 3309, NGC 3311, NGC 3312, NGC 3314, and NGC 3316. NGC 3311 is the brightest, but it appears as just a circular haze with a bright center.

Magnitude 12.8 NGC 3312 lies 5′ east-southeast of NGC 3311 and is the most visually interesting, but you'll need at least a 16-inch telescope to resolve its spiral arms, arching to the north and south.

NGC 3309 is another large elliptical that appears to be in contact with the western edge of NGC 3311. It looks similar but a bit smaller than its neighbor.

| OBJECT #157 | x Velorum |
|---|---|
| Constellation | Vela |
| Right ascension | 10h39m |
| Declination | –55°36′ |
| Magnitudes | 4.4/6.6 |
| Separation | 52″ |
| Type | Double star |

If you want to view a southern counterpart to Albireo (Beta [$\beta$] Cygni), this star is it. Either binoculars or a finder scope can split this binary, but a telescope/eyepiece combination that yields 50× seems ideal. The primary shines with a golden hue, and the secondary is blue.

| OBJECT #158 | IC 2602 |
|---|---|
| Constellation | Carina |
| Right ascension | 10h43m |
| Declination | −64°24′ |
| Magnitude | 1.9 |
| Size | 50′ |
| Type | Open cluster |
| Other names | The Southern Pleiades, the Theta Carinae Cluster, Caldwell 102 |

March's first object is nothing less than dazzling. You'll find it surrounding the bluish magnitude 2.7 Theta ($\theta$) Carinae, so it's sometimes called the Theta Carinae Cluster. More commonly, however, observers call it the Southern Pleiades because its discoverer, French astronomer Nicolas Louis de Lacaille, compared it to M45. Lacaille encountered IC 2602 March 3, 1752.

As with the actual Pleiades, this object usually looks better through binoculars because most telescopes spread out the stars too much. That said, if you own a short-focal-length refractor and an eyepiece that will provide at least a 1.5° field of view, you're in for a wonderful experience.

At low power, the Southern Pleiades appears like two clusters separated by a 0.3° gulf. The western half includes Theta Carinae and two arcs of stars that originate at Theta. One curves northward and the other southward. The eastern half of IC 2602 looks to me like a miniature version of the main part of the constellation Orion — with different brightnesses for the stars.

When you're satiated with the Southern Pleiades, nudge your telescope 0.7° south. There you'll find the magnitude 8 open cluster Melotte 101. A magnification around 75× is best for this object.

M.E. Bakich, *1,001 Celestial Wonders to See Before You Die*, Patrick Moore's Practical Astronomy Series,
DOI 10.1007/978-1-4419-1777-5_3, 

| OBJECT #159 | NGC 3338 |
|---|---|
| Constellation | Leo |
| Right ascension | 10h42m |
| Declination | 13°44′ |
| Magnitude | 10.9 |
| Size | 5.7′ by 3.4′ |
| Type | Spiral galaxy |

Our next target lies 7.9° west-southwest of magnitude 3.3 Chertan (Theta [θ] Leonis), or 1.1° west-southwest of the magnitude 5.5 star 52 Leonis. A small telescope shows only an evenly illuminated oval roughly 50% longer than it is wide stretching east-west. Through at least a 12-inch scope, however, you'll see several zones of differing brightnesses: the faint outer halo; then the slightly brighter central region (or inner halo); and finally the stellar nucleus. The magnitude 9.0 star SAO 99253 lies not quite 3′ west of NGC 3338.

| OBJECT #160 | NGC 3344 |
|---|---|
| Constellation | Leo Minor |
| Right ascension | 10h44m |
| Declination | 24°55′ |
| Magnitude | 9.9 |
| Size | 6.9′ by 6.4′ |
| Type | Spiral galaxy |
| Other name | The Sliced Onion Galaxy |

Look for this object 6.3° east-northeast of magnitude 3.4 Zeta (ζ) Leonis. An 8-inch telescope shows a central region that appears bright and concentrated. Double the aperture, and you'll marvel at this classic, face-on spiral. Its many arms wind tightly around the core, making the galaxy appear circular.

Two field stars lie superposed on the eastern half of the galaxy. The brightest, and farthest from the center, is magnitude 10.2 GSC 1977:2634. The other star lies closer in and glows at magnitude 11.5.

*Astronomy* magazine Contributing Editor Stephen James O'Meara dubbed NGC 3344 the Sliced Onion Galaxy, but not because of its face-on appearance. He likened observing this galaxy to peeling away the skin of an onion, with magnification as your carving tool.

**Object #161** M95 Adam Block/Mount Lemmon SkyCenter/University of Arizona

| OBJECT #161 | M95 (NGC 3351) |
|---|---|
| Constellation | Leo |
| Right ascension | 10h44m |
| Declination | 11°42′ |
| Magnitude | 9.7 |
| Size | 7.8′ by 4.6′ |
| Type | Barred spiral galaxy |

This terrific galaxy lies 3.6° northeast of magnitude 3.9 Rho ($\rho$) Leonis. Through an 8-inch telescope, M95 appears slightly elongated in a north-northeast to south-southwest orientation. The central region appears bright. A faint, outer ring surrounds the core. The ring is the brightest parts of M95's spiral arms. Through a 16-inch or larger telescope at high magnification, you'll immediately recognize this object's bar, which stretches east-west from one side of the ring to the other.

M95 belongs to the M96 Group of galaxies, sometimes called the Leo I Group, which lies at a mean distance of 38 million light-years. Don't associate the name with the dwarf galaxy Leo I (Object #136). That object has nothing to do with this group. Nine galaxies make up the Leo I Group. Along with M95 and M96 (Object #163), M105 (Object #166), NGC 3384 (Object #168), and NGC 3377 (Object #165) are the brightest members.

**Object #162** NGC 3359 Svend and Carl Freytag/Adam Block/NOAO/AURA/NSF

| OBJECT #162 | NGC 3359 |
|---|---|
| Constellation | Ursa Major |
| Right ascension | 10h46m |
| Declination | 63°13′ |
| Magnitude | 10.3 |
| Size | 7.2′ by 4.4′ |
| Type | Spiral galaxy |

Our next target lies 2.5° northwest of magnitude 1.8 Dubhe (Alpha [α] Ursae Majoris). This bright galaxy takes magnification well. Through a 10-inch telescope at 150×, you'll immediately notice the bar that runs north-south. The bar is brighter at its center and fades toward the ends. A 16-inch scope and 300× shows the beginnings of two spiral arms emanating from the bar's ends. The southern one curves to the east, while the northern one trends westward.

**Object #163** M96 Adam Block/NOAO/AURA/NSF

| OBJECT #163 | M96 (NGC 3368) |
|---|---|
| Constellation | Leo |
| Right ascension | 10h47m |
| Declination | 11°49′ |
| Magnitude | 9.2 |
| Size | 6.9′ by 4.6′ |
| Type | Spiral galaxy |

M96, the traveling companion of Object #161 (M95) lies only 0.7° east of it. Alternatively, you can find it 4.3° northeast of magnitude 3.9 Rho ($\rho$) Leonis. Through a 4-inch telescope, you'll see a bright, evenly illuminated oval 50% longer than it is wide stretching northwest to southeast. The tiny central region appears slightly brighter. This galaxy is one of the sky's brightest. As such, it takes magnification really well, but unless you observe it through a 20-inch or larger scope, you won't detect much in the way of detail.

| OBJECT #164 | NGC 3372 |
|---|---|
| Constellation | Carina |
| Right ascension | 10h44m |
| Declination | −59°52′ |
| Magnitude | 3.0 |
| Size | 2° |
| Type | Emission nebula |
| Other names | The Eta Carinae Nebula, Caldwell 92 |

Imagine a gas cloud nearly a light-year long expanding at 1.5 million mph (2.3 million km/h). Within the cloud sits a supermassive star radiating 5 million times the Sun's energy. This scenario describes the Eta Carinae Nebula and its 6th-magnitude central star, Eta Carinae.

Eta Carinae underwent a giant outburst about 150 years ago, when it became the second-brightest star in the sky. Although it released as much visible light as a supernova explosion would, the star survived the explosion and produced a nebula with two polar lobes and a large, thin equatorial disk. Astronomers call the bi-lobed nebula the Homunculus, derived from the Latin for "little man."

Observing the Eta Carinae Nebula offers a real treat for observers. Easily visible to the naked eye as the brightest spot in the southern Milky Way, NGC 3372 looks great through any size telescope.

At magnifications below 50×, the most prominent detail is a dark, V-shaped rift slicing through the nebula's center. Double the power, and you'll see a scalloped edge against the bright background.

Look toward NGC 3372's center for the Keyhole Nebula, a dark cloud with overlapping nebulosity. Double the magnification again, and find the yellowish region around Eta Carinae, which some amateurs have dubbed the Fried Egg. If the seeing is good, you may detect thin spokes of nebulosity that seem to emanate from the central star.

| OBJECT #165 | NGC 3377 |
|---|---|
| Constellation | Leo |
| Right ascension | 10h48m |
| Declination | 13°59′ |
| Magnitude | 10.4 |
| Size | 4.1′ by 2.6′ |
| Type | Elliptical galaxy |

Our next target lies 6.6° west-southwest of magnitude 3.3 Chertan (Theta [$\theta$] Leonis), or 0.4° east-southeast of the magnitude 5.5 star 52 Leonis. Through a 6-inch telescope at 150×, you'll see an oval twice as long as it is wide with a large central region and an extremely thin halo. Larger apertures and higher magnifications bring out the halo better, but not a lot better.

Through a 12-inch or larger scope, look for magnitude 13.6 NGC 3377A (also known as UGC 5889), which lies 6′ northwest of NGC 3377.

| OBJECT #166 | M105 (NGC 3379) |
|---|---|
| Constellation | Leo |
| Right ascension | 10h48m |
| Declination | 12°35′ |
| Magnitude | 9.3 |
| Size | 3.9′ by 3.9′ |
| Type | Elliptical galaxy |

Our next target travels in elite company. It sits only 0.8° north-northeast of M96 (Object #163). Beyond that, M105 doesn't offer observers much in the way of detail. The galaxy has a bright central region surrounded by a halo with an edge that's difficult to define. Although M105 appears circular at low magnifications, crank up the power past 250×, and you'll see that it's a fat oval, orienting northeast to southwest.

A member of the M96 Group of galaxies, NGC 3384, lies 7′ to the east-northeast. Non-member magnitude 11.8 NGC 3389 sits 10′ east-southeast of M105.

| OBJECT #167 | V Hydrae |
|---|---|
| Constellation | Hydra |
| Right ascension | 10h52m |
| Declination | −21°15′ |
| Magnitude range | 6.6–9.0 |
| Period | 531 days |
| Type | Variable star |

Our next target is the star V Hydrae. To find it, use binoculars, and look toward the center of the huge constellation Hydra. V Hydrae lies near that star pattern's border with Crater.

Locate the 4th-magnitude star Alkes (Alpha [α] Crateris). Sweep your binoculars 3.5° (that's about half the field of view of 7×50 binoculars) toward the south-southwest. Or, just center Alkes and look to about the 5 o'clock position at the edge of the field of view. V Hydrae should be sitting there. Once you locate V Hya through binoculars, switch to a telescope view at a magnification of 100×.

The special thing about V Hydrae is its color. Of all the stars I've seen — and I've seen a lot — this one's the reddest. Think ruby, crimson, or blood, and then deepen it a shade. I'm talking red.

Astronomers classify V Hydrae as a type of variable star known as a carbon star. A huge amount of carbon soot accumulates in the upper atmosphere of such stars. The carbon is black, not red, but the particles scatter short-wavelength light, which are the colors near the blue end of the spectrum. What's left for us to view is the red component of the star's light.

As the particles build up, the star fades in brightness and also gets redder. Eventually, the carbon absorbs enough radiation to escape the star, and the cycle starts again.

V Hydrae takes 531 days to complete one cycle. During that time, its magnitude drops from 6.6 to 9.0 and rises back again. That means at minimum, the star shines only 11% as bright as when it's at maximum.

| OBJECT #168 | NGC 3384 |
|---|---|
| Constellation | Leo |
| Right ascension | 10h48m |
| Declination | 12°37′ |
| Magnitude | 9.9 |
| Size | 5.4′ by 2.7′ |
| Type | Elliptical galaxy |

Once you've located M105 (Object #166), you probably have seen this galaxy. It sits a mere 7′ to the east-northeast of M105, which shines a scant 75% brighter. Through any size telescope, you'll see NGC 3384 as an oval twice as long as it is wide oriented northeast-southwest. The central region is large and bright, and the outer halo appears faint even through large scopes.

Because early nineteenth-century observers weren't perfect, you may run across references equating this galaxy with NGC 3371. They are the same object.

| OBJECT #169 | NGC 3412 |
|---|---|
| Constellation | Leo |
| Right ascension | 10h50m |
| Declination | 13°24′ |
| Magnitude | 10.4 |
| Size | 3.7′ by 2.2′ |
| Type | Barred spiral galaxy |

Our next object lies 1.3° southeast of the magnitude 5.5 star 52 Leonis. If you need a brighter starting point, choose magnitude 3.3 Chertan (Theta [$\theta$] Leonis). NGC 3412 sits 6° west-southwest of that star. Through an 8-inch telescope at 200×, you'll see a bright central region that spans half the galaxy's length. Around that, a reasonably bright halo gradually fades to black. At low power, this object is oval and evenly lit, about twice as long as it is wide. The magnitude 14.2 star GSC 852:749 lies 1′ north of NGC 3412's core.

| OBJECT #170 | The Broken Engagement Ring |
|---|---|
| Constellation | Ursa Major |
| Right ascension | 10h51m |
| Declination | 56°07′ |
| Type | Asterism |

You'll spot our next object, the Broken Engagement Ring, through binoculars. It lies only 1.5° west of one of the Big Dipper's Pointer stars: Merak (Beta [$\beta$] Ursae Majoris).

The Ring's brightest star also is its northernmost, magnitude 7.5 SAO 27788. I guess this star supposedly represents the ring's diamond. From it, the brightnesses drop off quite a bit. The next brightest star glows at magnitude 9.1, and the faintest is magnitude 9.9.

Generally, the Ring's stars lie south and west of SAO 27788. The Broken Engagement Ring spans 16′, which is half the diameter of the Full Moon. The best binocular views will be through units that give magnifications above 10×. Mount your binoculars on a camera tripod for a steadier view.

| OBJECT #171 | NGC 3414 |
|---|---|
| Constellation | Leo Minor |
| Right ascension | 10h51m |
| Declination | 27°58′ |
| Magnitude | 10.9 |
| Size | 3.5′ by 2.6′ |
| Type | Lenticular galaxy |

Our next target lies 6.8° west-southwest of magnitude 4.5 Xi ($\xi$) Ursae Majoris, or only 0.3° east of the magnitude 6.1 star 44 Leonis Minoris. This galaxy stretches into a rough oval shape oriented north-south. Apertures of less than 10′ show only an evenly illuminated face. Larger scopes delineate the small, bright central region from the slightly fainter halo.

When you're done observing NGC 3414, try for the much fainter NGC 3418. It lies 8′ to the north and glows at magnitude 13.5.

| OBJECT #172 | 54 Leonis |
|---|---|
| Constellation | Leo |
| Right ascension | 10h56m |
| Declination | 24°45′ |
| Magnitudes | 4.5/6.3 |
| Separation | 6.5″ |
| Type | Double star |

You'll love the color contrast here. Most observers see the primary as off-white and the secondary as deep blue. Some amateur astronomers, however, have reported the primary as a pale or robin's-egg blue.

| OBJECT #173 | NGC 3448 |
|---|---|
| Constellation | Ursa Major |
| Right ascension | 10h55m |
| Declination | 54°19′ |
| Magnitude | 11.6 |
| Size | 4.9′ by 1.4′ |
| Type | Irregular galaxy |

If your telescope has no go-to drive, start at 5th-magnitude 44 Ursae Majoris, and move 0.3° southeast. That's where you'll find NGC 3448.

Even the largest amateur telescopes won't reveal much detail in this galaxy. You'll see a central bulge and two smaller regions flanking it.

But this galaxy isn't worth looking at for its beauty or its brightness. Instead, it's the kind of object NGC 3448 is that makes it special. Astronomers classify it as an amorphous galaxy, a young object in which a large percentage of its mass has not yet formed stars. The Cigar Galaxy (M82) is the best example of this group.

A 12-inch scope under a dark, steady sky will reveal NGC 3448's companion galaxy, magnitude 14.2 UGC 6016. Astronomers have determined that a bridge of stellar material connects these two galaxies.

| OBJECT #174 | NGC 3486 |
|---|---|
| Constellation | Leo Minor |
| Right ascension | 11h00m |
| Declination | 28°58′ |
| Magnitude | 10.5 |
| Size | 6.6′ by 4.7′ |
| Type | Spiral galaxy |

This object lies 4.6° west-southwest of magnitude 4.5 Xi (ξ) Ursae Majoris. Through an 8-inch telescope at 200×, you'll see a wide, circular central region that's evenly lit surrounded by a fainter halo pockmarked with tiny dark areas. Through a 14-inch scope, you may detect two small, dark V-shaped intrusions at the northern and southern ends.

When you're through with NGC 3486, look for the magnitude 12.6 spiral NGC 3510, which lies 0.7° to the east.

| OBJECT #175 | NGC 3489 |
|---|---|
| Constellation | Leo |
| Right ascension | 11h00m |

| (continued) | |
|---|---|
| Declination | 13°54′ |
| Magnitude | 10.3 |
| Size | 3.2′ by 2.0′ |
| Type | Spiral galaxy |

Our next target lies 3.7° west-southwest of magnitude Chertan (Theta [θ] Leonis). Through a 6-inch telescope, this galaxy appears as a fat oval 50% longer than it is wide. It orients east-northeast to west-southwest. It appears evenly illuminated without outer structure.

| OBJECT #176 | NGC 3504 |
|---|---|
| Constellation | Leo Minor |
| Right ascension | 11h03m |
| Declination | 27°58′ |
| Magnitude | 10.9 |
| Size | 2.7′ by 2.1′ |
| Type | Barred spiral galaxy |

You'll find this object 4.8° southwest of magnitude 4.5 Xi (ξ) Ursae Majoris. Its oval shape, 50% longer than it is wide, is apparent through any size telescope, but detail is scant. The galaxy orients northwest to southeast.

NGC 3504 is the third in a line of four reasonably spaced galaxies that stretch 0.8° from the northeast to the southwest. Magnitude 13.9 NGC 3515 lies 0.4° northeast of NGC 3504. Midway between the two sits magnitude 12.3 NGC 3512. Finally, magnitude 14.6 NGC 3493 lies almost 0.5° southwest of NGC 3504. All four galaxies are spirals.

| OBJECT #177 | NGC 3507 |
|---|---|
| Constellation | Leo |
| Right ascension | 11h03m |
| Declination | 18°08′ |
| Magnitude | 11.8 |
| Size | 3.3′ by 2.9′ |
| Type | Spiral galaxy |

Our next object lies 3.5° southwest of magnitude 2.6 Zosma (Delta [δ] Leonis). This galaxy is a bit tough to observe because of the magnitude 10.5 star only 30″ northeast of its core. A slightly brighter star, magnitude 10.1 GSC 1433:1318, sits 3′ to the south.

Through an 8-inch telescope at 200×, the core appears doubled because of the superimposed star. The outer halo appears faint. Through a 16-inch scope, you'll see the dark lanes that divide the spiral arms curving through the halo.

| OBJECT #178 | NGC 3511 |
|---|---|
| Constellation | Crater |
| Right ascension | 11h03m |
| Declination | −23°05′ |
| Magnitude | 11.0 |
| Size | 5.5′ by 1.0′ |
| Type | Spiral galaxy |

Our next target, a nearly edge-on spiral and another of the sky's "needles," lies 2° west of Beta (β) Crateris. Elongated east-west, NGC 3511 appears four times as long as it is wide. At high magnifications through a 10-inch or larger telescope, the galaxy appears mottled. The oval core appears to be offset to the north of center.

Only 11′ to the south-southeast in the same telescopic field you can find the magnitude 11.5 barred spiral galaxy NGC 3513. Through a 12-inch scope at 300×, the slightly oval halo leads to a brighter central region. A more careful look reveals the thin central bar with its long axis oriented east-west.

| OBJECT #179 | NGC 3521 |
|---|---|
| Constellation | Leo |
| Right ascension | 11h06m |
| Declination | −0°02′ |
| Magnitude | 9.0 |
| Size | 12.5′ by 6.5′ |
| Type | Spiral galaxy |

Our next target is spiral galaxy NGC 3521. Because this object lies just 28 million light-years away, it appears bright and detailed.

You'll find this galaxy 4.5° northwest of magnitude 4.5 Phi ($\theta$) Leonis. It sits in a small region of Leo between Sextans and Virgo.

Through a 10-inch telescope, you'll see the bright, extended core surrounded by a diffuse halo. With a 16-inch scope, NGC 3521 will appear nearly twice as long as through the smaller instrument.

That much aperture also brings out fine — almost cotton-like — spiral structure, particularly near the edge of the halo and along the minor axis near the core. If your sky's transparency is good, also try to observe the long, dark dust lane that runs the length of the galaxy's western side.

| OBJECT #180 | NGC 3532 |
|---|---|
| Constellation | Carina |
| Right ascension | 11h06m |
| Declination | −58°40′ |
| Magnitude | 3.0 |
| Size | 55′ |
| Type | Open cluster |
| Other names | The Firefly Party Cluster, the Pincushion Cluster, Caldwell 91 |

This fabulous object sits in a terrific star field 4.7° south-southwest of magnitude 3.9 Pi ($\pi$) Centaurus. You'll spot it immediately with your naked eyes, but although it shines at 3rd magnitude, the individual stars it contains are too faint to see individually, so the impression is of a bright glow within the Milky Way.

A 4-inch telescope at 100× displays more than 100 stars. Use a larger aperture and higher magnification, and you'll quickly get lost in this cluster. With an eyepiece that gives a wide field of view, you'll notice numerous dark lanes dividing lines composed of dozens of stars.

If you expand your view past the cluster's tightly packed core, you'll spot many colorful stars. Two sit at the cluster's northeast end. Magnitude 6.9 SAO 238839 shines blue, while magnitude 6.2 SAO 238855 glows a deep red.

Even after I first observed NGC 3532 in 1986, I thought the deep-sky object with the best name was Thor's Helmet (Object #30). Now, I'm not so sure. Its hard to top a star cluster that got one of its common names because it resembled a large number of fireflies throwing a party!

| OBJECT #181 | IC 2631 |
|---|---|
| Constellation | Chamaeleon |
| Right ascension | 11h10m |
| Declination | −76°37′ |
| Size | 5.0′ |
| Type | Emission nebula |

Look for this target 2.7° northeast of magnitude 4.1 Gamma ($\gamma$) Chamaeleontis. IC 2631 resides on the northern end of the Chamaeleon I Dark Cloud, an active stellar nursery similar to the Rho Ophiuchi nebula. The Dark Cloud stretches more than 2° north to south with an extension westward that's nearly as long. Use your lowest-power, widest-field eyepiece to scan this region.

Through an 8-inch telescope at 200×, IC 2631 appears as a bright mist that surrounds the magnitude 9.0 star GSC 9410:2805.

**Object #182** M108 (top center) and M97 Yon Ough/Adam Block/NOAO/AURA/NSF

| OBJECT #182 | M108 (NGC 3556) |
|---|---|
| Constellation | Ursa Major |
| Right ascension | 11h12m |
| Declination | 55°40′ |
| Magnitude | 10.0 |
| Size | 8.1′ by 2.1′ |
| Type | Barred spiral galaxy |

Here's a great object for a large telescope. Not only is M108 a barred spiral galaxy, but it sits edge-on to our line of sight. M108 tilts east-northeastward (ok, or west-southwestward) and measures four times as long as it is wide. You can find it 1.5° east of magnitude 2.3 Merak (Beta [$\beta$] Ursae Majoris).

You'll immediately see the 12th-magnitude star that lies between us and M108's core. This stellar luminary has tricked many an observer into thinking they'd found a supernova.

Not many amateur telescopes will allow you to trace M108's spiral arms. If, however, you're lucky enough to observe this object with a 16-inch or larger scope, look for a darkening running along the galaxy's northeast side. That region appears less luminous because it's a massive dust lane in the star-forming areas of M108.

| OBJECT #183 | NGC 3557 |
|---|---|
| Constellation | Centaurus |
| Right ascension | 11h09m |
| Declination | −37°32′ |
| Magnitude | 10.6 |
| Size | 4′ by 3′ |
| Type | Elliptical galaxy |

Our next target lies 2.7° east of magnitude 4.6 Iota ($\iota$) Antliae. This galaxy appears oval, 50% longer than wide, oriented north-northeast to south-southwest. A thin halo you'll need 12 inches of aperture to see surrounds the wide central region.

Two spiral galaxies lie nearby. Magnitude 12.2 NGC 3564 lies 8′ to the east, and magnitude 12.3 NGC 3568 sits 11′ east-northeast of NGC 3557.

| OBJECT #184 | NGC 3572 |
|---|---|
| Constellation | Carina |
| Right ascension | 11h10m |
| Declination | −60°15′ |
| Magnitude | 6.6 |
| Size | 7′ |
| Type | Open cluster |

You'll find this object 2.7° west-southwest of the pair of 5th-magnitude stars Omicron$^1$ and Omicron$^2$ ($o^1$ and $o^2$) Centauri. Or will you? What I mean is that, within a 0.5°-wide area, astronomers have cataloged six bright open clusters.

NGC 3572 itself sits within the boundaries of the larger cluster magnitude 3.9 Collinder 240. The other clusters include magnitude 6.9 Hogg 10, which sits 7′ to the south-southeast of NGC 3572. Magnitude 8.1 Hogg 11 lies 11′ to NGC 3572's southeast. Magnitude 6.9 Trumpler 18 lies 0.4° to its south-southeast, and, finally, you'll find magnitude 8.2 NGC 3590 0.2° farther to the south-southeast.

This area explodes with stars when viewed through 15×70 or larger binoculars. A 4-inch telescope at 50× delineates the various clusters, but to see each one better, insert an eyepiece that magnifies 125×.

| OBJECT #185 | NGC 3585 |
|---|---|
| Constellation | Hydra |
| Right ascension | 11h13m |
| Declination | −26°45′ |
| Magnitude | 9.7 |
| Size | 6.9′ by 4.2′ |
| Type | Spiral galaxy |

Our next target lies 2° east-northeast of 5th-magnitude Chi ($\chi$) Hydrae. This spiral galaxy appears oval, not quite twice as long as it is wide, with its long axis oriented east-west. Its core brightens steadily to a stellar nucleus that can be glimpsed with some difficulty. Two stars lie equally distant from NGC 3585. Magnitude 8.6 SAO 179667 lies 8′ east, and magnitude 8.6 SAO 179663 sits 8′ southeast.

**Object #186** The Owl Nebula (M97) Gary White and Verlenne Monroe/Adam Block/NOAO/AURA/NSF

| OBJECT #186 | M97 (NGC 3587) |
|---|---|
| Constellation | Ursa Major |
| Right ascension | 11h15m |
| Declination | 55°01′ |
| Magnitude | 9.9 |
| Size | 194″ |
| Type | Planetary nebula |
| Other name | The Owl Nebula |

One of the best springtime planetary nebulae visible to Northern Hemisphere observers is the Owl Nebula (M97). That's amazing considering this object has quite a low surface brightness, so its details appear best to patient observers.

Astronomers long ago dubbed this object the "Owl" for the two dark circular regions visible in its disk. Each eye is slightly less than 1′ across, and the northwestern one appears a bit darker.

Pierre Méchain discovered this object February 16, 1781, but it was William Parsons, third earl of Rosse (1800–1867) who first called it the Owl Nebula. In March, 1848, he described, "Two stars considerably apart in the central region: dark penumbra around each spiral arrangements." During other observations, Rosse saw only one star and called the spiral form doubtful.

M97's orientation may be responsible for the owl's "eyes." Some astronomers have speculated this nebula is a torus (a sphere with a hollow cylinder from pole to pole) tilted to our line of sight so the cylinder's ends, where material is thinnest, form the eyes.

You'll need a big scope to spot the 16th-magnitude central star. It lies between the two eyes. Because the Owl Nebula has a low surface brightness, use an OIII filter and a magnification of about 100× to see it best. If you view through a 10-inch or larger telescope, look for the size discrepancy between the eyes. If your sky is dark, you may see the disk's outer 10% as a faint ring.

The Owl Nebula lies 2.3° southeast of Merak (Beta [$\beta$] Ursae Majoris).

| OBJECT #187 | NGC 3593 |
|---|---|
| Constellation | Leo |
| Right ascension | 11h14m |
| Declination | 12°49′ |
| Magnitude | 11.0 |
| Size | 5.2′ by 1.9′ |
| Type | Spiral galaxy |

You'll find this object 1.1° west-southwest of M95, or 0.6° south-southwest of the magnitude 5.3 star 73 Leonis. This lens-shaped galaxy stretches twice as long as it is wide in an east-west orientation. Through a 12-inch or larger telescope at 300× or above, you'll pick out the beginnings of NGC 3593's spiral arms at the east and west ends.

| OBJECT #188 | NGC 3607 |
|---|---|
| Constellation | Leo |
| Right ascension | 11h17m |
| Declination | 18°03′ |
| Magnitude | 9.9 |
| Size | 4.6′ by 4.1′ |
| Type | Spiral galaxy |

This target lies 2.6° south-southeast of magnitude 2.6 Zosma (Delta [$\delta$] Leonis). When you point your scope toward NGC 3607, you'll get three galaxies for the price of one. Magnitude 13.2 NGC 3605 lies 3′ to the southwest, and magnitude 11.7 NGC 3608 lies 6′ north. Both are elliptical galaxies.

Through an 8-inch telescope, NGC 3607 appears slightly oval in a northwest to southeast orientation. The central region appears broad and evenly illuminated, and the halo (you'll need a bigger scope to see it) is thin and faint.

| OBJECT #189 | NGC 3610 |
|---|---|
| Constellation | Ursa Major |
| Right ascension | 11h18m |
| Declination | 58°47′ |
| Magnitude | 10.7 |
| Size | 2.7′ by 2.3′ |
| Type | Elliptical galaxy |

Our next target lies 3.5° southeast of magnitude 1.8 Dubhe (Alpha [$\alpha$] Ursae Majoris). Generally, elliptical galaxies aren't known for their details. But crank up your aperture, and sometimes you'll get lucky.

Through a 30-inch Newtonian reflector, I saw three distinct brightness regions within this galaxy. The core appeared lenticular rather than stellar (an advantage of using a large telescope), about four times as long as wide. The region around the core was less bright, and NGC 3610's outer regions barely gave off enough light for me to observe them. I saw the faint galaxy PGC 213808, not quite 4′ to the southwest mainly because it's small and has a high surface brightness. It glows weakly at magnitude 16.9, however, so be advised.

| OBJECT #190 | NGC 3621 |
|---|---|
| Constellation | Hydra |
| Right ascension | 11h18m |
| Declination | −32°49′ |
| Magnitude | 8.9 |
| Size | 9.8′ by 4.6′ |
| Type | Spiral galaxy |
| Other name | The Frame Galaxy |

You'll find this object 3.3° west-southwest of magnitude 3.5 Xi (ξ) Hydrae. This object appears twice as long as it is wide, oriented north-northwest to south-southeast. The central area is wide and evenly illuminated. The halo, however, reveals mottling, which suggests spiral structure. Two 10th-magnitude stars lie at the southern edge of the galaxy.

NGC 3621's common name has nothing to do with how the galaxy looks. Rather, *Astronomy* magazine Contributing Editor David H. Levy gave it this name because a parallelogram of stars surrounds it, making the galaxy appear "framed." English astronomer Sir William Herschel (1738–1822) also mentioned the stellar parallelogram.

**Object #191** M65 Doc. G. and Dick Goddard/Adam Block/NOAO/AURA/NSF

| OBJECT #191 | M65 (NGC 3623) |
|---|---|
| Constellation | Leo |
| Right ascension | 11h19m |
| Declination | 13°06′ |
| Magnitude | 9.3 |
| Size | 8.7′ by 2.2′ |
| Type | Spiral galaxy |

Draw a line from magnitude 3.3 Chertan (Theta [θ] Leonis) to magnitude 4.0 Iota (ι) Leonis. At the midpoint of that line, you'll find the terrific galaxy M65. Alternatively, it lies 0.8° east-southeast of the magnitude 5.3 star 73 Leonis.

This galaxy belongs to the Leo Triplet — a triangular group that fits into a field of view 0.6° across. M65 anchors the Triplet's southwestern corner. The other members are M66 (Object #193) and NGC 3628 (Object #194).

M95 appears nearly four times as long as it does wide. Spiral structure is difficult to discern, but you'll see some mottling north and south of the galaxy's broad core.

| OBJECT #192 | NGC 3626 |
|---|---|
| Constellation | Leo |
| Right ascension | 11h20m |
| Declination | 18°21′ |
| Magnitude | 10.9 |
| Size | 2.7 by 1.9 |
| Type | Spiral galaxy |
| Other name | Caldwell 40 |

Our next target lies 2.6° southeast of magnitude 2.6 Zosma (Delta [$\delta$] Leonis). A small telescope won't show much more than this galaxy's oval shape, which orients north-northwest to south-southeast. Through an 8-inch scope at 300×, you'll see a faint outer halo surrounding a bright central region that sharpens to a tiny, starlike core.

| OBJECT #193 | M66 (NGC 3627) |
|---|---|
| Constellation | Leo |
| Right ascension | 11h20m |
| Declination | 12°59′ |
| Magnitude | 8.9 |
| Size | 8.2′ by 3.9′ |
| Type | Spiral galaxy |

From M65 (Object #191), move only 0.3° east-southeast to find an even brighter celestial wonder, M66. Through a 6-inch telescope, the galaxy looks hazy and twice as long as wide oriented north-south. A brighter central region dominates the view.

A 12-inch scope at 200× reveals a hint of the galaxy's spiral arms. Look for a hook-shaped feature pointing northeast from the southern end of the bar. The corresponding feature at the northern end is more difficult to see, so crank the magnification past 300×.

Larger telescopes show the spiral arms dotted with star-forming regions as well as a dark lane east of M66's central region. Through any scope, the galaxy has a well-defined edge.

**Object #194** King Hamlet's Ghost (NGC 3628) Adam Block/NOAO/AURA/NSF

| OBJECT #194 | NGC 3628 |
|---|---|
| Constellation | Leo |
| Right ascension | 11h20m |
| Declination | 13°35′ |
| Magnitude | 9.5 |
| Size | 14.0′ by 4.0′ |
| Type | Spiral galaxy |
| Other names | King Hamlet's Ghost, the Vanishing Galaxy |

Our next target rounds out the Leo Triplet. Although it glows more faintly than the Triplet's two Messier objects, NGC 3628 still ranks as one of the sky's brightest galaxies. From M65, nudge your telescope 0.6° northeast to center this fine object.

This edge-on starburst galaxy resembles M82 (Object #126). Astronomers theorize that NGC 3628 had an encounter with M66 that took place 800 million years ago, when the two galaxies were only 80,000 light-years apart.

Through a 12-inch telescope, you'll see a bright object four times as long as it is wide. A dust lane runs the galaxy's entire length. It crosses south of the center and is not parallel with the long axis. Through a 24-inch scope, the galaxy's ends appear thicker than the central region.

*Astronomy* magazine Contributing Editor Stephen James O'Meara gave NGC 3628 both of its common names. I like his reasoning for calling it the Vanishing Nebula. With each increase in magnification, he saw the galaxy blending more and more into the background. The reason he gives for that effect is that the galaxy's dust lane overwhelms the stars around it as he increases the telescope's power.

| OBJECT #195 | NGC 3631 |
|---|---|
| Constellation | Ursa Major |
| Right ascension | 11h21m |
| Declination | 53°10′ |
| Magnitude | 10.1 |
| Size | 5′ by 3.7′ |
| Type | Spiral galaxy |

You'll find this object 4.9° west of magnitude 2.4 Phecda (Gamma [$\gamma$] Ursae Majoris). NGC 3631 is a face-on spiral. Through an 8-inch telescope, you'll see a circular object with a brighter central region. A 14-inch scope at 300× and above shows the tightly wound spiral arms. As noticeable as the spiral arms is a dark lane that starts north of the core and winds around the eastern side, broadening as it goes.

| OBJECT #196 | NGC 3640 |
|---|---|
| Constellation | Leo |
| Right ascension | 11h21m |
| Declination | 3°14′ |
| Magnitude | 10.3 |
| Size | 4′ by 3.2′ |
| Type | Elliptical galaxy |

Look 2.8° south of magnitude 4.1 Sigma ($\sigma$) Leonis for our next object. If you observe through a 12-inch or larger telescope from a dark site, then NGC 3640 begins a 0.4°, slightly ragged line of five galaxies that stretches to the south-southeast. In order after NGC 3640, they are magnitude 13.1 NGC 3641, magnitude 14.1 NGC 3643, magnitude 14.3 NGC 3647, and magnitude 14.2 NGC 3644.

NGC 3640 appears slightly oval in an east-west orientation. It has a broad, bright central region and a faint, but reasonably thick, halo.

**Object #197** NGC 3642 Adam Block/Mount Lemmon SkyCenter/University of Arizona

| OBJECT #197 | NGC 3642 |
|---|---|
| Constellation | Ursa Major |
| Right ascension | 11h22m |
| Declination | 59°04′ |
| Magnitude | 10.8 |
| Size | 5.5′ by 4.7′ |
| Type | Spiral galaxy |

Our next target lies 3.5° southeast of magnitude 1.8 Dubhe (Alpha [α] Ursae Majoris). All telescopes show this galaxy as round. Through a 12-inch scope at 300×, the central region appears slightly irregular with a thin, faint halo surrounding it.

| OBJECT #198 | NGC 3665 |
|---|---|
| Constellation | Ursa Major |
| Right ascension | 11h24m |
| Declination | 38°45′ |
| Magnitude | 10.7 |
| Size | 4.3′ by 3.3′ |
| Type | Spiral galaxy |

This object sits 5.8° north of magnitude 3.5 Nu (ν) Ursae Majoris. It appears as an evenly illuminated oval, 50% longer than wide, and oriented north-northeast to south-southwest. Only the largest scopes will show the thin outer halo.

The magnitude 8.9 star SAO 62530 lies 14′ to the west-southwest. Then look 15′ southwest of NGC 3665 to find the magnitude 12.1 spiral NGC 3658.

| OBJECT #199 | NGC 3672 |
|---|---|
| Constellation | Crater |
| Right ascension | 11h25m |
| Declination | −9°48′ |
| Magnitude | 11.4 |
| Size | 3.9′ by 1.8′ |
| Type | Spiral galaxy |

Our next target lies 1° north of magnitude 4.8 Epsilon (ε) Crateris. Through a 14-inch telescope at 200×, the galaxy appears elongated twice as long as it is wide in a north-south orientation. Double the magnification at a true-dark site, and NGC 3672's outer edge appears scalloped. The most apparent spiral arm curves eastward from the northern tip. A dark indentation shows at the northeast edge.

| OBJECT #200 | NGC 3675 |
|---|---|
| Constellation | Ursa Major |
| Right ascension | 11h26m |
| Declination | 43°35′ |
| Magnitude | 10.0 |
| Size | 5.9′ by 3.1′ |
| Type | Spiral galaxy |

This object lies 3.1° east-southeast of magnitude 3.0 Psi (ψ) Ursae Majoris. Through an 8-inch telescope, NGC 3675 is an evenly lit disk twice as long as wide, which orients north-south. Through a 14-inch scope, you'll notice the western half appears a bit brighter.

| OBJECT #201 | NGC 3680 |
|---|---|
| Constellation | Centaurus |
| Right ascension | 11h25m |
| Declination | −43°15′ |
| Magnitude | 7.6 |
| Size | 7′ |
| Type | Open cluster |

To find our next target, look 9.1° northeast of magnitude 2.7 Mu (μ) Velorum. Through a 6-inch telescope at 100×, you'll spot roughly a dozen and a half magnitude 10.5 and fainter stars that divide into two groups, north and south, with a dark east-west lane between them.

| OBJECT #202 | NGC 3699 |
|---|---|
| Constellation | Centaurus |
| Right ascension | 11h28m |
| Declination | −59°57′ |
| Magnitude | 11.3 |
| Size | 67″ |
| Type | Planetary nebula |

This object sits 0.7° southwest of the magnitude 5.0 double star Omicron (o) Centauri in a really rich star field. Until recently, astronomers classified NGC 3699 as an HII region. They now recognize it as a bipolar planetary nebula.

And you'll see its double nature through a 12-inch telescope. Crank the magnification up to 250×, and the two disks, the brighter one to the north, appear mottled. A nebula filter really helps. This planetary isn't huge, but its high surface brightness allows you to use really high magnification, so don't be afraid to use your short-focal-length eyepieces.

| OBJECT #203 | N Hydrae |
|---|---|
| Constellation | Hydra |
| Right ascension | 11h32m |
| Declination | −29°16′ |
| Magnitudes | 5.8/5.9 |
| Separation | 9.2″ |
| Type | Double star |

You'll find N Hydrae 2.5° north of magnitude 3.5 Xi ($\xi$) Hydrae. The two stars of N are of essentially equal brightness. One shines white, while its companion is pale yellow.

| OBJECT #204 | NGC 3718 |
|---|---|
| Constellation | Ursa Major |
| Right ascension | 11h32m |
| Declination | 53°04′ |
| Magnitude | 10.6 |
| Size | 8.1′ by 4′ |
| Type | Spiral galaxy |

Look for this object 3.2° west of magnitude 2.4 Phecda (Gamma [$\gamma$] Ursae Majoris). Through a 10-inch telescope at 250×, this galaxy looks round at first glance. Take some time, however, to note faint, thick extensions to the north and south, which imply spiral arms. The central region is tiny but bright, surrounded by a large halo.

A nice double star with a separation of 33″ sits just on this galaxy's southern flank. The two components are magnitude 11.5 GSC 3825:806, which is the westernmost, and magnitude 11.7 GSC 577. Also, the magnitude 12.1 spiral galaxy NGC 3729 lies 12′ east-northeast.

| OBJECT #205 | NGC 3726 |
|---|---|
| Constellation | Ursa Major |
| Right ascension | 11h33m |
| Declination | 47°02′ |
| Magnitude | 10.4 |
| Size | 5.6′ by 3.8′ |
| Type | Spiral galaxy |

This tightly wound spiral is bright enough to classify as a small telescope object. Through a 4-inch scope, you'll notice a small oval of uniform brightness. An 8-inch instrument begins to reveal a mottled appearance. You're beginning to see the lanes and gaps that define the spiral arms.

Through a really large telescope, look for huge star-forming regions along NGC 3726's spiral arms. I've even used a nebula filter to get a better view of these areas when I viewed this galaxy through a 24-inch reflector.

To find this galaxy, look 2.3° west-southwest of magnitude 3.7 Chi ($\chi$) Ursae Majoris.

| OBJECT #206 | NGC 3735 |
|---|---|
| Constellation | Draco |
| Right ascension | 11h36m |
| Declination | 70°32′ |
| Magnitude | 10.6 |
| Size | 4.3′ by 3.7′ |
| Type | Spiral galaxy |

You'll find our next object 1.3° north-northeast of magnitude 3.8 Lambda (λ) Draconis quite close to the Dragon's border with Ursa Major. From our perspective, the spiral tilts 13° from edge-on, making it quite attractive.

Through a 10-inch telescope, the galaxy appears four times as long as it is wide. At high magnifications, NGC 3735 reveals a bright core that runs roughly two-thirds the length of this galaxy.

| OBJECT #207 | NGC 3766 |
|---|---|
| Constellation | Centaurus |
| Right ascension | 11h36m |
| Declination | −61°37′ |
| Magnitude | 5.3 |
| Size | 12′ |
| Type | Open cluster |
| Other names | The Pearl Cluster, Caldwell 97 |

You'll find our next target 1.5° north of Lambda (λ) Centauri, and what a sight it is. If your travel plans include the Southern Hemisphere in the near future, I strongly suggest you put NGC 3766 on your observing list. You'll thank me later.

Now, here's a bit of trivia about this object: It received its common name, the Pearl Cluster, February 15, 2006, from amateur astronomer and popularizer Ray Palmer. On that date, he founded the South Celestial Star Light Project to help fine-tune the names of Southern Hemisphere celestial objects. His reasoning is that astronomy must appeal to people, not be complicated or boring. People (especially young people), Palmer theorizes, will remember and relate to a sky object's popular name more than any of its catalog designations. I agree, and I'll be calling NGC 3766 the Pearl Cluster from now on.

At roughly 5th magnitude, you can see this cluster without optical aid, but you'll have to work at it because of the rich star field it's in. Binoculars, especially those that magnify 15× or more, will reveal dozens of stars. But the finest view comes through a telescope that magnifies between 75× and 100×.

Through a 4-inch scope, you can count 100 stars, the brightest of which shine at 7th magnitude. That number in itself provides a sweet view, but there's more. Riding seemingly in front of a pure-white carpet of diamonds are two pale rubies. One, magnitude 7.5 SAO 251483, lies midway between the cluster's center and its eastern edge. The other, magnitude 7.3 SAO 251470. lies the same distance from the center toward the west. Indeed, the Pearl Cluster is a Southern Hemisphere jewel.

**Object #208** Copeland's Septet Neil Jacobstein/Adam Block/NOAO/AURA/NSF

| OBJECT #208 | Hickson 57 |
|---|---|
| Constellation | Leo |
| Right ascension | 11h38m |
| Declination | 21°59′ |
| Size | 5′ by 2′ |
| Type | Galaxy group |
| Other name | Copeland's Septet |

You'll need serious aperture to get much out of our next object, but it's worth the trouble to transport and set up a large telescope just to say you've seen it. This group lies 5.7° east-northeast of magnitude 2.6 Zosma (Delta [δ] Leonis).

Scottish astronomer Ralph Copeland discovered this tiny group of galaxies in 1874. At the time, he worked for William Parsons, Third Earl of Rosse. Copeland almost certainly sighted what would come to be called his Septet through Parsons' 72-inch speculum-mirror reflector, the Leviathan of Parsonstown. In Copeland's descriptions for the *New General Catalogue*, he described five of the seven galaxies as "pretty bright."

The Septet comprises seven galaxies: magnitude 15.2 NGC 3745; magnitude 14.0 NGC 3746; magnitude 14.8 NGC 3748; magnitude 15.0 NGC 3750; magnitude 15.0 NGC 3751; magnitude 14.5 NGC 3753; and magnitude 14.3 NGC 3754. Please note that the listed magnitudes for these galaxies are all over the (sky) map, so you will find variations to the ones I've listed. Hickson 57 stretches some 6′ in a general north-south orientation.

Be patient when you observe Copeland's Septet, especially through telescopes smaller than 18′ in aperture. The key to spotting all seven is your site's seeing (atmospheric stability). If the stars near the zenith are twinkling at all, move on to the next object, and try for the Septet another night.

| OBJECT #209 | Collinder 249 |
|---|---|
| Constellation | Centaurus |
| Right ascension | 11h38m |
| Declination | −63°22′ |
| Size | 65′ by 40′ |
| Type | Open cluster |
| Other names | The Lambda Centauri Cluster, Caldwell 100 |

For our next target, find magnitude 3.1 Lambda (λ) Centauri. Beginning with that star and running toward the southeast throughout this whole area is the huge open star cluster Collinder 249. Its oval shape spans 1°. But this area contains a lot more than a star cluster. Through a 4-inch telescope from a dark site, you'll spot the HII region IC 2944 (often called the Running Chicken Nebula) concentrated around Lambda Centauri.

IC 2944 is famous for the dense, opaque dust clouds discovered by South African astronomer Andrew David Thackeray in 1950, now known as Thackeray's Globules. Astronomers find such regions in areas of intense star formation. They appear as shadows against the background nebula, which lies 5,900 light-years away. Ultraviolet radiation from nearby, recently formed, stars erodes the globules, and may ultimately dissipate them.

You won't see Thackeray's Globules visually, but through an 8-inch telescope and a low-power eyepiece equipped with a nebula filter, the region of IC 2944 appears bright. Some sources list its magnitude as 4.8. Just 12′ southeast of this nebula lies another, HII region, IC 2948. It appears larger, but fainter, than IC 2944. Some of you astrophotographers will recognize this area as also containing the Running Chicken Nebula.

**Object 210** Abell 1367 Svend and Carl Freytag/Adam Block/NOAO/AURA/NSF

| OBJECT #210 | Abell 1367 |
|---|---|
| Constellation | Leo |
| Right ascension | 11h45m |
| Declination | 19°50′ |
| Size | 100′ |
| Type | Galaxy cluster |
| Other name | The Leo Cluster |

This vast cluster sits mainly to the southwest of the magnitude 4.5 star 93 Leonis, although it extends a bit north and east of that star as well. Surveys have identified 542 galaxies within 1° of the cluster's center, and 1,682 within 2°. Abell 1367 sprawls over a region nearly 2.5° in diameter. It lies 330 million light-years away. At this distance, the galaxies' light appears faint, but a couple dozen members are bright enough to earn NGC designations.

For example, the centrally placed, nearly 1′ glow of magnitude 11.8 NGC 3842 shows up nicely through an 8-inch telescope at a dark site. Others include magnitude 12.7 NGC 3861, magnitude 12.7 NGC 3862, magnitude 13.3 NGC 3837, and magnitude 13.7 NGC 3840.

Upgrade to a 14-inch scope, insert an eyepiece that magnifies 300×, and scan the region, and you'll view 50 tiny galaxies. I've observed this cluster through a 30-inch reflector at a true-dark site. I can say honestly that, had I not stopped myself, I could have spent the entire night viewing Abell 1367. Sounds like a great future observing project to me!

| OBJECT #211 | NGC 3877 |
|---|---|
| Constellation | Ursa Major |
| Right ascension | 11h46m |
| Declination | 47°30′ |
| Magnitude | 11.0 |
| Size | 5.1′ by 1.1′ |
| Type | Spiral galaxy |

NGC 3877 is an easy galaxy to find. Just locate magnitude 3.7 Chi (χ) Ursae Majoris. If NGC 3877 isn't already in your field of view, simply pan south 0.3°.

This spiral tilts edge-on to us, although not perfectly so. You'll still be able to spot the angled disk. Few other details will be visible, however.

NGC 3877 is a member of the M109 group of galaxies. This small celestial family contains at least three dozen galaxies, with another dozen and a half possible members.

| OBJECT #212 | NGC 3887 |
|---|---|
| Constellation | Crater |
| Right ascension | 11h47m |
| Declination | −16°51′ |
| Magnitude | 10.6 |
| Size | 3.5′ by 2.4′ |
| Type | Barred spiral galaxy |

Our next target lies 1.6° north-northeast of magnitude 4.7 Zeta (ζ) Crateris. Through an 8-inch telescope at 200×, you'll see an oval haze 50% longer than wide with a slightly brighter center oriented north-northeast to south-southwest. A magnitude 12.0 star sits at the northern end.

| OBJECT #213 | NGC 3898 |
|---|---|
| Constellation | Ursa Major |
| Right ascension | 11h49m |
| Declination | 56°05′ |
| Magnitude | 10.8 |
| Size | 4.4′ by 2.6′ |
| Type | Spiral galaxy |

When you observe NGC 3898, look for its nearby companion, magnitude 12.0 NGC 3888, which sits 16′ to the southwest. On a 2009 observing trip, I saw both of these galaxies easily through a 30-inch Starmaster reflector. In a way, this pair reminded me of another, much brighter, galaxy duo, M81 and M82.

Like M81, NGC 3898 is brighter than its partner. But, like M82, NGC 3888 appears easier to see because it's smaller and has a higher surface brightness. Oh, and if you can use a 16-inch or larger aperture, look for magnitude 14.8 NGC 3889 between the two brighter galaxies.

To find NGC 3898, look 2.5° north-northwest of magnitude 2.4 Phecda (Gamma [γ] Ursae Majoris).

| OBJECT #214 | NGC 3918 |
|---|---|
| Constellation | Centaurus |
| Right ascension | 11h50m |
| Declination | −57°11′ |
| Magnitude | 8.1 |
| Size | 12″ |
| Type | Planetary nebula |
| Other name | The Blue Planetary |

This nice object lies 3.6° west-northwest of magnitude 2.8 Delta (δ) Crucis. Even small telescopes reveal the nebula's vivid blue, greenish-blue, or blue-green color (depending on your eyes' color sensitivity). The disk has incredible surface brightness and a sharp edge. Unfortunately, although you can crank up the magnification past 500× on this object, you won't see details other than the color.

| OBJECT #215 | NGC 3923 |
|---|---|
| Constellation | Hydra |
| Right ascension | 11h51m |
| Declination | −28°48′ |
| Magnitude | 9.6 |
| Size | 6.9′ by 4.8′ |
| Type | Elliptical galaxy |

Our next target sits 5.6° southwest of magnitude 4.0 Alchiba (Alpha [α] Corvi). Through an 8-inch telescope at 150×, you'll see an evenly illuminated oval that measures 50% longer than it is wide oriented northeast to southwest. The central region is broad and bright and a large oval halo surrounds it. If you view NGC 3923 through a light-bucket, look for magnitude 14.9 PGC 100033, which lies 8′ to the east.

**Object #216** NGC 3953 Tom and Gail Haynes/Adam Block/NOAO/AURA/NSF

| OBJECT #216 | NGC 3953 |
|---|---|
| Constellation | Ursa Major |
| Right ascension | 11h53m |
| Declination | 52°19′ |
| Magnitude | 9.8 |
| Size | 6.9′ by 3.6′ |
| Type | Barred spiral galaxy |

This bright galaxy lies 1.4° due south of magnitude 2.4 Phecda (Gamma [γ] Ursae Majoris). Although you can see it through a 4-inch telescope (and even through binoculars), many details emerge through a 12-inch or larger scope.

You'll see a north-south oriented oval that's twice as long as it is wide. At magnifications above 300×, the center is bright and appears extended in a rough east-west direction. Is that the bar? The northern edge outshines the southern and looks rounder. The southern edge thins rapidly. Images confirm that this is because of a spiral arm that curves westward.

| OBJECT #217 | NGC 3962 |
|---|---|
| Constellation | Crater |
| Right ascension | 11h55m |
| Declination | −13°58′ |
| Magnitude | 10.7 |
| Size | 2.6′ by 2.2′ |
| Type | Elliptical galaxy |

Look 3.2° north of magnitude 5.7 Eta (η) Crateris. Through a 12-inch telescope, the galaxy is ever-so-slightly oval in a north-south orientation. Its central region appears bright and extended, and only with extreme magnification will you spot the thin halo.

| OBJECT #218 | M109 (NGC 3992) |
|---|---|
| Constellation | Ursa Major |
| Right ascension | 11h58m |
| Declination | 53°23′ |
| Magnitude | 9.8 |
| Size | 7.6′ by 4.3′ |
| Type | Barred spiral galaxy |

If you're star-hopping to M109, first center magnitude 2.4 Phecda (Gamma [γ] Ursae Majoris) in your telescope. From Phecda, move 0.6° to the east-southeast.

Although M109 has a low surface brightness, it is still a great example of a barred spiral galaxy. An 8-inch telescope under superb conditions will reveal the bar. Be sure to use a magnification of 200× or above.

The bar extends from either side of a bright, non-stellar nucleus. It measures about 1′ wide.

| OBJECT #219 | NGC 3994 |
|---|---|
| Constellation | Ursa Major |
| Right ascension | 11h58m |
| Declination | 32°17′ |
| Magnitude | 12.7 |
| Size | 0.9′ by 0.5′ |
| Type | Spiral galaxy |

Now here's a small group of galaxies best suited for a large-aperture telescope. Before an early-2009 observing outing, the last time I observed this faint trio was in 1977. I gotta get out more. To find this object, look 7.5° northwest of magnitude 4.3 Gamma (γ) Comae Berenices.

NGC 3995 is the most interesting because it reveals faint hints of non-symmetrical spiral arms. Although NGC 3994 has the same magnitude as NGC 3995, it appeared brighter because of its smaller size. NGC 3991 glows somewhat fainter at magnitude 13.0. One of the cool things about the galaxies in this group is their nearly similar orientations.

| OBJECT #220 | NGC 3998 |
|---|---|
| Constellation | Ursa Major |
| Right ascension | 11h58m |
| Declination | 55°27′ |
| Magnitude | 10.6 |
| Size | 2.7′ by 2.3′ |
| Type | Spiral galaxy |

Remember the original *Star Trek* episode "The Trouble with Tribbles"? When I observe NGC 3998, it reminds me of a Tribble because it's oval, it has a soft edge, and I got the sense of three-dimensionality from viewing it. Call me crazy. Another fairly bright galaxy, NGC 3990, sits 3′ west of NGC 3998. Between the two, look for MGC +09–20–046 through a 16-inch or larger telescope. I've seen it through a 30-inch scope as a (still) faint oval I could observe with direct vision.

This object lies 1.9° north-northeast of magnitude 2.4 Phecda (Gamma [$\gamma$] Ursae Majoris).

| OBJECT #221 | Corvus the Crow |
|---|---|
| Right ascension | 12h24m |
| Declination | −18° |
| Size | 183.8 square degrees |
| Type | Constellation |

Here's a great star group to look for if you're just starting out: the small but easy-to-see Corvus. This constellation only has three bordering star patterns: Virgo the Maiden lies to the north and east; Crater the Cup lies to the west; and Hydra the Water Snake sits to the south.

In size, little Corvus ranks 70th out of the 88 constellations that cover the sky. It occupies only 0.45% of the sky.

The best date to see Corvus is March 28. That's when the constellation lies opposite the Sun. It rises at sunset, sets at sunrise, and is visible in the sky all night. The worst date to look for Corvus is September 27. That's when its stars line up with the Sun.

Although Corvus has no Messier object within its borders, it's a near miss. Just to the constellation's north, in Virgo, is one of the best galaxies in the sky. It's the Sombrero Galaxy (M104).

You can use Corvus' top two stars as pointers to a bright blue luminary. Draw a line from magnitude 2.6 Gienah (Gamma [$\gamma$] Corvi) through magnitude 2.9 Algorab (Delta [$\delta$] Corvi), and extend that line 4 times the distance between the two stars, and you'll arrive at Spica, the brightest star in Virgo.

| OBJECT #222 | NGC 4027 |
|---|---|
| Constellation | Corvus |
| Right ascension | 12h00m |
| Declination | −19°16′ |
| Magnitude | 11.2 |
| Size | 3.8′ by 2.3′ |
| Type | Barred spiral galaxy |

Our next target lies 4.2° west-southwest of magnitude 2.6 Gienah (Gamma [$\gamma$] Corvi). In a 16-inch scope, the galaxy's comma shape is evident. The central region is irregular and broadly concentrated. A faint arc reaches from the west end and curves to the north. Scientists think NGC 4027's one-armed spiral shape resulted from an interaction with nearby NGC 4027A. Through a 14-inch or larger scope, you'll spot this magnitude 15.0 spiral 4′ to the south. It measures nearly 1′ across.

| OBJECT #223 | Epsilon (ε) Chamaeleontis |
|---|---|
| Constellation | Chamaeleon |
| Right ascension | 12h00m |
| Declination | −78°13′ |
| Magnitudes | 5.4/6.0 |
| Separation | 0.4″ |
| Type | Double star |

So, you want to test the optics in your 12-inch telescope? Well, here's the star. With a sub-arcsecond separation, Epsilon Cha requires large aperture and high power. A clean split shows most observers a bluish-white or white primary and a yellow-white secondary.

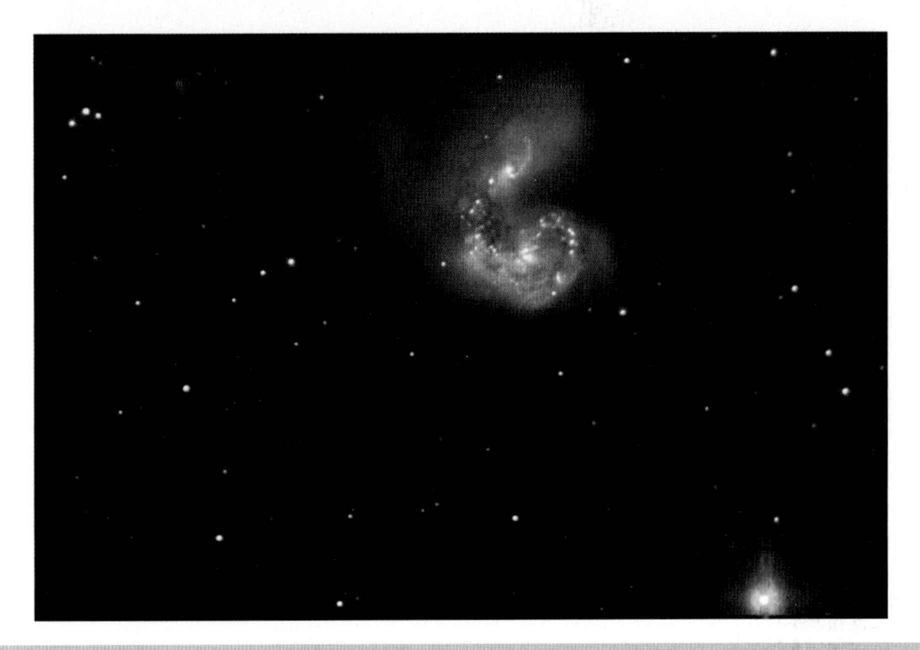

**Object #224** The Antennae (NGC 4038 + NGC 4039) Bob and Bill Twardy/Adam Block/NOAO/AURA/NSF

| OBJECT #224 | NGC 4038–9 |
|---|---|
| Constellation | Corvus |
| Right ascension | 12h02m |
| Declination | −18°52′ |
| Magnitude | 10.5 |
| Size | 5.4′ by 3.9′ |
| Type | Interacting galaxies |
| Other names | The Antennae, the Ringtail Galaxies, Caldwell 60 (NGC 4038), Caldwell 61 (NGC 4039) |

Astronomers refer to NGC 4038 and its companion NGC 4039 as the Antennae because of their bright tails. These debris trails are the result of violent tidal interactions between the two massive galaxies. Along the length of these tails are concentrations of stars as large as dwarf galaxies. At a distance of about 60 million light-years, the Antennae represent the nearest example of a galaxy collision. You'll find this pair 3.6° west-southwest of magnitude 2.6 Gienah (Gamma [$\gamma$] Corvi).

Through a 4-inch telescope, you'll see two incredibly faint, cottony smudges. The larger and brighter smudge, NGC 4038, sits to the northwest. A 12-inch telescope at 200× shows both galaxies' cores as ovals roughly twice as long as they are wide. Double the magnification, and you'll see a patchy surface with bright and dark knots and traces of at least one tidal tail. To study these features that give the Antennae their name, you'll need to view through at least a 20-inch scope. The tails' surface brightness is uniform, so you'll be able to trace their entire length.

**Object #225** NGC 4051 George Seitz/Adam Block/NOAO/AURA/NSF

| OBJECT #225 | NGC 4051 |
|---|---|
| Constellation | Ursa Major |
| Right ascension | 12h03m |
| Declination | 44°31′ |
| Magnitude | 10.0 |
| Size | 5.2′ by 3.9′ |
| Type | Spiral galaxy |

Our next target lies 4.4° southeast of magnitude 3.7 Chi ($\chi$) Ursae Majoris. A small telescope shows an oval 50% as long as it is wide oriented from the northwest to the southeast. A 14-inch telescope at 300× shows the galaxy to be riddled with tiny dark markings. The central region is the center of a faint S shape. The brightest extension of the S begins at the southeastern edge and winds to the north.

| OBJECT #226 | NGC 4052 |
|---|---|
| Constellation | Crux |
| Right ascension | 12h02m |

| (continued) | |
|---|---|
| Declination | −63°12′ |
| Magnitude | 8.8 |
| Size | 7′ |
| Type | Open cluster |

Our next target makes a nearly equilateral triangle with magnitude 4.3 Theta[1] ($\theta^1$) and magnitude 4.7 Theta[2] ($\theta^2$) Crucis. NGC 4052 appears west of them in the field of view of a medium-power eyepiece. Through an 8-inch telescope at 150×, you'll spot 50 stars ranging from 11th to 13th magnitude. One nice east-west line of more than half a dozen stars stretches south of the center. Be warned. This cluster isn't easy to pick out from the surrounding rich Milky Way.

| OBJECT #227 | NGC 4062 |
|---|---|
| Constellation | Ursa Major |
| Right ascension | 12h04m |
| Declination | 31°54′ |
| Magnitude | 11.2 |
| Size | 4.0′ by 1.8′ |
| Type | Spiral galaxy |

This object lies 6.1° northwest of magnitude 4.3 Gamma ($\gamma$) Comae Berenices. Although you can spot NGC 4062 through an 8-inch telescope, details are lacking. Even through a 30-inch reflector under a jet-black sky, I detected no spiral detail in this galaxy. What I did notice was an ever-so-gradual drop in brightness from a wide nucleus to its edges. It seemed the longer I looked, the farther into space the outer diaphanous halo stretched.

| OBJECT #228 | NGC 4105 |
|---|---|
| Constellation | Hydra |
| Right ascension | 12h07m |
| Declination | −29°46′ |
| Magnitude | 10.4 |
| Size | 3.7′ by 1.7′ |
| Type | Elliptical galaxy |

Our next object lies 5° south of magnitude 4.0 Alchiba (Alpha [$\alpha$] Corvi). NGC 4105 is the brighter and westernmost of a close pair of small galaxies. The center of the overlapping magnitude 11.4 galaxy NGC 4106 lies only 1′ away. The high surface brightness of this pair allows you to use high magnification.

NGC 4105 appears just slightly out of round with a bright center. NGC 4106, a near twin of NGC 4105, measures 1.6′ by 1.3′.

Two fainter galaxies lie equidistant nearby. Magnitude 13.1 IC 3005 lies 17′ south-southeast, and magnitude 13.5 IC 2996 sits 17′ southwest.

| OBJECT #229 | NGC 4103 |
|---|---|
| Constellation | Crux |
| Right ascension | 12h07m |
| Declination | −61°15′ |
| Magnitude | 7.4 |
| Size | 6′ |
| Type | Open cluster |

This easy-to-see target lies along Crux's western edge 2° west-southwest of magnitude 3.6 Epsilon (ε) Crucis. Through a 6-inch telescope at 150×, you'll see a tight grouping of some 50 stars randomly distributed across the field of view. Near the center, half a dozen pairs of stars are easy to see.

| OBJECT #230 | Canes Venatici |
|---|---|
| Right ascension | 13h04m |
| Declination | 40°30′ |
| Size | 465.19 square degrees |
| Type | Constellation |

Our next object may be a bit difficult to see for beginners. It's the constellation Canes Venatici the Hunting Dogs. This constellation only has three bordering star patterns: Ursa Major the Great Bear lies to the north and west; Coma Berenices the Hair of Berenice forms part of its southern border; and Boötes the Herdsman lies to the east and also completes the southern border.

In size, Canes Venatici ranks 38th out of the 88 constellations that cover the sky. It occupies a respectable 1.13% of the sky.

The best time to see Canes Venatici is early April. That's when the constellation lies opposite the Sun. It rises at sunset, sets at sunrise, and is visible in the sky all night. The worst time to look for Canes Venatici is in early October. That's when its stars line up with the Sun.

Canes Venatici bustles with deep-sky objects. It contains five Messier objects within its borders: globular cluster M3, and spiral galaxies M51, M63, M94, and M106.

Finding Canes Venatici isn't easy. It's rated as the 5th faintest constellation in the sky. From a dark site, look just below the Big Dipper's handle. The brightest star in this region — and that's not saying much — is Cor Caroli (Alpha [α] Canum Venaticorum), which shines at magnitude 2.8. Canes Venatici's only other reasonably bright star, Chara (Beta [β] Canum Venaticorum) lies 5° to the west-northwest and glows weakly at magnitude 4.3.

| OBJECT #231 | NGC 4111 |
|---|---|
| Constellation | Canes Venatici |
| Right ascension | 12h07m |
| Declination | 43°04′ |
| Magnitude | 10.7 |
| Size | 4.4′ by 0.9′ |
| Type | Spiral galaxy |

On the western border of Canes Venatici, you'll find one of the constellation's many classic edge-on galaxies — and another of my celestial "needles." Just look 5.2° west-northwest of magnitude 4.2 Chara (Beta [β] Canum Venaticorum). Through an 8-inch telescope, this galaxy appears featureless, five times as long as it is wide oriented northwest to southeast, and evenly illuminated. The central region appears oval.

The magnitude 8.1 star SAO 44039 sits 30″ to the northeast. While you're in the area, look for several nearby galaxies: Magnitude 13.0 NGC 4117 and magnitude 14.7 NGC 4118 lie 8′ east-northeast, and magnitude 14.1 NGC 4109 sits 5′ south-southwest.

| OBJECT #232 | NGC 4125 |
|---|---|
| Constellation | Draco |
| Right ascension | 12h08m |
| Declination | 65°11′ |
| Magnitude | 9.7 |
| Size | 6.1′ by 5.1′ |
| Type | Elliptical galaxy |

To find our next object, move 5.5° southeast from magnitude 3.8 Lambda (λ) Draconis. NGC 4125 also forms an equilateral triangle (extending upward) with the top stars in the bowl of the Big Dipper, Dubhe and Megrez (Alpha [α] and Delta [δ] Ursae Majoris, respectively). The galaxy lies 8° from each of those bright stars.

Through an 8-inch telescope, you'll see a bright, broadly concentrated oval core with a surrounding haze. Larger apertures extend the length of the elliptical galaxy, and some observers have even noted a nearly stellar nucleus.

| OBJECT #233 | NGC 4143 |
|---|---|
| Constellation | Canes Venatici |
| Right ascension | 12h10m |
| Declination | 42°32′ |
| Magnitude | 10.7 |
| Size | 2.9′ by 1.9′ |
| Type | Spiral galaxy |

The small spiral galaxy NGC 4143 lies 4.6° west-southwest of magnitude 4.2 Chara (Beta [β] Canum Venaticorum). It lies at the southern edge of the Ursa Major galaxy cluster. This gathering contains some 80 galaxies.

NGC 4143 has a high surface brightness, so you can really crank up the magnification. Unfortunately, even at 350× all you'll see is a bright central region surrounded by a faint, oval halo twice as long as it is wide.

**Object #234** M98 Adam Block/NOAO/AURA/NSF

| OBJECT #234 | M98 (NGC 4192) |
|---|---|
| Constellation | Coma Berenices |
| Right ascension | 12h14m |

| (continued) | |
|---|---|
| Declination | 14°54′ |
| Magnitude | 10.1 |
| Size | 9.1′ by 2.1′ |
| Type | Spiral galaxy |

This fine target lies 7.2° east of magnitude 2.1 Denebola (Beta [$\beta$] Leonis). Through an 8-inch telescope at 200×, you'll see a fat spindle four times as long as it is wide oriented north-northwest to south-southeast. The central region is broad and slightly brighter than the arms. Through a 16-inch telescope and an eyepiece with a nebula filter, you'll see knots in the spiral arms marking the positions of large star-forming regions.

| OBJECT #235 | NGC 4214 |
|---|---|
| Constellation | Canes Venatici |
| Right ascension | 12h16m |
| Declination | 36°20′ |
| Magnitude | 9.6 |
| Size | 8′ by 6.6′ |
| Type | Irregular galaxy |

Our next object lies 8.3° west of magnitude 2.9 Cor Caroli (Alpha [$\alpha$] Canum Venaticorum). Through an 8-inch telescope at 200×, this galaxy orients northwest to southeast. The central region is long and bright, and the halo is large. Through larger apertures and higher magnifications, both appear irregular.

| OBJECT #236 | The Y of Virgo |
|---|---|
| Constellation | Virgo |
| Right ascension | 12h42m |
| Declination | −1°27′ |
| Type | Asterism |

Our next object may be a bit out of sequence, as far as its right ascension goes, but I wanted to introduce it before we get into individual deep-sky treats within the constellation itself. The Y is a large asterism in Virgo. It's made up of six stars, the faintest of which still shines brighter than 4th magnitude, so you'll see it easily even from the suburbs.

Start at Virgo's brightest star, Spica (Alpha [$\alpha$] Virginis). This blue-white 1st-magnitude luminary marks the bottom of the Y. From Spica, move 14.5° northwest to magnitude 2.7 Porrima (Gamma [$\gamma$] Virginis).

From Porrima, the Y branches in two directions, toward the north-northeast and the west-northwest. And if you're a fan of strange star names, you'll enjoy these. The north-northeast leg contains Minelauva (Delta [$\delta$] Virginis) and Vindemiatrix (Epsilon [$\varepsilon$] Virginis). The west-northwest branch contains Zaniah (Eta [$\eta$] Virginis) and Zavijava (Beta [$\beta$] Virginis).

**Object #237** The Silver Streak Galaxy (NGC 4216) Ken Siarkiewicz/Adam Block/NOAO/AURA/NSF

| OBJECT #237 | NGC 4216 |
|---|---|
| Constellation | Virgo |
| Right ascension | 12h16m |
| Declination | 13°09′ |
| Magnitude | 10.0 |
| Size | 7.8′ by 1.6′ |
| Type | Spiral galaxy |
| Other names | The Silver Streak Galaxy, the Weaver's Shuttle Galaxy |

To find NGC 4216, start at the hindquarters of Leo the Lion. From Chertan (Theta [$\theta$] Leonis), draw a line through Denebola (Beta [$\beta$] Leonis), and extend it nearly as far eastward as the distance between those two stars. Your destination is the western edge of the Coma-Virgo Cluster of galaxies. From a dark site, a 10-inch telescope will reveal several hundred galaxies here, so use a high-quality star chart, and be sure of your identifications.

NGC 4216 itself appears as a streak of light four times as long as it is wide. The core is bright, but to see its bulge requires a 12-inch or larger scope. Look for 12th-magnitude NGC 4206 lies 12′ to the southwest of NGC 4216. It has a similar appearance to its brighter neighbor.

The name "Weaver's Shuttle" comes to this galaxy from the astronomy popularizer George F. Chambers (1841–1915), who revised Admiral William Smyth's *Cycle of Celestial Objects* in 1881, Chambers described NGC 4216 this way: "This is a very curious object, in shape resembling a weaver's shuttle, and lying across the parallel; the upper branch is the faintest, and the centre exhibits a palpable nucleus, which in my instrument brightens at intervals, as the eye rallies."

| OBJECT #238 | NGC 4236 |
|---|---|
| Constellation | Draco |
| Right ascension | 12h17m |
| Declination | 69°28′ |
| Magnitude | 9.6 |
| Size | 21.0′ by 7.5′ |

| (continued) | |
|---|---|
| Type | Barred spiral galaxy |
| Other name | Caldwell 3 |

You'll find our next target 1.5° west of magnitude 3.9 Kappa (κ) Draconis. Although NGC 4236's listed magnitude is bright, it's so huge that it has a low surface brightness, so you might be hard pressed to see it through any telescope less than 10 inches in aperture.

It appears as a ghostly mist six times as long as it is wide. Larger telescopes reveal a number of faint knots that are star-forming regions similar to our galaxy's Orion Nebula. Through a 14-inch telescope, look for the brightest of these HII regions at the south end of the spindle.

**Object #239** The Silver Needle Galaxy (NGC 4244) Joe Naughton and Steve Stafford/Adam Block/NOAO/AURA/NSF

| OBJECT #239 | NGC 4244 |
|---|---|
| Constellation | Canes Venatici |
| Right ascension | 12h18m |
| Declination | 37°49′ |
| Magnitude | 10.4 |
| Size | 17.0′ by 2.2′ |
| Type | Spiral galaxy |
| Other names | The Silver Needle Galaxy, Caldwell 26 |

To observers, spring in the Northern Hemisphere means it's time to hunt galaxies. And while the giant constellations Virgo, Ursa Major, and Leo hold hundreds of worthy targets, don't overlook Canes Venatici. Although its stars are faint (Canes Venatici is only the 84th brightest constellation), it contains many bright galaxies.

Four of these made Messier's list — M51, M63, M94, and M106. And although the Silver Needle Galaxy doesn't stand out like these luminaries, it has a lot to offer observers.

NGC 4244's disk appears almost edge-on, tilting only 5° to our line of sight. Through the eyepiece, this galaxy has a relatively even illumination — the core appears only slightly brighter than the rest.

What sets the Silver Needle Galaxy apart, however, is its length-to-width ratio and its appearance through small telescopes.

Once you view this object through a 4-inch or smaller telescope from a dark site, you'll understand why its name contains the word "needle." Through larger scopes, the needle-like quality disappears. To the northeast, the galaxy fades gradually into a broad haze. On the opposite side of the core, the surface appears more mottled and irregular.

**Object #240** M99 Adam Block/NOAO/AURA/NSF

| OBJECT #240 | M99 (NGC 4254) |
|---|---|
| Constellation | Coma Berenices |
| Right ascension | 12h19m |
| Declination | 14°25′ |
| Magnitude | 9.9 |
| Size | 4.6′ by 4.3′ |
| Type | Spiral galaxy |
| Other names | The Pinwheel Nebula, St. Catherine's Wheel |

Our next object is a galaxy, but not the more famous Pinwheel Galaxy, which is M51 in Canes Venatici. The so-called Pinwheel Nebula (a nineteenth -century term for any diffuse deep-sky object) lies at the southwestern edge of Coma Berenices. The nearest bright star is Denebola (Beta [$\beta$] Leonis). Once you find that magnitude 2.1 luminary, M99 lies 7.2° due east.

When you first observe M99, you might think the galaxy looks a bit "off." In fact, even through a 10-inch telescope, M99 appears to have only one spiral arm, which winds under the galaxy's southern edge. This arm appears brighter because it's full of large star-forming regions. Through the telescope, you'll recognize them as the bright "clumpy" areas.

You'll need at least a 14-inch telescope and magnifications above 250× to see M99's two other spiral arms. They're thinner than the southern arm, but the one that extends northward is the easier of the two to see. The core these arms attach to is large and evenly illuminated, and it takes up about a quarter of M99's overall diameter.

I don't know who gave M99 its other common name, St. Catherine's Wheel, but that's not the kind of mental picture I want to bring to the eyepiece. Named after the Christian martyr St. Catherine of Alexandria, the wheel was a medieval torture device used to break the bones of anyone placed upon it.

**Object #241** M106 Adrian Zsilavec and Michelle Qualls/Adam Block/NOAO/AURA/NSF

| OBJECT #241 | M106 (NGC 4258) |
|---|---|
| Constellation | Canes Venatici |
| Right ascension | 12h19m |
| Declination | 47°18′ |
| Magnitude | 8.4 |
| Size | 20.0′ by 8.4′ |
| Type | Spiral galaxy |

One of the best and brightest galaxies on this or any other list (including Messier's) also ranks as one of the least observed. It may be that because M106 falls near the end of Messier's list, observers think it's not a worthy target. Perhaps because it doesn't have a catchy name like the Sombrero Galaxy (M104), it seems less appealing. Nothing could be further from the truth.

M106's equatorial plane tilts to our line of sight, as does that of the Andromeda Galaxy (M31), so many features resemble those seen in that object. This orientation may explain why M106's dust lanes appear so prominent.

M106 lies in the northwest corner of Canes Venatici, and it makes a triangle with magnitude 3.7 Chi ($\chi$) Ursae Majoris and magnitude 2.4 Phecda (Gamma [$\gamma$] Ursae Majoris). It lies 5.5° east of Chi and 7.5° southeast of Gamma. First impressions are of a bright core surrounded by an oval haze. Surrounding the nucleus is a stretched-out inner disk a third of the galaxy's size.

Through a 10-inch or larger telescope, you'll begin to see the mottled texture and spiral structure. The strikingly linear northern arm appears more pronounced, while the southern arm looks more diffuse. Be patient when observing this arm, and wait for moments of good seeing to help you pull out detail.

| OBJECT #242 | NGC 4278 |
|---|---|
| Constellation | Coma Berenices |
| Right ascension | 12h20m |
| Declination | 29°17′ |
| Magnitude | 10.1 |
| Size | 3.8′ |
| Type | Elliptical galaxy |

Our next target lies 1.8° northwest of magnitude 4.3 Gamma ($\gamma$) Comae Berenices. Through a 10-inch telescope at magnifications above 250×, this galaxy resembles an unresolved globular cluster. The brightness decreases from an intense core through the bright central region to the halo and off to the blackness of space.

Look for another elliptical galaxy, magnitude 13.1 NGC 4283, which lies 3′ to the east-northeast. It measures 1.4′ across.

**Object #243** NGC 4298 + NGC 4302 Jeff Hapeman/Adam Block/NOAO/AURA/NSF

| OBJECT #243 | NGC 4302 |
|---|---|
| Constellation | Coma Berenices |
| Right ascension | 12h22m |
| Declination | 14°36′ |
| Magnitude | 11.6 |
| Size | 4.7′ by 0.9′ |
| Type | Spiral galaxy |

To find our next object, you'll have to dig deeply into the heart of the Coma-Virgo Cluster of galaxies. From the magnitude 5.1 star 6 Comae Berenices, nudge your telescope 1.4° to the east-southeast. Oriented north-south, NGC 4302 appears more than four times as long as it is wide. Its surface brightness is uniform except for a slightly brighter wide center.

Only 2′ east of NGC 4302 lies magnitude 11.4 NGC 4298. It measures 3.2′ by 1.9′ and has a broad central concentration.

**Object #244** M61 Adam Block/NOAO/AURA/NSF

| OBJECT #244 | M61 (NGC 4303) |
|---|---|
| Constellation | Virgo |
| Right ascension | 12h22m |
| Declination | 4°28′ |
| Magnitude | 9.7 |
| Size | 6.0′ by 5.9′ |
| Type | Spiral galaxy |

To find M61, first locate the stars 16 and 17 Virginis. The former shines at magnitude 5.0, and the latter at magnitude 6.5. M61 lies between these two stars, only a bit closer to 17 Virginis. This is a face-on spiral galaxy, however its arms wind tightly around the core, so this object doesn't look nearly as good as M101. That said, a 12-inch telescope will allow you to see the stubby extensions of two arms. Through really big scopes at high magnification, look for a thick bar that runs north-south through this object.

In his 1844 classic, *Cycle of Celestial Objects*, William Smyth describes M61 as "A large pale-white nebula between the Virgin's shoulders. This is a well defined object, but so feeble as to excite surprise that Messier detected it with his 3½-foot telescope in 1779."

| OBJECT #245 | M100 (NGC 4321) |
|---|---|
| Constellation | Coma Berenices |
| Right ascension | 12h23m |
| Declination | 15°49′ |
| Magnitude | 9.3 |
| Size | 6.2′ by 5.3′ |
| Type | Spiral galaxy |

In the Northern Hemisphere's spring, the Milky Way encircles us along the horizon, turning the rest of the sky into a large, clear window for viewing distant galaxies. The north galactic pole, that point in the sky farthest from the plane of our galaxy, resides in the constellation Coma Berenices.

Despite being a mid-sized constellation, it contains no fewer than eight Messier objects — including M100 — and dozens of bright, nearby galaxies of the Coma-Virgo galaxy cluster.

The Coma-Virgo cluster, sometimes shortened to the Virgo cluster, lies approximately 60 million light-years away. Astronomers estimate between 1,500 and 2,000 galaxies populate this cluster, which forms the heart of the Local Supercluster. The Local Group, of which our Milky Way is part, is another collection of galaxies within the Local Supercluster. The Virgo cluster's mass measures some 1.2 quadrillion ($1.2 \times 10^{15}$) suns and it spans more than 14 million light-years.

Pierre Méchain discovered M100 March 15, 1781. It ranks as one of the brightest galaxies in the Coma-Virgo cluster, and it is a great target for amateur telescopes. It lies 8.3° east of magnitude 2.1 Denebola (Beta [$\beta$] Leonis), or 1.9° east-northeast of the magnitude 5.1 star 6 Comae Berenices.

Through an 8-inch scope at low to medium power, look for a haze about 4′ long. You won't see this galaxy's spiral arms until you crank the magnification past 200×, and then only on the best nights.

The arms appear as brighter regions just to the east and west of the nucleus. Through a 12-inch telescope, you'll trace the spiral structure twice as far from the core. Two faint dwarf galaxies lie to the north and east. Magnitude 13.9 NGC 4322, to the north, appears to be M100's true companion, while magnitude 13.3 NGC 4328 lies in the foreground to the east.

**Object #246** M40 Anthony Ayiomamitis

| OBJECT #246 | M40 |
|---|---|
| Constellation | Ursa Major |
| Right ascension | 12h22m |
| Declination | −28°05′ |
| Magnitude | 9.0/9.6 |
| Separation | 53″ |
| Type | Double star |
| Other name | Winnecke 4 |

M40 is the most unusual entry in Charles Messier's catalog. The object is a widely spaced double star. And, it's not even that interesting. The two stars glow faintly at 9th magnitudes and the separation is nearly an arcminute wide. The primary star appears light yellow, and the secondary is a deep-yellow, almost bordering on pale orange.

Despite its ho-hum nature, M40 still is a Messier object, and so it deserves at least one look during your observing lifetime. You'll find it 1.4° northeast of magnitude 3.3 Megrez (Delta [δ] Ursae Majoris).

| OBJECT #247 | NGC 4349 |
|---|---|
| Constellation | Crux |
| Right ascension | 12h25m |
| Declination | −61°54′ |
| Magnitude | 7.4 |
| Size | 15′ |
| Type | Open cluster |

You'll find this object 1.3° north-northwest of magnitude 0.8 Acrux (Alpha [α] Crucis). Through a 6-inch telescope at 150×, you'll count 75 stars. The cluster's brightest star, magnitude 8.4 SAO 251883, sits at the southeastern end. A straight line of a half dozen 11th- and 12th-magnitude stars sits nearer the center and stretches from the northwest to the southeast.

**Object #248** NGC 4361 Adam Block/Mount Lemmon SkyCenter/University of Arizona

| OBJECT #248 | NGC 4361 |
|---|---|
| Constellation | Corvus |
| Right ascension | 12h25m |
| Declination | –18°47′ |
| Magnitude | 10.9 |
| Size | 45″ |
| Type | Planetary nebula |

Our next target lies 2.6° south-southwest of magnitude 2.9 Algorab (Delta [$\delta$] Corvi). Through an 8-inch telescope at 250×, you'll see a slightly stretched irregular disk oriented northeast to southwest with a diffuse edge. The 13th-magnitude central star is just visible, and is easy through larger apertures.

| OBJECT #249 | NGC 4365 |
|---|---|
| Constellation | Virgo |
| Right ascension | 12h24m |
| Declination | 7°19′ |
| Magnitude | 9.6 |
| Size | 6.9′ by 5′ |
| Type | Elliptical galaxy |

This object lies 5.2° southwest of magnitude 4.9 Rho ($\rho$) Virginis. Trek carefully here because many other galaxies lie in this region. Through an 8-inch telescope at 200×, you'll see a fat oval oriented northeast to southwest. The central region is bright and the surrounding halo is wide.

Magnitude 12.9 NGC 4370 lies 10′ to the northeast. If you have a large scope, try for magnitude 14.3 NGC 4366, which sits midway between the two galaxies.

| OBJECT #250 | M84 (NGC 4374) |
|---|---|
| Constellation | Virgo |
| Right ascension | 12h25m |
| Declination | 12°53′ |
| Magnitude | 9.1 |
| Size | 5.1′ by 4.1′ |
| Type | Elliptical galaxy |

It's easy to confuse M84 with nearby M86. Both lie midway between magnitude 2.1 Denebola (Beta [$\beta$] Leonis) and magnitude 2.9 Vindemiatrix (Epsilon [$\varepsilon$] Virginis). M84 lies a bit more westerly. Also, under high magnifications, M84 appears smaller and slightly fainter than its neighbor.

Although this galaxy was bright enough for Charles Messier to see, it has few distinguishing characteristics. Its core is large and definitely nonstellar. Look for a fainter halo that surrounds the wide core.

| OBJECT #251 | Melotte 111 |
|---|---|
| Constellation | Coma Berenices |
| Right ascension | 12h25m |
| Declination | 26°06′ |
| Magnitude | 1.8 |
| Size | 275′ |
| Type | Open cluster |
| Other names | The Coma Berenices star cluster, Ariadne's Hair, Thisbe's Veil |

Coma Berenices is a poorly defined constellation containing only three stars brighter than magnitude 4.5. In this star group's northwest corner, you'll find the yellow star Gamma ($\gamma$) Comae Berenices.

Gamma sits 170 light-years away, and it rests in front of the Coma Berenices Cluster, which lies about 100 light-years farther from us.

This loose open cluster also carries the designation Melotte 111. British astronomer Philibert Jacques Melotte (1880–1961) placed it in *A catalog of star clusters shown on the Franklin-Adams chart plates*, which appeared in the *Memoirs of the Royal Astronomical Society*, volume 60, published in 1915. Still, it wasn't until 1938 that astronomers recognized the Coma Berenices Cluster as a true star cluster.

Melotte 111 contains some 40 stars between magnitudes 5 and 10. About a dozen rise above naked-eye visibility. Because this object spans more than 4°, you'll need optics with a wide field of view to see all the stars simultaneously. Start by using binoculars with front lenses that are 50 mm or larger in aperture, and then move to your telescope and select your lowest-power eyepiece.

This cluster's two other common names originated in antiquity. Greek mathematician Eratosthenes (c. 276 B.C. – c 195 B.C.) wrote that the stars represented the hair of the mythological figure Ariadne, daughter of King Minos of Crete.

"Thisbe's Veil" comes from Ovid's *Metamorphoses*. The myth involves Pyramis and Thisbe, two lovers who committed suicide because of a misunderstanding. Jupiter honored their mutual devotion, however, and placed Thisbe's Veil in the sky, where it now resides in Coma Berenices.

| OBJECT #252 | M85 (NGC 4382) |
|---|---|
| Constellation | Coma Berenices |
| Right ascension | 12h25m |
| Declination | 18°11′ |
| Magnitude | 9.1 |
| Size | 7.5′ by 5.7′ |
| Type | Spiral galaxy |

Lenticular galaxy M85 lies in Coma Berenices and is one of the sky's brightest galaxies. Even so, you'll want to get away from city lights, which hinder galaxy observing more than any other kind of observing.

A lenticular galaxy like M85 has, as the word implies, a lens shape, so it kind of looks like an edge-on spiral. It doesn't have a spiral galaxy's arm structure, however. Also, it's not a place where stars are forming because the nebulae turned into stars millions of years ago.

To find M85, first locate the magnitude 4.7 star 11 Comae Berenices. From there, head 1.2° east-northeast, and you'll sweep up M85.

A small scope's view is disappointing. It shows only a luminous core surrounded by an oval halo. Through an 8-inch telescope look for a 13th-magnitude star less than 1′ north of the core. Use a 12-inch or larger scope, and you'll see the brightness difference as you move out from M85's central region. You'll also start to see a subtle color — yellow. It seems this galaxy mainly contains old yellow stars.

When you've had your fill of M85, look just 8′ east for the magnitude 10.9 barred spiral galaxy NGC 4394. This galaxy has a bright core, and its bar runs its full length.

| OBJECT #253 | NGC 4372 |
|---|---|
| Constellation | Musca |
| Right ascension | 12h26m |
| Declination | −72°39′ |
| Magnitude | 7.3 |
| Size | 10′ |
| Type | Globular cluster |
| Other name | Caldwell 108 |

Only 0.7° southwest of magnitude 3.8 Gamma ($\gamma$) Muscae lies NGC 4372, one of the sky's least concentrated globular clusters. At a distance of 15,000 light-years, its brightest stars glow at 12th magnitude and appear through an 8-inch telescope more like an open cluster — a loose collection of mainly 12th

through 14th-magnitude stars randomly distributed across the field of view. The cluster shows almost no central concentration. Just 5′ northwest of the core lies the magnitude 6.6 star SAO 256939. The best strategy to observing NGC 4372's faint stars is to move SAO 256939 out of the field of view.

| OBJECT #254 | NGC 4395 |
|---|---|
| Constellation | Canes Venatici |
| Right ascension | 12h26m |
| Declination | 33°33′ |
| Magnitude | 10.0 |
| Size | 13.2′ by 11′ |
| Type | Spiral galaxy |

You'll find our next target 7.8° southwest of magnitude 2.8 Cor Caroli (Alpha [α] Canum Venaticorum). This galaxy is huge: It covers 20% of the area of the Full Moon. Because it has a low surface brightness, you won't see much more than the faint central glow through an 8-inch telescope. Step up to a 14-inch scope and use a magnification around 150×. You'll see the starlike core, and, toward the southeast, two star-forming regions in NGC 4395 that carry their own NGC numbers.

The fainter region, NGC 4400 lies 2′ away. From the galaxy's core. Slightly brighter NGC 4401 lies 2′ to the east-southeast of NGC 4395's center.

| OBJECT #255 | M86 (NGC 4406) |
|---|---|
| Constellation | Virgo |
| Right ascension | 12h26m |
| Declination | 12°57′ |
| Magnitude | 8.9 |
| Size | 12.0′ by 9.3′ |
| Type | Elliptical galaxy |

M86 proves that not all elliptical galaxies are circular. In fact, this bright object appears oval even at low magnifications. Modern galaxy classification schemes lean toward M86 as a lenticular (lens-shaped) galaxy, rather than a strict elliptical. Cranking up the power will reveal M86's starlike core.

To find M84, look midway between Denebola (Beta [β] Leonis) and Vindemiatrix (Epsilon [ε] Virginis). Just don't confuse it with slightly fainter and rounder M84, which lies 0.2° away.

| OBJECT #256 | NGC 4414 |
|---|---|
| Constellation | Coma Berenices |
| Right ascension | 12h26m |
| Declination | 31°13′ |
| Magnitude | 10.3 |
| Size | 4.4′ by 3′ |
| Type | Spiral galaxy |

Our next target lies 3° north of magnitude 4.3 Gamma (γ) Comae Berenices. It's a bright galaxy you can see easily even through a 4-inch telescope. It appears oval shaped, 50% longer than it is wide, and elongated from the north-northwest to the south-southeast. Its central region is tiny, but brighter than the surrounding halo. Through a 10-inch scope, the halo fades gradually.

| OBJECT #257 | NGC 4429 |
|---|---|
| Constellation | Virgo |
| Right ascension | 12h27m |
| Declination | 11°06′ |
| Magnitude | 10.2 |

| (continued) | |
|---|---|
| Size | 5.8′ by 2.8′ |
| Type | Spiral galaxy |

Our next object lies in the heart of the Virgo Cluster of galaxies 1.6° west-northwest of the magnitude 6.3 star 20 Virginis. Two stars flank this galaxy. Magnitude 9.1 SAO 100102 lies 2′ to the north-northeast, and magnitude 9.2 SAO 100103 lies 5′ to the south-southeast.

NGC 4429's central region has even illumination and is broad, taking up approximately one-third of the galaxy's length. The halo is brighter than for most galaxies. Be sure to use at least an 8-inch telescope and a magnification above 250× to see it clearly.

**Object #258** The Eyes (NGC 4435 + NGC 4438) Adam Block/NOAO/AURA/NSF

| OBJECT #258 | NGC 4435 + NGC 4438 |
|---|---|
| Constellation | Virgo |
| Right ascension | 12h28m |
| Declination | 13°01′ |
| Magnitude | 10.0 |
| Size | 8.5′ by 3′ |
| Type | Spiral galaxy |
| Other designation | The Eyes |

If you have a 12-inch or larger telescope, pay particular attention to NGC 4435 and NGC 4438, a pair of galaxies called the Eyes, or Markarian's Eyes in honor of Armenian astronomer Beniamin Markarian (1913–1985). This duo lies 0.4° east of M86. Because of immense gravitational forces, these galaxies are tearing themselves apart. Astronomers estimate the two passed within 16,000 light-years of each other some time in the past. NGC 4438 is the more distorted galaxy. Try to spot this object's irregular outer regions.

| OBJECT #259 | Alpha (α) Crucis |
|---|---|
| Constellation | Crux |
| Right ascension | 12h27m |
| Declination | −63°06′ |
| Magnitudes | 1.4/1.9 |
| Separation | 4″ |
| Type | Double star |
| Other name | Acrux |

Although any size telescope will split Acrux, this luminary makes a startlingly beautiful double star split at high power through an 8-inch or larger telescope. Both stars are 1st magnitude and each is blue, albeit ever-so-slightly different shades.

**Object #260** NGC 4449 John and Christie Connors/Adam Block/NOAO/AURA/NSF

| OBJECT #260 | NGC 4449 |
|---|---|
| Constellation | Canes Venatici |
| Right ascension | 12h28m |
| Declination | 44°06′ |
| Magnitude | 9.6 |
| Size | 5.5′ by 4.1′ |
| Type | Irregular galaxy |
| Other name | Caldwell 21 |

I think you'll like our next object, irregular galaxy NGC 4449. You'll find it 2.9° north-northwest of magnitude 4.2 Chara (Beta [β] Canum Venaticorum). The galaxy's high surface brightness makes it easy to observe.

Astronomers classify NGC 4449 as a magellanic type galaxy because it appears similar to the Large Magellanic Cloud, a satellite galaxy of the Milky Way visible from the Southern Hemisphere. Both galaxies have large stellar bars running through them.

Through an 8-inch telescope, you'll see NGC 4449's unusual rectangular shape. It has a bright, concentrated nucleus that also looks rectangular. Crank the magnification past 250×, and examine the irregular halo outside this galaxy's core.

If your seeing is good, an 11-inch scope will help you spot several concentrations of star-forming activity. The main one lies to the north, but a smaller one is just south of NGC 4449's core. Larger telescopes will bring out more detail in the galaxy's central region.

| OBJECT #261 | NGC 4450 |
|---|---|
| Constellation | Coma Berenices |
| Right ascension | 12h29m |
| Declination | 17°05′ |
| Magnitude | 10.1 |
| Size | 5.0′ by 3.4′ |
| Type | Spiral galaxy |

You'll find our next object just south of the midway point of a line that connects the magnitude 4.7 star 11 Comae Berenices with the magnitude 5.7 star 25 Comae Berenices. The galaxy is relatively bright and oval, measuring 50% longer than it is wide with a north-south orientation. A thick halo surrounds an equally stretched central region.

| OBJECT #262 | M49 (NGC 4472) |
|---|---|
| Constellation | Virgo |
| Right ascension | 12h30m |
| Declination | 8°00′ |
| Magnitude | 8.4 |
| Size | 8.1′ by 7.1′ |
| Type | Elliptical galaxy |

M49 is a bright galaxy that appears slightly oval. The core occupies the central two-thirds of this object. A fainter outer region envelops the core. Because this galaxy shines so brightly, you can crank up the power and reveal this outer halo.

| OBJECT #263 | NGC 4473 |
|---|---|
| Constellation | Coma Berenices |
| Right ascension | 12h30m |
| Declination | 13°26′ |
| Magnitude | 10.2 |
| Size | 4.5′ by 2.5′ |
| Type | Elliptical galaxy |

Our next object lies buried among the galactic detritus of the Coma-Virgo cluster of galaxies. Look for NGC 4473's elliptical shape 3.7° east-southeast of the magnitude 5.1 star 6 Comae Berenices. Scan the area carefully because many other galaxies reside nearby.

Through a 10-inch telescope, look for a bright oval twice as long as it is wide in an east-west orientation. Its broad central region spans some 50% of the galaxy's length. You'll need high magnification to spot any part of the faint outer halo.

| OBJECT #264 | Delta (δ) Corvi |
|---|---|
| Constellation | Corvus |
| Right ascension | 12h30m |
| Declination | −16°31′ |
| Magnitudes | 3.0/9.2 |

| (continued) | |
|---|---|
| Separation | 24.2″ |
| Type | Double star |
| Other name | Algorab |

Algorab is the "pointer" star in Corvus that lies closest to Spica (Alpha [α] Virginis). The primary and secondary shine blue-white and orange, respectively. Luckily, the wide separation keeps the main star from overwhelming its companion, so the colors are nicely apparent.

Regarding this star's name, according to the J. K. Rowling character Regulus Black in the *Harry Potter* series, "Algorab means raven in Arabian." Ok, maybe so. But a better explanation seems to be the one put forth by Richard Hinckley Allen, who wrote the seminal *Star-Names and Their Meanings* (G. E. Stechert, 1899). Allen indicated that Algorab was a modern name that first appeared in 1803 in the *Palermo Catalogue*, a listing of 7,646 stars compiled by Italian astronomer Giuseppe Piazzi (1746–1826).

| OBJECT #265 | ADS 8573 |
|---|---|
| Constellation | Corvus |
| Right ascension | 12h30m |
| Declination | −13°24′ |
| Magnitudes | 6.5/8.6 |
| Separation | 2.2″ |
| Type | Double star |

You'll find this pair 3.1° due north of magnitude 3.0 Algorab (Delta [δ] Corvi). The brighter component shines yellow, and the secondary gives off an orange light. It's a nice contrast, but use a magnification above 150× to see the stars well.

Regarding this binary's designation, "ADS" refers to any entry in the *General Catalogue of Double Stars Within 120° of the North Pole*. American astronomer Robert Grant Aitken (1864–1951) compiled this list of 17,180 double stars. The Carnegie Institution published it as a two-volume set in 1932.

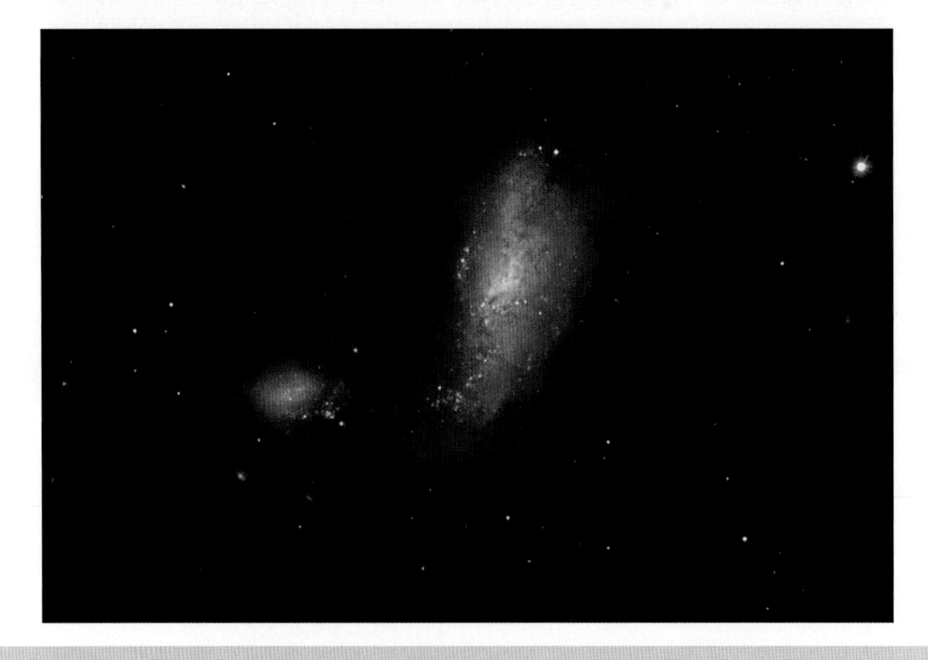

**Object #266** The Cocoon Galaxy (NGC 4490) Michael Gariepy/Adam Block/NOAO/AURA/NSF

| OBJECT #266 | NGC 4490 |
|---|---|
| Constellation | Canes Venatici |
| Right ascension | 12h31m |
| Declination | 41°38′ |
| Magnitude | 9.8 |
| Size | 6.4′ by 3.3′ |
| Type | Barred spiral galaxy |
| Other name | The Cocoon Galaxy |

You'll find this nice target 0.7° west-northwest of magnitude 4.2 Chara (Beta [β] Canum Venaticorum), in the same field of view of a low-power eyepiece. The galaxy appears as an irregularly bright oval halo (the "cocoon") that envelops a bright central region. Its spiral arms remain invisible to all but the largest amateur instruments.

But don't stop now. This object is a two-in-one treat. Look just 3′ north of NGC 4490's western end for its magnitude 12.5 companion, irregular galaxy NGC 4485. This pair interacts gravitationally, which explains why the brighter member looks like it does.

**Object #267** Virgo A (M87) Adam Block/NOAO/AURA/NSF

| OBJECT #267 | M87 (NGC 4486) |
|---|---|
| Constellation | Virgo |
| Right ascension | 12h31m |
| Declination | 12°24′ |
| Magnitude | 8.6 |
| Size | 7.1′ by 7.1′ |
| Type | Elliptical galaxy |
| Other names | Virgo A, the Smoking Gun |

To astronomers, M87 is a treasure-trove of science. It's a colossal object with a mass in excess of 3 trillion Suns and a diameter that may reach half a million light-years. M87 also possesses a huge array of globular clusters, perhaps numbering in the tens of thousands.

Visually, however, you could do better. Through any telescope up to 20 inches in aperture, M87 appears circular. Its core stands out as bright, and it spans about one-third of the galaxy's overall size.

Tear your gaze away from M87 for a moment to spot magnitude 11.4 NGC 4478 and magnitude 12.0 NGC 4476, which lie 8.5′ and 12.5′ west-southwest of their brilliant host, respectively.

To see the jet emanating from M87's nucleus, you'll have to use a 30-inch or larger telescope under a pristine sky. Through smaller-aperture scopes, you'll see M87 as perfectly round. A bright core, one-third of the galaxy's size, blazes in the center. Look for two companion galaxies, NGC 4476 and NGC 4478, about 10′ west.

Astronomers refer to M87 as Virgo A because letter designations refer to strong sources of radio waves. M87 was the first radio source discovered in Virgo, so it received the letter A. This galaxy's other moniker, the "Smoking Gun" comes from astronomers comparing M87's jet (as seen on images) to the front of the barrel of a recently fired gun.

| OBJECT #268 | Gamma (γ) Crucis |
|---|---|
| Constellation | Crux |
| Right ascension | 12h31m |
| Declination | −57°07′ |
| Magnitudes | 1.6/6.4 |
| Separation | 111″ |
| Type | Double star |
| Other name | Gacrux |

This is the one of the easiest binaries in the sky to find and to split. The orange primary is a red giant star. Even at the lowest power, you'll spot the blue secondary nearly 2′ to the north-northeast.

| OBJECT #269 | NGC 4494 |
|---|---|
| Constellation | Coma Berenices |
| Right ascension | 12h31m |
| Declination | 25°46′ |
| Magnitude | 9.8 |
| Size | 4.6′ by 4.4′ |
| Type | Elliptical galaxy |

Our next target is a bright galaxy that appears ever so slightly oval at high magnifications. Through a 12-inch or larger telescope, you'll see the faint, thin halo that surrounds the central region. Through smaller scopes, the galaxy just appears as an evenly illuminated disk.

To find NGC 4494, look 1.2° north of the magnitude 5.5 star 21 Comae Berenices.

**Object #270** M88 Adam Block/Mount Lemmon SkyCenter/University of Arizona

| OBJECT #270 | M88 (NGC 4501) |
|---|---|
| Constellation | Coma Berenices |
| Right ascension | 12h32m |
| Declination | 14°25′ |

| (continued) | |
|---|---|
| Magnitude | 9.6 |
| Size | 6.1′ by 2.8′ |
| Type | Spiral galaxy |

Our next target is a great one. You'll find it 8° west-northwest of magnitude 2.9 Vindemiatrix (Epsilon [ε] Virginis). As you move toward M88's position, you'll see the faint glow of dozens — if not hundreds — of galaxies. Luckily, M88 is one of the sky's brightest.

Through a 6-inch telescope, the galaxy appears as an oval haze more than twice as long as it is wide surrounding a brighter central region. Through a 12-inch scope at a magnification of 300×, you'll see some of this galaxy's spiral structure. Through larger instruments, look for a pair of 14th-magnitude stars separated by 20″ near the southeastern end of the halo.

| OBJECT #271 | NGC 4517 |
|---|---|
| Constellation | Virgo |
| Right ascension | 12h33m |
| Declination | 0°07′ |
| Magnitude | 10.4 |
| Size | 9.9′ by 1.4′ |
| Type | Spiral galaxy |

Sitting nearly alone 2.7° northwest of Porrima (Gamma [γ] Virginis), NGC 4517 looks like a shiny splinter drifting through space. This galaxy measures seven times as long as wide in apparent size, and only through the biggest scopes will you see the slightly brighter central region. A 10th-magnitude star (sorry, it's not a supernova) sits just northeast of the core.

| OBJECT #272 | NGC 4526 |
|---|---|
| Constellation | Virgo |
| Right ascension | 12h34m |
| Declination | 7°42′ |
| Magnitude | 9.6 |
| Size | 7′ by 2.5′ |
| Type | Spiral galaxy |
| Other names | The Hairy Eyebrow Galaxy |

You'll find our next object 7.7° west-southwest of magnitude 2.9 Vindemiatrix (Epsilon [ε] Virginis). *Astronomy* magazine Contributing Editor Stephen James O'Meara christened this object the Hairy Eyebrow Galaxy because of how the dense dust lane visible on Hubble Space Telescope images forms a bushy brow.

Through any size telescope, this lenticular galaxy reveals scant details. Through 10-inch and larger scopes at magnifications above 250×, you'll separate the wide central region and the fainter outer halo.

Two relatively bright stars flank this galaxy. Look 7′ to the east-northeast for magnitude 6.8 SAO 119479. Then, look 7′ west for magnitude 7.0 SAO 119466.

| OBJECT #273 | IC 3568 |
|---|---|
| Constellation | Camelopardalis |
| Right ascension | 12h33m |
| Declination | 82°33′ |
| Magnitude | 10.6 |
| Size | 10″ |
| Type | Planetary nebula |
| Other names | The Baby Eskimo Nebula, the Sliced Lime Nebula, the Theoretician's Planetary |

Our next object lies in the far-northern sky, only 8° from the celestial pole. The tiny inner core appears bright. You'll need at least a 10-inch telescope and 200× to see the 13th-magnitude central star. A slightly fainter shell surrounds the core. You'll also spot a 12th-magnitude star just 15″ to the west. Whatever scope you view this object through, be sure to use as high a magnification as the sky will allow.

The common names given to this planetary are just as interesting as the view of the object itself, but my favorite is the Theoretician's Planetary. This comes from what Bruce Balick, an astronomer at the University of Washington who has written a number of stories for *Astronomy* magazine, said about this object in 1996: "If IC 3568 did not exist, it might have been created by theoreticians."

**Object #274** The Lost Galaxy (NGC 4535) Adam Block/Mount Lemmon SkyCenter/University of Arizona

| OBJECT #274 | NGC 4535 |
|---|---|
| Constellation | Virgo |
| Right ascension | 12h34m |
| Declination | 8°12′ |
| Magnitude | 10.0 |
| Size | 7.0′ by 6.4′ |
| Type | Spiral galaxy |
| Other names | The Lost Galaxy, McLeish's Object |

Here's a little-know showpiece through large amateur telescopes. NGC 4535 is a barred spiral with a large, bright central region. This area looks rectangular because of the bar. Closer examination at magnifications above 300× reveals two faint spiral arms that emanate from the bar. Oh, and one note: The 13th-magnitude star enmeshed in the northern spiral arm is not a supernova.

Robert Burnham, writing in his *Celestial Handbook*, says that early twentieth-century American amateur astronomer Leland S. Copeland called NGC 4536 the Lost Galaxy "from its hazy phantom-like appearance in the amateur telescope."

McLeish's Object was a name given in honor of astronomer David McLeish, who made a number of discoveries at Cordoba Observatory in Argentina. And, please note, McLeish's Interacting Object is a completely different galaxy. McLeish discovered that object, now cataloged as IRAS 20048–6621, in 1948.

| OBJECT #275 | NGC 4536 |
|---|---|
| Constellation | Virgo |
| Right ascension | 12h35m |
| Declination | 2°11′ |
| Magnitude | 10.6 |
| Size | 6.4′ by 2.6′ |
| Type | Spiral galaxy |

This galaxy is an unusual spiral. NGC 4536's arms extend nearly straight out from the core. Near the core, they appear thick and bright. One-third of the way from the core to each arm's end, however, their brightnesses and thicknesses decrease dramatically. A 7th-magnitude star (SAO 119485) sits just 13′ east-northeast of this galaxy.

| OBJECT #276 | 24 Comae Berenices |
|---|---|
| Constellation | Coma Berenices |
| Right ascension | 12h35m |
| Declination | 18°23′ |
| Magnitudes | 5.2/6.7 |
| Separation | 20.3″ |
| Type | Double star |

This binary is another nice one to use as a test of color acuity. Of course, there's no right answer, but what colors do you see? Many observers see yellow and blue, while others see both as white. Still others detect an orange and green pair.

**Object #277** M91 Thomas and Gail Haynes/Adam Block/NOAO/AURA/NSF

| OBJECT #277 | M91 (NGC 4548) |
|---|---|
| Constellation | Coma Berenices |
| Right ascension | 12h35m |
| Declination | 14°30′ |
| Magnitude | 10.2 |
| Size | 5.0′ by 4.1′ |
| Type | Barred spiral galaxy |

The easiest way to find our next target is to start at bright M88 (Object #270), and move 0.8° east. Through a 6-inch telescope, you'll see a nearly rectangular object a bit longer than it is wide. Its central region is broad and bright. Through a 12-inch scope at 200×, you'll easily see the bar. Look even more closely, and you might spot one of the spiral arms. The brighter one extends southward from the east end of the bar.

| OBJECT #278 | M89 (NGC 4552) |
|---|---|
| Constellation | Virgo |
| Right ascension | 12h36m |
| Declination | 12°33′ |
| Magnitude | 9.8 |
| Size | 3.4′ by 3.4′ |
| Type | Elliptical galaxy |

Two-thirds of a degree south of M90, you'll find M89, a small, round elliptical galaxy. Only through large amateur telescopes will you see that the outer regions of this galaxy form a faint ring that fades rapidly with increasing distance from M89's center.

**Object #279** NGC 4559 Jeff Hapeman/Adam Block/NOAO/AURA/NSF

| OBJECT #279 | NGC 4559 |
|---|---|
| Constellation | Coma Berenices |
| Right ascension | 12h36m |
| Declination | 27°58′ |
| Magnitude | 10.0 |
| Size | 12.0′ by 4.9′ |
| Type | Spiral galaxy |
| Other name | Caldwell 36 |

Our next target lies 2° east-southeast of magnitude 4.3 Gamma ($\gamma$) Comae Berenices. NGC 4559 is a member of the Coma I galaxy group, a loose association containing some 30 members. The group lies approximately 30 million light-years away.

Through an 8-inch telescope, NGC 4559 appears strongly elliptical, three times as long as it is wide. Its dead center appears starlike and the surrounding central region is only a bit fainter. The outer halo, however, isn't easy to see unless you move up to a 14-inch scope. Then, the irregular brightness will suggest to you the position of the galaxy's spiral arms.

**Object #280** The Needle Galaxy (NGC 4565) Bruce Hugo and Leslie Gaul/Adam Block/NOAO/AURA/NSF

| OBJECT #280 | NGC 4565 |
|---|---|
| Constellation | Coma Berenices |
| Right ascension | 12h36m |
| Declination | 25°59′ |
| Magnitude | 9.6 |
| Size | 14.0′ by 1.8′ |
| Type | Spiral galaxy |
| Other names | The Needle Galaxy, Caldwell 3 |

Point your telescope at the Needle Galaxy, and you'll see the sky's finest edge-on spiral. Don't confuse the Needle Galaxy with the Silver Needle Galaxy (NGC 4244), which lies 12.5° north-northwest.

Through an 8-inch telescope, you'll see a streak roughly 10′ long but only 1.5′ thick oriented northwest to southeast. As you increase your scope's aperture, the apparent length of the Needle Galaxy also increases. A 16-inch scope shows NGC 4565's full extent.

A dust lane runs the entire length of this object, masking much of the arms' brightness. The central region features a small bulge, and that's the easiest place to detect the dust lane. The dark streak is offset a bit to the north of center because NGC 4565 is inclined 3.5° from edge-on.

If your telescope's aperture is 12 inches or more, try for the magnitude 13.5 galaxy NGC 4562. From the Needle Galaxy, this faint object lies 13′ southwest.

Although we see NGC 4565's edge, astronomers speculate that a face-on view of the Needle Galaxy would resemble M100 (Object #245). At its distance of 31 million light-years, NGC 4565 spans more than 150,000 light-years. It resides within the Coma-Sculptor Cloud of galaxies.

| OBJECT #281 | The Stargate |
|---|---|
| Constellation | Corvus |
| Right ascension | 12h36m |
| Declination | −12°01′ |
| Type | Asterism |

Our next object is a small telescope target with many names, but the one amateur astronomers use most is the Stargate. You'll find the asterism of the Stargate midway between magnitude 3.0 Delta ($\delta$) Corvi and magnitude 4.7 Chi ($\chi$) Virginis.

The Stargate is a triangle of stars within a larger stellar triangle. Neither triangle is huge. The large triangle measures roughly 5′ on a side. The small triangle spans less than 1′ on a side. Some observers have reported seeing the Stargate through binoculars. You'll need a unit with a magnification of at least 10× and with front lenses 50 millimeters across or larger. I've seen it through 16×70 binoculars.

If you have trouble resolving the smaller triangle through your binoculars, point a telescope at this group. Use an eyepiece that gives a magnification of about 50×. That power will show both triangles well.

The brightest of the Stargate's outer three luminaries shines at magnitude 6.6. Stars of magnitudes 6.7 and 9.9 complete the big triangle. The small triangle's stars glow more subtly, at magnitudes 8.0, 9.7, and 10.6.

| OBJECT #282 | NGC 4567/8 |
|---|---|
| Constellation | Virgo |
| Right ascension | 12h37m |
| Declination | 11°15′ |
| Magnitude | 11.3/10.9 |
| Size | 3.1′ by 2.2′ and 4.3′ by 2.0′ |
| Type | Spiral galaxies |
| Other name | The Siamese Twins |

For a wonderful example of interacting galaxies, turn your telescope toward the Siamese Twins. Under a dark sky, even a 6-inch telescope will reveal their overall "V" shape. To see any detail, however, you'll need a 12-inch or larger instrument.

To tell these two galaxies apart, remember that NGC 4568 appears slightly brighter and a bit longer than its companion.

M.E. Bakich, *1,001 Celestial Wonders to See Before You Die*, Patrick Moore's Practical Astronomy Series, DOI 10.1007/978-1-4419-1777-5_4, © Springer Science+Business Media, LLC 2010

**Object #283** M90 Paul and Daniel Koblas/Adam Block/NOAO/AURA/NSF

| OBJECT #283 | M90 (NGC 4569) |
|---|---|
| Constellation | Virgo |
| Right ascension | 12h37m |
| Declination | 13°10′ |

| (continued) | |
|---|---|
| Magnitude | 9.5 |
| Size | 10.5′ by 4.4′ |
| Type | Spiral galaxy |

M90 may be one of the least interesting spiral galaxies you'll ever observe. That's too bad, because we tend to expect more from Messier objects.

You will see an object that measures two times as long as wide. M90's spiral arms wind tightly around it, however, so unless your scope's mirror measures two feet across, be content to just check this bright galaxy off your list, and move on.

| OBJECT #284 | NGC 4589 |
|---|---|
| Constellation | Draco |
| Right ascension | 12h37m |
| Declination | 74°12′ |
| Magnitude | 10.7 |
| Size | 3.0′ by 2.7′ |
| Type | Elliptical galaxy |

Our next object lies 4.4° due north of magnitude 3.9 Kappa ($\kappa$) Draconis. Through a 12-inch telescope, a bright central disk comes into view, surrounded by a much darker oval halo 2′ by 1′ across. Can you also see the spindle of the magnitude 14.0 spiral galaxy NGC 4572? It glows dimly only 7′ west-northwest of NGC 4589.

| OBJECT #285 | M58 (NGC 4579) |
|---|---|
| Constellation | Virgo |
| Right ascension | 12h38m |
| Declination | 11°49′ |
| Magnitude | 9.6 |
| Size | 5.5′ by 4.6′ |
| Type | Barred spiral galaxy |

Just about any size telescope will show M58's slightly oval structure. Through a 16-inch or larger scope, however, you should be able to pick out the brighter central bar. Around the bar, a fainter halo region represents the galaxy's tightly wound spiral arms.

| OBJECT #286 | Markarian's Chain |
|---|---|
| Constellations | Coma Berenices and Virgo |
| Right ascension | 12h40m |
| Declination | 13° |
| Magnitudes | various |
| Sizes | various |
| Type | Galaxy group |

Nineteenth-century observers called the area bounded by the stars Arcturus, Spica, and Denebola the "realm of the nebulae." They weren't describing nebulae in the current sense, however. Their nebulae were galaxies, which looked nebulous through the small telescopes most observers used.

Across the heart of this region stretches Markarian's Chain, a group of galaxies bounded on the west by the large lenticular (lens-shaped) galaxies M84 and M86. From those two giants, the chain swings to the northeast. Where it ends depends on your perspective. Some observers end the chain at NGC 4477. Others swing from that galaxy to the northwest and include NGC 4459. The chain is named for Armenian astronomer Beniamin Egishevich Markarian (1913–1985), who discovered an energetic class of galaxies in the 1960s.

The key to observing this region is having a good finder chart. Orient it so it matches your eyepiece's field of view. The finder chart will show a lot more area and objects, of course. What you want is to be able to move your scope left, right, up, and down and have those directions correspond to your chart.

| OBJECT #287 | M68 (NGC 4590) |
|---|---|
| Constellation | Hydra |
| Right ascension | 12h40m |
| Declination | −26°45′ |
| Magnitude | 7.6 |
| Size | 12′ |
| Type | Globular cluster |

Our next target lies just outside the naked-eye limit. You'll spot M68 3.5° south-southeast of magnitude 2.7 Beta (*β*) Corvi. Although you can see this cluster as a fuzzy glow through binoculars, you'll get a whole new perspective by pointing a small telescope at this object.

At low powers, look at the wide central region. It spans half of M68's diameter. Also, if your eyepiece gives a wide field of view, check out the star field M68 lies in. The individual stars around the cluster all seem to be about the same brightness, which adds to the easy visibility of the globular.

At a dark site, crank up the magnification to 200× and beyond through your 4-inch scope to resolve a dozen or so of M68's stars. Notice that the core doesn't appear round or evenly illuminated. If you use a 6-inch or larger scope, you can see "through" the cluster's brighter stars to a haze of fainter background points of light.

**Object #288** The Sombrero Galaxy (M104) Morris Wade/Adam Block/NOAO/AURA/NSF

| OBJECT #288 | M104 (NGC 4594) |
|---|---|
| Constellation | Virgo |
| Right ascension | 12h40m |
| Declination | –11°37′ |
| Magnitude | 8.0 |
| Size | 7.1′ by 4.4′ |
| Type | Spiral galaxy |
| Other name | The Sombrero Galaxy |

Deep within Virgo's heart lies a showpiece spiral galaxy guaranteed to delight amateur astronomers and the general public alike. The Sombrero Galaxy is undoubtedly one of the finest objects you can see through a small telescope.

M104 was the first galaxy to have a large redshift detected. Redshift refers to the direction of motion away from us caused by the universe's expansion. In 1912, American astronomer Vesto M. Slipher (1875–1969) discovered the Sombrero Galaxy was moving away from us at a speed of 2.2 million mph (3.6 million km/h).

The Sombrero Galaxy's lens shape and the dark dust lane that splits it are easy to spot. The galaxy's two sections have unequal brightnesses — the north outshines the south because M104 inclines 6° to our line of sight. The dust lane, therefore, appears to cross south of center.

Through a 4-inch telescope, you may detect the dust lane only near the Sombrero's center. The core is bright, and a large halo surrounds it, even extending above and below the sections of the spiral arms nearest the nucleus.

| OBJECT #289 | NGC 4605 |
|---|---|
| Constellation | Ursa Major |
| Right ascension | 12h40m |
| Declination | 61°37′ |
| Magnitude | 10.9 |
| Size | 6.0′ by 2.4′ |
| Type | Spiral galaxy |
| Other names | The Faberge Egg Galaxy, the Frankenstein Galaxy |

You won't need a big telescope to observe this galaxy. It's bright with a huge central region. The outer halo is quite faint and small, but, beyond that, details are lacking. One thing you can look for, however, is that NGC 4605's oval shape isn't perfect — it seems to bulge outward toward the south.

*Astronomy* magazine Contributing Editor Stephen James O'Meara gave this object its two common names. Because he saw slender waves of light and a star-studded face, O'Meara said it looked like a Faberge Egg with enameling.

He also called it the Frankenstein Galaxy because he observed it looks like it was badly pieced together with different body parts.

| OBJECT #290 | Struve 1669 |
|---|---|
| Constellation | Corvus |
| Right ascension | 12h41m |
| Declination | –13°01′ |
| Magnitudes | 6.0/6.1 |
| Separation | 5.4″ |
| Type | Double star |

Crank the power to 100× or more, and target this equally matched pair. The slightly brighter primary shines yellow. Its companion is white. You'll find this tight binary 4.5° northeast of Algorab (Delta [δ] Corvi).

This double star's name derives from German astronomer Friedrich Georg Wilhelm von Struve (1793–1864). He published a catalog of his binary star discoveries in 1827, and then measured 2,714 double stars and cataloged them in 1837.

| OBJECT #291 | Gamma (γ) Virginis |
|---|---|
| Constellation | Virgo |
| Right ascension | 12h42m |
| Declination | −1°27′ |
| Magnitudes | 3.5/3.5 |
| Separation | 0.5″ |
| Type | Double star |
| Common names | Porrima; Arich |

Gamma Virginis is one of the most famous double stars in the sky. Much research has been aimed at this binary, and astronomers have even written poems about it. When you observe it, make a note to revisit the stars every year or two. Currently, this is a close binary that requires a 12-inch or larger telescope. But the pair is widening, and by 2020 splitting it will be within reach of even small scopes.

According to Richard Hinckley Allen in *Star-Names and Their Meanings* (1899), this star's name is Latin, and it refers to Postvorta, one of two ancient goddesses of prophecy. Several sources refer to "Arich" as a traditional name for this star. Allen, however, doesn't mention this variant.

Admiral Smyth in *Cycle of Celestial Objects* devotes eight pages (which include four tables and three diagrams) to this double star. He described the pair as, "A 4, silvery white; B 4, pale yellow, but though marked by Piazzi of equal magnitude with A, it has certainly less brilliance; and the colours are not always of the same intensity, but whether owing to atmospherical or other causes, remains undecided."

| OBJECT #292 | M59 (NGC 4621) |
|---|---|
| Constellation | Virgo |
| Right ascension | 12h42m |
| Declination | 11°39′ |
| Magnitude | 9.6 |
| Size | 4.6′ by 3.6′ |
| Type | Elliptical galaxy |

Although M59 does grace Charles Messier's famous list, don't get your hopes up as you point your telescope toward it. Note its oval glow and uniform illumination that only begins to fade close to the galaxy's edge. Got it? O.K., time to move on.

**Object #293** The Whale Galaxy (NGC 4631) John Vickery and Jim Matthes/Adam Block/NOAO/AURA/NSF

| OBJECT #293 | NGC 4631 |
|---|---|
| Constellation | Canes Venatici |
| Right ascension | 12h42m |
| Declination | 32°32′ |
| Magnitude | 9.8 |
| Size | 17′ × 3.5′ |
| Type | Spiral galaxy |
| Common names | The Whale Galaxy, the Herring Galaxy, Caldwell 32 |

At first glance, the Whale Galaxy, with its bulging core and asymmetrical material distribution doesn't look like an edge-on spiral galaxy. Yet this object is one of the sky's brightest edge-on galaxies.

William Herschel discovered NGC 4631 March 20, 1787. Since then, observers have enjoyed views of this object through all sizes of telescopes. Through 4- to 6-inch scopes, you'll see an imperfect lens shape that's bigger and brighter on one side.

Larger scopes will reveal NGC 4631's companion — dwarf spheroidal galaxy NGC 4627, which sits 2.5′ northwest. NGC 4627's gravity has distorted the Whale's once-classic spiral structure. In fact, the largest telescopes reveal a faint bridge of material connecting the two galaxies.

For the Whale Galaxy, the close passage of the smaller galaxy really has stirred things up. Its central region is a maelstrom of star formation. Huge clumps of stars, visible through 12-inch or larger telescopes, lie all along the spiral arms. If you're lucky enough to observe NGC 4631 through a scope with an aperture bigger than 16′, look for dark areas made of dust and cold gas amidst the bright patches.

| OBJECT #294 | NGC 4609 |
|---|---|
| Constellation | Crux |
| Right ascension | 12h42m |
| Declination | −63°00′ |

| (continued) | |
|---|---|
| Magnitude | 6.9 |
| Size | 6′ |
| Type | Open cluster |
| Other name | Caldwell 98 |

Because NGC 4609 sits only 1.8° east of Crux's brightest star, magnitude 0.8 Acrux (Alpha [α] Crucis), and because it lies 5′ northwest of the only naked-eye star within the Coalsack, you'll have no trouble locating this object.

The star, magnitude 5.3 SAO 252002, also goes by the designation BZ Crucis because of its slight variable nature. Equidistant from BZ Crucis, but on the side opposite NGC 4609, is the magnitude 10.3 open cluster Hogg 15.

Through a 4-inch telescope at 75×, you'll see NGC 4609 break into about a dozen stars. Larger apertures reveal a second layer of about 20 fainter stars.

| OBJECT #295 | NGC 4636 |
|---|---|
| Constellation | Virgo |
| Right ascension | 12h43m |
| Declination | 2°41′ |
| Magnitude | 9.4 |
| Size | 5.9′ by 4.6′ |
| Type | Elliptical galaxy |

To find our next target, look 3.3° west-southwest of magnitude 3.4 Minelauva (Delta [δ] Virginis). Through an 8-inch telescope, this object has an oval shape with a northwest to southeast orientation. Its central region is small but bright, and the outer halo quickly fades to the blackness of the surrounding space.

| OBJECT #296 | M60 (NGC 4649) |
|---|---|
| Constellation | Virgo |
| Right ascension | 12h44m |
| Declination | 11°33′ |
| Magnitude | 8.8 |
| Size | 7.1′ by 6.1′ |
| Type | Elliptical galaxy |

This bright, but under-observed galaxy lies 1.4° north-northeast of magnitude 4.9 Rho (ρ) Virginis. Through medium-sized telescopes, M60 easily shows as a double galaxy. Its companion, NGC 4647, glows 3 magnitudes fainter, but it's still well within the light-grasp of a 6-inch scope.

M60 itself appears just slightly out-of-round, but otherwise featureless. M59 lies less than half a degree west of M60. A low-power telescope/eyepiece combination will show them both.

In *Cycle of Celestial Objects*, Smyth describes a contemporary's view of this type of object. and gives a reason for astronomers to observe them: "The hypothesis of Sir John Herschel, upon double nebulae, is new and attracting. They may be stellar systems each revolving round the other: each a universe, according to ancient notions. But as these revolutionary principles of those vast and distant firmamental clusters cannot for ages yet be established, the mind lingers in admiration, rather than comprehension of such mysterious collocations. Meantime our clear duty is, so industriously to collect facts, that much of what is now unintelligible, may become plain to our successors, and a portion of the grand mechanism now beyond our comprehension, revealed."

**Object #297** The Hockey Stick (NGC 4656-7) Doug Matthews/Adam Block/NOAO/AURA/NSF

| OBJECT #297 | NGC 4656-7 |
|---|---|
| Constellation | Canes Venatici |
| Right ascension | 12h44m |
| Declination | 32°10′ |
| Magnitude | 10.4 |
| Size | 14′ × 3′ |
| Type | Irregular galaxy |
| Other names | The Hockey Stick, the Crowbar Galaxy, the Hook Galaxy |

Our next deep-sky object is the fine Hockey Stick Galaxy. Not many bright stars lie nearby, so to find it you'll need to head 6.6° south-southwest of magnitude 2.8 Cor Caroli. That's the Alpha (α) star in Canes Venatici.

English astronomer Sir William Herschel discovered this galaxy in 1787. The core is the brightest part of the galaxy, followed by the blade, which lies to the northeast. NGC 4656's disk widens and gets fainter toward the southwest.

As you view this unusual galaxy, it's interesting to note that astronomers gave the stick and the bright knot in the blade separate NGC numbers. The long stick is NGC 4656, while the knot is NGC 4657.

Because you're already in the area, you owe it to yourself to take a look 0.5° northwest of the Hockey Stick Galaxy. There you'll find the magnitude 9.0 Whale Galaxy (NGC 4631). Gravitational interaction between these two objects is what tore the Hockey Stick Galaxy apart and gave it its unusual appearance.

| OBJECT #298 | Y Canum Venaticorum |
|---|---|
| Constellation | Canes Venatici |
| Right ascension | 12h45m |
| Declination | 45°26′ |
| Magnitude | 4.8 |

| (continued) | |
|---|---|
| Period | 158 days |
| Type | Variable star |
| Other name | La Superba |

Our next target is one of the reddest stars in the sky. It's official astronomical designation is Y Canum Venaticorum, but observers usually call it La Superba. It received its common name from Italian astronomer Angelo Secchi. The star's color so impressed him that he christened it "the superb one," or La Superba.

Astronomers classify La Superba as a semi-regular variable star. That means most of the time the star's brightness varies between a peak of magnitude 4.8 and a low of magnitude 6.3. It takes 160 days to go from one peak to the next . . . usually.

La Superba has a surface temperature near the minimum for stars, about 2,800 K. (That's about 4,600° F.) Compare that to the temperature at our Sun's surface, which is 5,800 K (10,000° F).

La Superba also is a carbon star. Carbon compounds like soot accumulate in the star's upper atmosphere. The particles scatter light near the blue end of the spectrum. What's left for us to view is the red component of the star's light. As the particles build up, the star fades in brightness and also gets redder. Eventually, the carbon absorbs enough radiation to escape the star, and the cycle starts again.

To find La Superba, look a bit more than 7° north-northwest of magnitude 2.8 Cor Caroli (Alpha [α] Canum Venaticorum). A sky-distance of 7° equals the field of view of many binoculars. And although you can find La Superba through binoculars, its color appears best through small telescopes.

| OBJECT #299 | Beta (β) Muscae |
|---|---|
| Constellation | Musca |
| Right ascension | 12h46m |
| Declination | −68°06′ |
| Magnitudes | 3.7/4.0 |
| Separation | 1.3″ |
| Type | Double star |

You'll need a superb 4-inch telescope to split this star. Use a magnification above 200×. Both components appear pale blue or blue-white.

| OBJECT #300 | NGC 4665 |
|---|---|
| Constellation | Virgo |
| Right ascension | 12h45m |
| Declination | 3°03′ |
| Magnitude | 10.3 |
| Size | 3.5′ |
| Type | Spiral galaxy |

A famous astronomer long ago made a mistake designating this object. And it may not be the only one. German-born English astronomer Sir William Herschel assigned it two entries in his catalog. Those became NGC 4664 and NGC 4665. NGC 4624 may also refer to this galaxy.

You'd think a galaxy with that many numbers would look terrific. Unfortunately, this isn't the second coming of the Andromeda Galaxy. Through a 10-inch telescope, you'll see the bright, stretched-out central region oriented north-south. Not quite 2′ southwest of NGC 4665's core lies the magnitude 10.7 star GSC 293:1166.

| OBJECT #301 | NGC 4697 |
|---|---|
| Constellation | Virgo |
| Right ascension | 12h49m |

| (continued) | |
|---|---|
| Declination | –5°48′ |
| Magnitude | 9.2 |
| Size | 7.2′ by 4.7′ |
| Type | Elliptical galaxy |
| Other name | Caldwell 52 |

Our next object, NGC 4697, ranks as one of the sky's brightest galaxies. You'll it 5.3° west of magnitude 4.4 Theta ($\theta$) Virginis. A small telescope will reveal its hazy nature and oblong shape.

Step up to an 11-inch scope, and you'll see much more detail. Regions of this galaxy outside the core show a threefold variance in brightness, getting fainter as you move away from the core.

The bright central region's shape may remind you of a spiral galaxy. In fact, astronomers categorize NGC 4697 as a lenticular galaxy — one with characteristics of an elliptical but with a toehold on the first rung of a spiral galaxy's evolutionary ladder.

If you're using a 16-inch or larger telescope, you might notice an ultra-faint object 6′ west-northwest of NGC 4697. That's PGC 170203. This spiral galaxy has a dismal magnitude of 15.1, so I'll forgive you if you don't spend much time observing it.

| OBJECT #302 | NGC 4699 |
|---|---|
| Constellation | Virgo |
| Right ascension | 12h49m |
| Declination | –8°40′ |
| Magnitude | 9.6 |
| Size | 3.8′ by 2.8′ |
| Type | Spiral galaxy |

You'll find our next target 1.6° west-northwest of magnitude 4.8 Psi ($\psi$) Virginis. Any telescope will show this galaxy as an oval object 50% longer than it is wide, oriented north-northeast to south-southwest. Through 8-inch and larger scopes, you'll see the outer halo that marks where the tightly wound spiral arms reside. The central region appears wide and bright. The magnitude 10.7 star GSC 5535:1227 lies 5′ to the east.

| OBJECT #303 | DY Crucis |
|---|---|
| Constellation | Crux |
| Right ascension | 12h47m |
| Declination | –59°42′ |
| Magnitude range | 8.4–9.8 |
| Type | Variable star |
| Other name | Ruby Crucis |

Our next target is one of the sky's reddest stars. It's also incredibly easy to find if you can observe south of about latitude 25° north. First find magnitude 1.3 Mimosa (Beta [$\beta$] Crucis). Train your telescope on that brilliant star, crank the magnification past 100×, and look just 2′ to its west.

Ruby Crucis also goes by the designations GSC 8659:1394, TYC 8659–1394–1, and NSV 19481. The only tough part about this observation is to move Mimosa out of the field of view so you can see Ruby Crucis without all that glare. Oh, and the nearer you can view it to its minimum brightness, the redder it will appear.

| OBJECT #304 | Abell 3526 |
|---|---|
| Constellation | Centaurus |
| Right ascension | 12h49m |
| Declination | –41°18′ |

| (continued) | |
|---|---|
| Size | 180′ |
| Type | Galaxy cluster |
| Other name | The Centaurus Cluster |

Astronomers catalog our next object as the Centaurus Galaxy Cluster, but, actually, it's two galaxy clusters we see in the same direction. Most of the galaxies belong to the cluster Cen 30. Its brightest galaxy is NGC 4696 and it lies 160 million light-years away.

The other cluster, Cen 45, has NGC 4709 at its center. This much looser cluster lies at a distance of 220 million light-years.

You'll find this combination object 7.6° southwest of magnitude 2.8 Iota (ι) Centaurus. Through a 16-inch telescope, you'll spot roughly 20 galaxies in a 2° area. Magnitude 11.9 NGC 4696 appears oval, elongated east-west, and measures 4.7′ by 3.3′. Magnitude 11.1 NGC 4709 lies 15′ to the east-southeast. It looks nearly circular, 2.3′ by 2′, with a bright core.

| OBJECT #305 | 32 Camelopardalis |
|---|---|
| Constellation | Camelopardalis |
| Right ascension | 12h49m |
| Declination | 83°25′ |
| Magnitudes | 5.3/5.8 |
| Separation | 21.6″ |
| Type | Double star |

This binary lies in the far northwestern section of Camelopardalis roughly 7° from Polaris (Alpha [α] Ursae Minoris). The slightly brighter primary is blue, and its companion glows white.

| OBJECT #306 | NGC 4710 |
|---|---|
| Constellation | Coma Berenices |
| Right ascension | 12h50m |
| Declination | 15°10′ |
| Magnitude | 11.0 |
| Size | 3.9′ by 1.2′ |
| Type | Spiral galaxy |

You'll find this object 5.4° west-southwest of magnitude 4.3 Alpha (α) Comae Berenices. Through a 12-inch telescope at low power, NGC 4710 is another of the sky's needles. This one orients north-northeast to south-southwest. Crank up the magnification past 250×, and you'll spot the bright central bulge with the edge-on spiral arms on either side. Finally, look 19′ southwest of NGC 4710 for the magnitude 13.8 spiral galaxy IC 3806.

| OBJECT #307 | NGC 4725 |
|---|---|
| Constellation | Coma Berenices |
| Right ascension | 12h50m |
| Declination | 25°30′ |
| Magnitude | 9.4 |
| Size | 11.0′ by 8.3′ |
| Type | Spiral galaxy |

You'll find this object 5.9° east-southeast of magnitude 4.3 Gamma (γ) Comae Berenices. Through a 6-inch telescope, the galaxy appears as a bright oval region surrounding an intense core. The galaxy orients northeast to southwest.

Through a 14-inch scope, you'll spot the spiral arms. The one attached to the northeast side glows a bit brighter. You'll also spot two fainter galaxies in the same field of view. Magnitude 12.5 NGC 4712 lies 0.2° west. Magnitude 12.2 NGC 4747 lies 0.4° northeast.

**Object #308** M94 Adam Block/NOAO/AURA/NSF

| OBJECT #308 | M94 (NGC 4736) |
|---|---|
| Constellation | Canes Venatici |
| Right ascension | 12h51m |
| Declination | 41°07′ |
| Magnitude | 8.2 |
| Size | 13.0′ by 11.0′ |
| Type | Spiral galaxy |

When does a face-on spiral galaxy not look like a spiral galaxy? When it's M94. This notable object — the brightest galaxy in Canes Venatici — has tightly wrapped spiral arms. Through small telescopes, most observers might classify it as an elliptical galaxy.

Pierre Méchain discovered M94 March 22, 1781. Messier observed M94, determined its position, and added it to his catalog just 2 days later. He called it a "nebula without star." Look for it 3° north-northwest of magnitude 2.9 Cor Caroli (Alpha [α] Canum Venaticorum).

British astronomer and author Admiral William H. Smyth (1788–1865) gave a more detailed description, but it certainly doesn't describe a galaxy: "A comet-like nebula; a fine, pale-white object with evident symptoms of being a compressed cluster of small stars. It brightens toward the middle."

Through an 8-inch scope, you'll see the tiny nucleus surrounded by a bright disk that measures only 30″ across. A much fainter oval halo surrounds the disk. Increase your telescope's aperture to 16′, and you'll begin to see the tightly wound spiral arms close to the nucleus.

| OBJECT #309 | NGC 4731 |
|---|---|
| Constellation | Virgo |
| Right ascension | 12h51m |
| Declination | –6°24′ |
| Magnitude | 11.5 |
| Size | 6.6′ by 4.2′ |
| Type | Spiral galaxy |

Our next target is not a bright galaxy — only magnitude 11.5 — but it has several features I think you'll find worth your observing time. NGC 4731 lies 3.3° east-northeast of magnitude 4.7 Chi ($\chi$) Virginis.

This dim galaxy appears highly distorted into an "S" shape because it doesn't travel through space alone. You'll easily spot its brighter companion. Look only 0.8° to the northwest for magnitude 9.2 NGC 4697 (Object #301). Gravitational interaction with this elliptical galaxy has nearly destroyed NGC 4731's spiral arms.

Through a 10-inch telescope, observe NGC 4731's long, relatively bright central bar. If your observing site is dark enough, crank up the power past 200×, and look at the wide, irregular spiral arms that originate from each side of the bar.

The western arm appears somewhat brighter. Tiny bright patches within both arms signal the existence of hotspots of star formation. Through a 20-inch or larger telescope, use a nebular filter to increase the contrast of those regions and the galaxy's already formed stars.

At an estimated distance of 65 million light-years, NGC 4731 sits on the far side of the Virgo Cluster of galaxies.

| OBJECT #310 | NGC 4762 |
|---|---|
| Constellation | Virgo |
| Right ascension | 12h53m |
| Declination | 11°14′ |
| Magnitude | 10.3 |
| Size | 9.1′ by 2.2′ |
| Type | Barred spiral galaxy |

This object lies 2.3° west of magnitude 2.9 Vindemiatrix (Epsilon [$\varepsilon$] Virginis). More than four times as long as it is wide, NGC 4762 appears as a white line through medium-sized telescopes. If you want to show someone an edge-on galaxy, here's your chance.

You won't see a central bulge through any size scope. All you will notice is that the core appears ever-so-slightly brighter than the arms.

| OBJECT #311 | The Coalsack |
|---|---|
| Constellation | Crux |
| Right ascension | 12h53m |
| Declination | –63°18′ |
| Size | 400′ by 300′ |
| Type | Dark nebula |
| Common names | Magellan's Spot, the Black Magellanic Cloud, and Caldwell 99 |

Although Crux the Southern Cross is the smallest of the 88 constellations, it's also the brightest when measured by bright stars per unit area. Any dark object, therefore — especially one as large as the Coalsack — will stand out.

The Coalsack has the greatest impact as a naked-eye object. When you use optics, the field of view shrinks, thus lessening the contrast with the surrounding bright Milky Way star field. Also, binoculars and telescopes reveal the Coalsack is not devoid of stars, and that takes away from its appeal.

The Coalsack falls mainly within the boundaries of Crux, but parts of it lie in Musca and Centaurus. And although we can't give a discovery date — Southern Hemisphere residents had seen it for thousands of years — Spanish explorer Vincente Yánez Pinzón (c. 1460–after 1523) reported observing it in 1499.

| OBJECT #312 | NGC 4753 |
|---|---|
| Constellation | Virgo |
| Right ascension | 12h52m |
| Declination | −1°12′ |
| Magnitude | 9.9 |
| Size | 6′ by 2.8′ |
| Type | Spiral galaxy |

Our next target lies 2.7° east of magnitude 3.5 Porrima (Gamma [γ] Virginis). Through even a 4-inch telescope, this galaxy appears football shaped with the long axis stretching east-west. It's twice as long as it is wide with a large, bright central region and a fainter, but noticeable halo.

| OBJECT #313 | NGC 4755 |
|---|---|
| Constellation | Crux |
| Right ascension | 12h54m |
| Declination | −60°20′ |
| Magnitude | 4.2 |
| Size | 10′ |
| Type | Open cluster |
| Other names | The Jewel Box, the Kappa Crucis Cluster, Caldwell 94 |

Many amateur astronomers consider the Jewel Box Cluster the sky's finest open cluster. It's not the biggest or the brightest, or even the most populous. The reason NGC 4755 stops so many observers in their tracks is its colorful stars.

Almost all open clusters contain hot, recently formed stars. Most of them are blue but appear white through a telescope. The Jewel Box, however, contains more than half a dozen stars of various shades of blue, yellow, and orange.

French astronomer Nicolas Louis de Lacaille (1713–1762) discovered NGC 4755 during his trip to South Africa in 1751–1752. English astronomer Sir John Herschel's (1792–1871) eloquent description led to other astronomers coining the popular name "Jewel Box."

NGC 4755's alternate proper name, the Kappa Crucis Star Cluster, does not derive from a single star because no star in the cluster is bright enough to have garnered such attention. "Kappa," instead, refers to the entire cluster.

A 6-inch telescope and an eyepiece that yields 50× may be the best combination with which to view NGC 4755. Through this setup, you'll see nearly a dozen stars that exhibit color, twenty additional white stars, and a faint backdrop composed of some 200 cluster members.

| OBJECT #314 | Alpha (α) Canum Venaticorum |
|---|---|
| Constellation | Canes Venatici |
| Right ascension | 12h56m |
| Declination | 38°19′ |
| Magnitudes | 2.9/5.5 |
| Separation | 19″ |
| Type | Double star |
| Other name | Cor Caroli |

This stellar pair makes a nice sight through small telescopes in the Northern Hemisphere's spring under even moderate magnification. The brighter primary glows blue-white, and that provides a nice color-contrast to the yellow secondary. This double is a great target for star parties.

Prior to 1725, this star was known as the Lion's liver. In that year, however, British Astronomer Royal Edmond Halley (1656–1742) designated it Cor Caroli, in honor of Charles II (1630–1685). According to Richard Hinckley Allen, Halley did this at the suggestion of the court physician Sir Charles Scarborough, who said that the star had shone with special brilliancy on the eve of the king's return to London on May 29, 1660.

| OBJECT #315 | M64 (NGC 4826) |
|---|---|
| Constellation | Coma Berenices |
| Right ascension | 12h57m |
| Declination | 21°41′ |
| Magnitude | 8.5 |
| Size | 9.2′ by 4.6′ |
| Type | Spiral galaxy |
| Common names | The Blackeye Galaxy, the Sleeping Beauty Galaxy |

English astronomer Edward Pigott (1753–1825) discovered M64 March 23, 1779. Messier independently found it a year later and added it to his catalog. William Herschel discovered this galaxy's dark dust feature, which he compared it to a black eye. The name caught on.

The dust lane is prominent, but only when viewed through a 10-inch or larger telescope. The lane sits north of the nucleus, separating the core from the northern spiral arm. The arms wrap tightly around the core, so you'll need a 16-inch scope to see any detail in them. If you view through a scope of this aperture, look for a halo formed by the arms' outermost regions.

| OBJECT #316 | NGC 4815 |
|---|---|
| Constellation | Musca |
| Right ascension | 12h58m |
| Declination | –64°57′ |
| Magnitude | 8.6 |
| Size | 3.0′ |
| Type | Open cluster |

Our next object sits at the southern edge of the Coalsack (Object #311). Specifically, you can find it 1.1° west-northwest of magnitude 5.7 Theta ($\theta$) Muscae.

NGC 4815 is an open cluster, but even through a 12-inch telescope you'll only resolve the brightest 15 or so members. Two bright stars dominate the cluster. The easternmost is magnitude 9.6 GSC 8997:563. A bit more than 1′ west-northwest lies magnitude 10.0 GSC 8997:72.

| OBJECT #317 | NGC 4856 |
|---|---|
| Constellation | Virgo |
| Right ascension | 12h59m |
| Declination | –15°03′ |
| Magnitude | 10.4 |
| Size | 4.3′ by 1.2′ |
| Type | Spiral galaxy |

This object lies near Virgo's western border with Corvus. Through an 8-inch telescope at 200×, you'll see a disk with a small, bright central region. The galaxy stretches three times as long as it is wide in a northeast to southwest orientation. For those of you using 14-inch or larger scopes, crank the power past 350×, and look for a magnitude 13.1 star just barely east of the core.

With that aperture, you'll be able to catch some fainter galactic quarry nearby. First, target the magnitude 13.1 spiral galaxy NGC 4877. It measures 2.3′ by 0.9′ and lies 21′ to the southeast of NGC 4856. Between the two galaxies lie two stars, magnitude 9.5 GSC 6112:285 and magnitude 9.2 SAO 157648. Then look 6′ northeast of NGC 4856 for PGC 44645. This magnitude 14.9 spiral will really test your eyesight. It measures 1.6′ by only 0.4′ thick.

| OBJECT #318 | The Spring Triangle |
|---|---|
| Right ascension | 13 h |
| Declination | 9°30′ |
| Type | Asterism |

Our next objects is a naked-eye asterism called the Spring Triangle. This giant geometrical figure is visible in the spring all night long from any location in the Northern Hemisphere.

Three dazzling stars mark this asterism. The brightest is Arcturus (Alpha [α] Boötis), which shines at magnitude −0.04 near the bottom of Boötes the Herdsman. Orange Arcturus is the fourth-brightest nighttime star overall and the brightest north of the celestial equator.

Next in brightness is Spica (Alpha Virginis), the luminary of Virgo the Maiden. Spica is the very definition of a 1st-magnitude star, but its brightness isn't constant. Its apparent magnitude varies between 0.92 and 1.04 over a period of just more than 4 days. Unlike Arcturus, Spica shines with a blue-white intensity that betrays its hot surface temperature of more than 20,000° Fahrenheit (11,400 K). Arcturus' orange surface is cooler, on the order of 7,300° F (4,300 K).

The third Spring Triangle star is Denebola (Beta [β] Leonis), the star that marks the tail of Leo the Lion. And although Denebola, at magnitude 2.1, is the sky's 59th-brightest star, it's only 36% as bright as Spica, and it emits just 14% the light output of Arcturus.

| OBJECT #319 | NGC 4833 |
|---|---|
| Constellation | Musca |
| Right ascension | 13h00m |
| Declination | −70°53′ |
| Magnitude | 7.8 |
| Size | 13.5′ |
| Type | Globular cluster |
| Other name | Caldwell 105 |

You'll find our next treat 0.7° north-northwest of magnitude 3.6 Delta (δ) Muscae. NGC 4833 is easy to spot through binoculars or a finder scope. This is about as loosely concentrated a globular cluster as you can see. Because of this feature, you'll see about 30 of its outer stars randomly strewn across the field of view through an 8-inch telescope at 200×. The central area appears more concentrated and oval, elongated east-west.

| OBJECT #320 | Abell 1656 |
|---|---|
| Constellation | Coma Berenices |
| Right ascension | 13h00m |
| Declination | 27°59′ |
| Size | 319′ |
| Type | Galaxy cluster |
| Other name | The Coma Galaxy Cluster |

Our next target is for those of you with large telescopes. It's the Coma Galaxy Cluster, also designated Abell 1656. You'll find this group 2.7° due west of magnitude 4.2 Beta (β) Comae Berenices.

For those of you using go-to drives, "Abell 1656" may not be in your database. Instead, target either of this cluster's brightest galaxies, magnitude 11.9 NGC 4874 or magnitude 11.5 NGC 4889.

Abell 1656 spans a whopping 4°. Within that neighborhood, hundreds of member galaxies lie in range of a large amateur telescope. The Coma Galaxy Cluster's richest region, however, is the center, which measures 0.5° across and covers the same area as the Full Moon.

Even through a large scope, you won't pull out much detail from individual members. The exceptions are magnitude 12.8 NGC 4911 and magnitude 12.5 NGC 4921. Both are spirals and respond well to magnification above 300×.

But, perhaps the main point of viewing the Coma Galaxy Cluster is just to see it. After all, this is a group of nearly 1,000 galaxies that lies more than 300 million light-years away.

| OBJECT #321 | NGC 4889 |
|---|---|
| Constellation | Coma Berenices |
| Right ascension | 13h00m |
| Declination | 27°59′ |
| Magnitude | 11.5 |
| Size | 2.8′ by 2′ |
| Type | Elliptical galaxy |
| Other name | Caldwell 35 |

Look for our next target 2.6° west of magnitude 4.3 Beta ($\beta$) Comae Berenices. A small telescope shows an oval glow 50% longer than it is wide, oriented northwest to southeast. Within the glow lies a slightly brighter central region. Don't be distracted by the magnitude 7.2 star SAO 82595, which lies only 9′ to the northwest.

Larger telescopes really don't reveal much more detail in NGC 4889. However, the larger your scope, the more faint smudges — each one a system of billions of stars — you'll see in this incredible area of sky.

| OBJECT #322 | NGC 4945 |
|---|---|
| Constellation | Centaurus |
| Right ascension | 13h05m |
| Declination | –49°28′ |
| Magnitude | 8.8 |
| Size | 23.0′ by 5.9′ |
| Type | Barred spiral galaxy |
| Other name | Caldwell 83 |

This treat sits only 0.3° east of magnitude 4.8 Xi$^1$ ($\xi^1$) Centauri. This is a huge, bright, nearly edge-on galaxy oriented northeast to southwest that looks great through any size telescope. The galaxy shows even illumination across its surface except at its ends. The northeastern one glows brighter. Through a 12-inch or larger scope at 300×, look for a dark indentation near the northeastern end. Our next object (NGC 4976) lies 0.5° east. A fainter galaxy, magnitude 12.5 NGC 4945A, sits 0.3° southeast of NGC 4945.

| OBJECT #323 | NGC 4976 |
|---|---|
| Constellation | Centaurus |
| Right ascension | 13h09m |
| Declination | –49°30′ |
| Magnitude | 10.1 |
| Size | 5.6′ by 3′ |
| Type | Elliptical galaxy |

Our next target lies 0.5° east of NGC 4945 (Object #322). Through a 6-inch telescope, you'll see an evenly illuminated oval 50% longer than it is wide oriented north-northwest to south-southeast. A larger scope may differentiate the thin outer halo from the central region, but that's about it.

**Object #324** NGC 5005 Ray and Emily Magnani/Adam Block/NOAO/AURA/NSF

| OBJECT #324 | NGC 5005 |
|---|---|
| Constellation | Canes Venatici |
| Right ascension | 13h11m |
| Declination | 37°03′ |
| Magnitude | 9.8 |
| Size | 5.8′ by 2.8′ |
| Type | Spiral galaxy |
| Other name | Caldwell 29 |

You'll find our next treat 3 east-southeast of magnitude 2.8 Cor Caroli (Alpha [α] Canum Venaticorum). Through an 8-inch telescope, you'll pick out the bright stellar core surrounded by a fainter oval disk. A 16-inch scope at 300× still won't fully resolve the tightly wound spiral arms, but it will let you see the uneven brightness that marks the positions of immense dust lanes.

| OBJECT #325 | M53 (NGC 5024) |
|---|---|
| Constellation | Coma Berenices |
| Right ascension | 13h13m |
| Declination | 18°10′ |
| Magnitude | 7.7 |
| Size | 12.6′ |
| Type | Globular cluster |

This nice small telescope target is globular cluster M53 in Coma Berenices. To find it, look a little less than 1° northeast of Coma Berenices' brightest star. That's magnitude 4.3 Diadem (Alpha [α] Comae Berenices).

M53 sits 60,000 light-years from both the Sun and the Milky Way's center. It lies in our galaxy's halo region. Through a 4-inch scope under a dark sky, you'll see several dozen faint stars.

Many stars concentrate in the wide core of this cluster. Few field stars lie around M53, so you'll have no trouble defining where the cluster's stars end.

| OBJECT #326 | NGC 5033 |
|---|---|
| Constellation | Canes Venatici |
| Right ascension | 13h13m |
| Declination | 36°36′ |
| Magnitude | 10.2 |
| Size | 10.5′ by 5.1′ |
| Type | Spiral galaxy |

Our next target lies 1.7° east-northeast of magnitude 5.2 star 14 Canum Venaticorum. It appears twice as long as it is wide and elongated north-northwest to south-southeast. The broad bright central region overwhelms the faint spiral structure through anything less than a 14-inch telescope. Through large scopes at 350× and above, you'll detect the faint spiral arms that wrap loosely around the inner glow. They show up best just to the east and west of the galaxy's center.

**Object #327** The Sunflower Galaxy (M63) Adam Block/Mount Lemmon SkyCenter/University of Arizona

| OBJECT #327 | M63 (NGC 5055) |
|---|---|
| Constellation | Canes Venatici |
| Right ascension | 13h16m |
| Declination | 42°02′ |
| Magnitude | 8.6 |
| Size | 13.5′ by 8.3′ |
| Type | Spiral galaxy |
| Other name | The Sunflower Galaxy |

As spring begins to take hold in the Northern Hemisphere, the Sunflower Galaxy blooms near the Big Dipper. Target it, and images of summertime fields will dance through your head.

Pierre Méchain discovered this galaxy June 14, 1779. It marked the first of his deep-sky finds. He reported it to his friend Messier, who immediately included it in his catalog.

M63 lies 5.7° from the Whirlpool Galaxy (M51), and both of these objects belong to the M51 galaxy group. NGC 5195 (the Whirlpool Galaxy's companion) plus five galaxies fainter than magnitude 12.3 round out the collection.

M63 reveals a wealth of detail to a careful observer. Through small telescopes, the nucleus appears stellar, and a 3′ long oval halo surrounds it. Through a 10-inch telescope, the halo shows clumpy structure formed by stellar associations and star-forming regions within M63's spiral arms.

Outside the central region, the arms fade with increasing distance from the nucleus. Through even large amateur telescopes, you'll only see hints of the many spiral arms that radiate outward. Astronomers refer to M63 and similar galaxies as "flocculent" spirals — those having only patchy, localized spiral structure.

| OBJECT #328 | NGC 5053 |
|---|---|
| Constellation | Coma Berenices |
| Right ascension | 13h16m |
| Declination | 17°42′ |
| Magnitude | 9.9 |
| Size | 10.5′ |
| Type | Globular cluster |

In the same low-power telescopic field as M53 (Object #325), you'll find one of the Milky Way's least concentrated globular clusters, NGC 5053. Its brightest stars barely top magnitude 14, so, although you can spot the cluster through a small scope, you'll need an 8-inch telescope to resolve its stars as individual points. This unusual globular looks like an open cluster. It has only a few dozen widely spaced stars strewn across its width. Look closely and see if you can pick out its roughly triangular shape.

| OBJECT #329 | NGC 5068 |
|---|---|
| Constellation | Virgo |
| Right ascension | 13h19m |
| Declination | −21°02′ |
| Magnitude | 9.8 |
| Size | 7.3′ by 6.4′ |
| Type | Spiral galaxy |

Our next target lies 2.1° due north of magnitude 3.0 Gamma (γ) Hydrae. Through a 10-inch telescope it appears irregularly round. The southern half slightly outshines the northern section. Through telescopes with 18-inch apertures or more, this object resolves to a superb face-on spiral with many bright knots within the halo that surrounds the extended central region.

Two other galaxies worthy of your time lie nearby. Magnitude 11.4 NGC 5087 lies 0.5° northeast, while magnitude 10.5 NGC 5084 sits 0.8° to the south-southeast. Both are spirals.

| OBJECT #330 | NGC 5102 |
|---|---|
| Constellation | Centaurus |
| Right ascension | 13h22m |
| Declination | −36°38′ |
| Magnitude | 8.8 |
| Size | 9.8′ by 4.0′ |
| Type | Spiral galaxy |
| Other name | Iota's Ghost |

This object sits only 0.3° east-northeast of magnitude 2.8 Iota (ι) Centauri. NGC 5102 appears relatively bright because it lies less than 11 million light-years away. Unfortunately, that distance also means the light from this galaxy spreads out quite a bit, so it doesn't show as many details as other similarly sized objects. Through an 8-inch telescope, look for a bright central region surrounded by a large oval halo twice as long as it is wide. Oh, and you'll get your best views if you move bright Iota Centauri out of the field of view.

| OBJECT #331 | Zeta (ζ) Ursae Majoris |
|---|---|
| Constellation | Ursa Major |
| Right ascension | 13h24m |
| Declination | 54°56′ |

| (continued) | |
|---|---|
| Magnitudes | 2.4/4.0 |
| Separation | 12′/14.4″ |
| Type | Double star |
| Other name | Mizar + Alcor |

At the bend of the Big Dipper's handle, you'll find Mizar (Zeta [ζ] Ursae Majoris) and Alcor (80 Ursae Majoris). The two stars are separated by 12′, an easy split for most observers just using their naked eyes.

Telescopically, Mizar itself splits into two components separated by 14″. Mizar's companion shines at magnitude 4.0. This was the first star astronomers telescopically identified as a double. Italian astronomer Giovanni Battista Riccioli (1598–1671) made this discovery at Bologna in 1650.

Originally, Zeta carried the name "Merak" (or, "Mirak"), which repeated the common name for Merak (Beta [β] Ursae Majoris). Merak means "the loin." Joseph Justus Scalinger (1540–1609) changed the name to Mizar.

| OBJECT #332 | NGC 5128 |
|---|---|
| Constellation | Centaurus |
| Right ascension | 13h26m |
| Declination | –43°01′ |
| Magnitude | 6.7 |
| Size | 31.0′ by 23.0′ |
| Type | Irregular galaxy |
| Other names | Centaurus A, the Hamburger Galaxy, Caldwell 77 |

Seeing Centaurus A high in the sky is one of the thrills of Southern Hemisphere observing. Observers call it the Hamburger Galaxy because two stellar regions (the bun) surround a dark dusty lane (the burger). Unfortunately, most northern viewers get only a taste of this object's details. For example, from Tucson, Arizona, NGC 5128 climbs to a maximum altitude of 15°. Viewing any object through that much of Earth's atmosphere presents a distorted view. For best results, head farther south.

Australian astronomer James Dunlop (1793–1848) discovered NGC 5128 and published the observation within a list of 629 objects titled "A catalogue of nebulae and clusters of stars in the southern hemisphere, observed at Parramatta in New South Wales," which appeared in the *Philosophical Transactions of the Royal Society*, Volume 118, 1828.

NGC 5128's appearance arises from a galactic collision. The main body of Centaurus A — a giant elliptical galaxy — is absorbing a smaller spiral galaxy. The two objects collided more than 200 million years ago, causing huge bouts of star formation.

Through small telescopes, NGC 5128 appears round with a wide, dark lane cutting the galaxy in half. Use a 12-inch or larger scope, and you'll see a thin wedge of light shining through the lane's western end. That lane widens on both ends.

**Object #333** Omega Centauri (NGC 5139) Adam Block/NOAO/AURA/NSF

| OBJECT #333 | NGC 5139 |
|---|---|
| Constellation | Centaurus |
| Right ascension | 13h27m |
| Declination | –47°29′ |
| Magnitude | 3.5 |
| Size | 36.3′ |
| Type | Globular cluster |
| Other names | Omega Centauri, Caldwell 80 |

Centaurus contains the nearest star system to our own, the spectacular double star Alpha ($\alpha$) Centauri and notable galaxies like Centaurus A. For amateur astronomers, however, the biggest draw to this constellation is the sky's top globular cluster — Omega Centauri.

NGC 5139 does not have the common name "Omega" because of its shape, as in the case of the Omega Nebula (M17). Rather, it appeared as a "star" labeled with the Greek letter Omega ($\omega$) on German cartographer Johann Bayer's 1603 star atlas *Uranometria*. Bayer labeled stars with Greek letters to designate their brightnesses within constellations. Because Bayer interpreted a historical listing of NGC 5139 as a star, he assigned it Omega.

Omega Centauri is a wonder to behold through binoculars or telescopes of any size. The cluster appears slightly larger than the Full Moon, and, because it's rotating relatively quickly, its shape is slightly out-of-round. Through an 8-inch telescope, you'll see 1,000 stars, each a pinprick of light. At high power, the stars appear nearly uniformly distributed across the field of view. Through scopes with apertures larger than 16 inches, crank up the magnification, and look for individual red supergiants within this cluster.

**Object #334** The Whirlpool Galaxy (M51) Jon and Bryan Rolfe/Adam Block/NOAO/AURA/NSF

| OBJECT #334 | M51 (NGC 5194) |
|---|---|
| Constellation | Canes Venatici |
| Right ascension | 13h30m |
| Declination | 47°12′ |
| Magnitude | 8.4 |
| Size | 8.2′ by 6.9′ |
| Type | Spiral galaxy |
| Other names | The Whirlpool Galaxy, Lord Rosse's Nebula, the Question Mark |

Even in a publication of the 1,000 best celestial wonders, the Whirlpool Galaxy stands out. It won't disappoint you when you view it through a small telescope; see it through a big scope, however, and it will knock your socks off. It lies 3.6° southwest of magnitude 1.9 Alkaid (Eta [$\eta$] Ursae Majoris).

Messier discovered the object that would become his catalog's 51st entry while observing a comet October 13, 1773. Méchain discovered the Whirlpool's smaller companion, NGC 5195, March 21, 1781. This object's spiral structure first revealed itself through William Parsons' (The Third Earl of Rosse) 6-foot reflector at Parsonstown, Ireland, in 1845.

NGC 5195 lies some distance behind the plane of M51's disk. Although photographs appear to show a connecting arm between the two galaxies, this is an illusion. Computer models indicate NGC 5195 passed close to the Whirlpool's disk some 70 million years ago and then plunged through its plane.

You'll see M51's spiral arms through 8-inch or larger telescopes. Through a 12-inch scope, they appear patchy with much greater detail. Look for the thin, dark dust lanes that follow the arms' inner edges. Also, try to spot the apparent connecting arm between M51 and NGC 5195. Although M51 is larger, NGC 5195's core appears brighter.

| OBJECT #335 | NGC 5189 |
|---|---|
| Constellation | Musca |
| Right ascension | 13h34m |

| (continued) | |
|---|---|
| Declination | −65°59′ |
| Magnitude | 9.9 |
| Size | 153″ |
| Type | Planetary nebula |
| Other name | The Spiral Planetary Nebula |

Our next target lies 2.7° east-southeast of magnitude 5.7 Theta ($\theta$) Muscae. NGC 5189 is a planetary nebula that shows five sets of "ansae." This Latin term for "handles" describes small nodules emanating out from the central star. In fact, so many observers' initial impressions are of a spiral galaxy that this object earned the moniker the "Spiral Planetary Nebula."

Many observers have noted the resemblance of NGC 5189 to a barred spiral galaxy. A thin, bright bar traverses the planetary and surrounds its 13th-magnitude central star. Through a 12-inch telescope at 300×, you'll see a nebulous arm wrapping to the north from the west end of the bar and curling around the magnitude 11.0 star GSC 9003:1874. To spot the southern arm, you'll need at least a 20-inch scope.

**Object #336** The Southern Pinwheel Galaxy (M83) Allan Cook/Adam Block/NOAO/AURA/NSF

| OBJECT #336 | M83 (NGC 5236) |
|---|---|
| Constellation | Hydra |
| Right ascension | 13h37m |
| Declination | −29°52′ |
| Magnitude | 7.5 |
| Size | 15.5′ by 13.0′ |
| Type | Spiral galaxy |
| Other name | The Southern Whirlpool Galaxy |

Dubbing a galaxy the "Southern Whirlpool" puts a lot of pressure on it to perform visually. For M83, that's no problem. Some observers rate this galaxy the finest barred spiral visible to northern observers.

Lacaille discovered M83 February 23, 1752. It was the third galaxy discovered. Only the Andromeda Galaxy (M31) and its companion M32 preceded it. Messier added it to his catalog March 18, 1781. You can find it 7.2° west-southwest of magnitude 3.3 Pi (π) Hydrae.

M83 is one of the brightest members of a collection of some 14 galaxies called the M83 group. The other notable object in this group is Centaurus A.

The Southern Whirlpool Galaxy appears nearly face-on, so you'll see its spiral structure through telescopes with apertures as small as 6 inches. The core is small and round, and the bar extends to the northeast and southwest. Both spiral arms are easy to see, but the one that wraps southward from the bar's northeastern end shows up better. Through 12-inch and larger scopes, you'll see large clumps of stars and star-forming regions along the arms.

| OBJECT #337 | NGC 5247 |
|---|---|
| Constellation | Virgo |
| Right ascension | 13h38m |
| Declination | −17°53′ |
| Magnitude | 9.9 |
| Size | 5.4′ by 4.9′ |
| Type | Spiral galaxy |

Look for our next object 6.9° northeast of magnitude 3.0 Gamma (γ) Hydrae. Although images of this galaxy show wonderfully arcing spiral arms curving around the nucleus, they're too faint to see through most amateur telescopes. Through smaller scopes at magnifications above 200×, you'll see a bright central region surrounded by a slightly oval haze that orients northeast to southwest.

| OBJECT #338 | The Kite |
|---|---|
| Constellation | Boötes |
| Right ascension | 14h40m |
| Declination | 29°15′ |
| Type | Asterism |

Our next object is a northern-sky springtime asterism many amateur astronomers call the Kite. Now, while many people see a kite in this part of the sky, I see an ice cream cone. (You can form either pattern from the same stars.) The cone is made of six stars in the constellation Boötes the Herdsman. Let's start with the easiest to find — Arcturus (Alpha [α] Boötis) — the night sky's fourth-brightest star.

Finding Arcturus is easy. First find the Big Dipper. Then follow the curve of the Dipper's handle until you encounter a bright orange star. That's Arcturus, and it's the bottom of the ice cream cone.

The thin sugar-cone tips slightly to the north-northeast, so from Arcturus, head up to Epsilon (ε) and Delta (δ) Boötis to make the cone's left side, and Rho (ρ) and Gamma (γ) Boötis for the right. Beta (β) Boötis marks the top of the scoop of ice cream.

This used to be a two-scoop cone. Boötes, however, lies near its highest point in the sky during the hottest days of summer. So, sometime long ago, the second scoop melted, slipped off, and now lies just to the east of the cone — as the constellation Corona Borealis.

**Object #339** NGC 5248 Dale Niksch/Adam Block/NOAO/AURA/NSF

| OBJECT #339 | NGC 5248 |
|---|---|
| Constellation | Boötes |
| Right ascension | 13h38m |
| Declination | 8°53′ |

| (continued) | |
|---|---|
| Magnitude | 10.3 |
| Size | 6.2′ by 4.6′ |
| Type | Spiral galaxy |
| Other name | Caldwell 45 |

Deep in Boötes' southwestern corner lies that constellation's brightest galaxy, NGC 5248. It lies in a relatively barren region of sky 8.9° east-southeast of magnitude 2.9 Vindemiatrix (Epsilon [$\varepsilon$] Virginis).

Even through a small telescope, you'll notice the bright core. Through a 10-inch or larger scope at magnifications above 200×, you can pick out the short spiral arms. Each contains a much brighter region of star formation that appears as a curved line.

Through a 14-inch instrument, look for two 15th-magnitude satellite galaxies of NGC 5248. UGC 8575 sits 0.5° west of the main object, and UGC 8629 lies 0.5° to the southeast.

| OBJECT #340 | MyCn 18 |
|---|---|
| Constellation | Musca |
| Right ascension | 13h40m |
| Declination | −67°23′ |
| Magnitude | 12.9 |
| Size | 25″ |
| Type | Planetary nebula |
| Other name | The Hourglass Nebula |

Admittedly, our next object is a faint one. You'll need at least a 16-inch telescope and a magnification above 400× to see much detail in it, but this object is worth the hunt if you're in the Southern Hemisphere. Look for the Hourglass Nebula (sometimes called the Engraved Hourglass Nebula) 2.4° east of magnitude 4.8 Eta ($\eta$) Muscae. Oh, and while you can look for it by name, don't expect to see an hourglass. That moniker only came about after the Hubble Space Telescope imaged MyCn 18.

Visually, this object appears as two ultra-faint smoke rings just beginning to merge. The merger point appears brightest, but it's also tiny.

The designation "MyCn" comes from Margaret W. Mayall (1902–1996) and Annie Jump Cannon (1863–1941), of whose list of 39 emission-line objects the Hourglass Nebula is number 18. They discovered MyCn 18 in 1940.

| OBJECT #341 | NGC 5253 |
|---|---|
| Constellation | Centaurus |
| Right ascension | 13h40m |
| Declination | −31°39′ |
| Magnitude | 10.2 |
| Size | 5.1′ by 1.3′ |
| Type | Irregular galaxy |

You'll find our next target 7.3° northwest of magnitude 2.1 Menkent (Theta [$\theta$] Centauri). Alternatively, you can locate it by first finding M83. From that bright object, move 1.9° south-southeast. This peculiar dwarf galaxy lies nearby (11 million light-years), but you won't see many details.

Astronomers believe NGC 5253 may have been a dwarf elliptical galaxy until an ancient encounter with M83. Through an 8-inch telescope, a bright central region dominates the view. A 12-inch scope reveals several tiny, bright knots with the brightest at the northeastern end.

**Object #342** M3 Bill Uminski and Cyndi Kristopeit/Adam Block/NOAO/AURA/NSF

| OBJECT #342 | M3 (NGC 5272) |
|---|---|
| Constellation | Canes Venatici |
| Right ascension | 13h42m |
| Declination | 28°23′ |
| Magnitude | 6.2 |
| Size | 16.2′ |
| Type | Globular cluster |

This terrific small telescope target is a springtime globular cluster. To find M3, start at the brilliant star Arcturus (Alpha [α] Boötis). Draw a line 25° long up to the northwest until you hit Cor Caroli (Alpha Canum Venaticorum).

M3 lies near the midpoint of this line. No other bright deep-sky object lies nearby, so you'll have no trouble zeroing in on M3. And, here's a test for you sharp-eyed observers: Try to spot M3 without optical aid from a dark site. At magnitude 6.2, it's a tough naked-eye catch, but many observers have seen it, so it's not impossible.

M3 looks great even through a 4-inch telescope.The cluster has a wide, bright center that accounts for about half of this object's width. Surrounding the center are dozens of stars whose density gradually decreases with their distance from M3's core.

Start with a magnification around 100× and increase the power if the steadiness of the air warrants it. M3 isn't small — its overall size is half that of the Full Moon — but it is dense. Through ever-larger scopes, you'll resolve more and more stars in this amazing cluster.

I like Admiral William H. Smyth's description of M3 from his 1844 observing classic, *Cycle of Celestial Objects*. He writes, "A brilliant and beautiful globular congregation of not less than 1,000 small stars, between the southern Hound and the knee of Boötes; it blazes splendidly towards the centre, and has outliers in all directions, except the *sf* [south following], where it is so compressed that, with its stragglers, it has something of the figure of the luminous oceanic creature called *Medusa pellucens*."

| OBJECT #343 | NGC 5286 |
|---|---|
| Constellation | Centaurus |
| Right ascension | 13h46m |
| Declination | –51°22′ |
| Magnitude | 7.2 |
| Size | 9.1′ |
| Type | Globular cluster |
| Other name | Caldwell 84 |

Our next target lies 2.3° north-northeast of magnitude 2.3 Epsilon (ε) Centauri. In addition to the globular cluster, you'll also immediately see the brilliant foreground star SAO 241157, also known as M Centauri, which lies 4′ from the cluster's center. Despite the pairing, the star has nothing to do with the cluster, which lies 200 times as far away.

NGC 5286 contains many stars, but its brightest shine at magnitude 13.5, so you'll have trouble resolving it through telescopes with apertures less than about 14 inches. The central region appears concentrated, but the outer stars are difficult to see because of the brightness of M Centauri.

| OBJECT #344 | NGC 5281 |
|---|---|
| Constellation | Centaurus |
| Right ascension | 13h47m |
| Declination | –62°54′ |
| Magnitude | 5.9 |
| Size | 5′ |
| Type | Open cluster |
| Other name | The Little Scorpion Cluster |

This bright cluster lies 3.3° southwest of magnitude 0.6 Hadar (Beta [β] Centauri). A 4-inch telescope at 100× reveals three dozen stars in a tiny area. The cluster's brightest member is magnitude 6.6 SAO 252442, which lies just north of center. This star forms the top of a slightly curving line of a half dozen points that arcs gently toward the southwest.

*Astronomy* magazine Contributing Editor Stephen James O'Meara christened this cluster the Little Scorpion because he saw that animal patterned in its stars, complete with claws and a raised tail.

| OBJECT #345 | NGC 5308 |
|---|---|
| Constellation | Ursa Major |
| Right ascension | 13h47m |
| Declination | 60°58′ |
| Magnitude | 11.4 |
| Size | 3.7′ by 0.7′ |
| Type | Spiral galaxy |

For our next target, look 3.9° south-southwest of magnitude 3.7 Thuban (Alpha [α] Draconis). At magnifications under about 200×, this galaxy looks like a luminous splinter. Cranking the power up thickens its appearance and reveals a slightly fatter nucleus. Through a 12-inch telescope, observers have reported a dark lane. Using a 30-inch Newtonian reflector, I saw a slightly more luminous "line" extending a bit from the nucleus in both directions along NGC 5308's long axis.

| OBJECT #346 | NGC 5322 |
|---|---|
| Constellation | Ursa Major |
| Right ascension | 13h49m |
| Declination | 60°11′ |

| (continued) | |
|---|---|
| Magnitude | 10.1 |
| Size | 6′ by 4.1′ |
| Type | Elliptical galaxy |

The next object lies 4.5° south-southwest of magnitude 3.7 Thuban (Alpha [α] Draconis). Through an 8-inch telescope, you'll see an east-west oriented oval half again as long as it is wide. The central region is bright and wide, and a thin halo you'll need high magnification to see surrounds it.

| OBJECT #347 | NGC 5350 |
|---|---|
| Constellation | Canes Venatici |
| Right ascension | 13h53m |
| Declination | 40°22′ |
| Magnitude | 11.3 |
| Size | 3.1′ by 2.5′ |
| Type | Barred spiral galaxy |

Toward the eastern edge of Canes Venatici lies the NGC 5353 galaxy group, also known as Hickson 68. NGC 5350 is its largest member. Find it 7.8° west-northwest of Seginus (Gamma [γ] Boötis).

Through a 12-inch telescope at 300×, you'll note that NGC 5350 has a subtle bar that traverses the center from east to west.

An interacting pair of galaxies, magnitude 11.1 NGC 5353 and magnitude 11.5 NGC 5354, lie 4′ south-southwest of NGC 5350. NGC 5353 measures 2.2′ by 1.1′, while NGC 5354 spans 1.4′. These three galaxies join with magnitude 13.0 NGC 5355 and magnitude 13.7 NGC 5358 to form this compact galaxy group. The magnitude 6.5 star HIP 67778 lies 3′ west-southwest of NGC 5350.

| OBJECT #348 | NGC 5316 |
|---|---|
| Constellation | Centaurus |
| Right ascension | 13h54m |
| Declination | −61°52′ |
| Magnitude | 6.0 |
| Size | 13′ |
| Type | Open cluster |

Our next target lies 1.9° southwest of magnitude 0.6 Hadar (Beta [β] Centauri). Now, at 6th magnitude, this cluster should be visible to observers at a dark site. That's a bit problematic because of the rich Milky Way star field it lies in.

Through a 4-inch telescope at 150×, you'll see three dozen 9th- and 10th-magnitude stars. A 10-inch scope at the same magnification brings into view a second tier of fainter stars, raising the total number past 50.

| OBJECT #349 | NGC 5315 |
|---|---|
| Constellation | Circinus |
| Right ascension | 13h54m |
| Declination | −66°31′ |
| Magnitude | 9.8 |
| Size | 5″ |
| Type | Planetary nebula |

To find this tiny but bright nebula, look 5.2° west-southwest of magnitude 3.2 Alpha (α) Circini. Through a 12-inch scope it has a faint bluish color and begins to show a disk at magnifications above 200×. A nebula filter such as an OIII will really help. The central star isn't all that tough to see. It glows at magnitude 14.2.

| OBJECT #350 | NGC 5367 |
|---|---|
| Constellation | Centaurus |
| Right ascension | 13h58m |
| Declination | −39°59′ |
| Size | 2.5′ by 2.5′ |
| Type | Reflection nebula |

Our next object lies 2° northwest of magnitude 4.4 Chi ($\chi$) Centauri. Through a 10-inch telescope, an evenly illuminated haze 3′ across surrounds an associated star that shines at magnitude 9.8. To the northeast lies a detached region 2′ across. Don't use a nebula filter on this object because the nebula is reflected starlight, which comprises all wavelengths.

**Object #351** M101 Adam Block/Mount Lemmon SkyCenter/University of Arizona

| OBJECT #351 | M101 (NGC 5457) |
|---|---|
| Constellation | Ursa Major |
| Right ascension | 14h03m |
| Declination | 54°21′ |
| Magnitude | 7.9 |
| Size | 26.0′ by 26.0′ |
| Type | Spiral galaxy |

Only one thing prevents spiral galaxy M101 from making every observer's top ten list — its surface brightness. Covering slightly more area than the Full Moon, M101's light spreads out so much that only large amateur telescopes (those 12 inches and larger in aperture) do it justice.

M101 still represents one of the sky's "grand design" spiral galaxies — one with prominent and clearly defined spiral arms. Usually, the arms mostly or completely surround such galaxies. Only about 10% of all spiral galaxies fall into the grand design category.

From a dark site through a large telescope, look for M101's multiple spiral arms. The core is concentrated but broad, not starlike. Many star-forming regions and stellar associations lie along the spiral arms. In fact, at least five (NGC 5447, NGC 5455, NGC 5461, NGC 5462, and NGC 5471) are bright enough to have their own NGC numbers. Of these, NGC 5447 is the most prominent. Find it 6′ southwest of M101's core. Several other objects within M101 once carried similar designations, but astronomers no longer recognize them. Use a nebula filter to tell the difference between star-forming regions and associations. The filter will dim the stars within the associations, but not the nebular gas of the star-forming regions.

To get a rough idea of the position of M101, make an equilateral triangle with the end two stars of the Big Dipper's handle. Alternatively, it lies 1.5° east-northeast of the magnitude 5.7 star 86 Ursae Majoris.

M101 is one of the most beautiful face-on spirals in the sky. From a dark site, a 12-inch telescope will show its multiple spiral arms surrounding a non-stellar core.

Through a 16-inch or larger telescope, use a nebula filter to cut down the brightness of stars within M101. This technique will help you see the glowing hydrogen clouds better.

Of M101, Smyth, writing in *Cycle of Celestial Objects* (1844), said, "It is one of those globular nebulae that seem to be caused by a vast agglomeration of stars, rather than by a mass of diffused luminous matter; and though the idea of too dense a crowd may intrude, yet the paleness tells of its inconceivable distance, and probable discreteness."

| OBJECT #352 | IC 972 |
|---|---|
| Constellation | Virgo |
| Right ascension | 14h04m |
| Declination | −17°15′ |
| Magnitude | 13.9 |
| Size | 43″ |
| Type | Planetary nebula |
| Other name | Abell 37 |

At nearly 14th magnitude, you might be inclined to pass IC 972 by for less difficult targets. That's fine if you're viewing through a 4-inch scope, but if you have a 10-inch or larger instrument, have a look at the faint outer layers of a once Sun-like star. Because of its small size, IC 972 has a reasonable surface brightness. Better known as Abell 37, this object appears uniformly illuminated with a sharp edge.

| OBJECT #353 | NGC 5466 |
|---|---|
| Constellation | Boötes |
| Right ascension | 14h06m |
| Declination | 28°32′ |
| Magnitude | 9.0 |
| Size | 11.0′ |
| Type | Globular cluster |

This object sits roughly 5° due east of the magnificent globular M3. NGC 5466 shines 3 magnitudes fainter than M3, but it's definitely worth a look. This object is one of the least-dense globulars you can observe.

Through a 12-inch telescope at 150×, you'll resolve two dozen stars, but be aware: they're faint. NGC 5466's individual stars glow at 14th magnitude. The star SAO 83172 lies in the same field of view as the globular. It appears bright compared to NGC 5466. The star sits 20′ east-southeast of the cluster and shines at magnitude 6.9.

| OBJECT #354 | Circinus Galaxy |
|---|---|
| Constellation | Circinus |
| Right ascension | 14h13m |
| Declination | −65°20′ |
| Magnitude | 10.1 |
| Size | 8.7′ by 2.8′ |
| Type | Spiral galaxy |

Our next target, while not a telescopically fascinating object, is well worth a look. You'll find it 3.1° west of magnitude 4.2 Alpha (α) Circini. This Seyfert galaxy wasn't discovered until 1977 because it lies only 4° from the Milky Way's plane, which really dims the light of distant galaxies.

With a diameter of more than 300,000 light-years, the galaxy is enormous. Interestingly, it is also isolated. The Circinus Galaxy is not a member of the Local Group or any nearby galaxy group, nor does it have any known companions.

Through a 12-inch telescope at 100×, it shows a medium-bright core surrounded by a faint halo. It appears nearly three times as long as it is wide. Also, please note that the listed brightnesses for this object vary by as much as 2 magnitudes.

| OBJECT #355 | Kappa (κ) Boötis |
|---|---|
| Constellation | Boötes |
| Right ascension | 14h16m |
| Declination | 51°47′ |
| Magnitudes | 4.6/6.6 |
| Separation | 13.4″ |
| Type | Double star |

The components are blue-white and white, or blue and white, depending on whose eyes you believe. You'll find this nice pair 1.8° west of magnitude 4.0 Theta (θ) Boötis.

| OBJECT #356 | NGC 5523 |
|---|---|
| Constellation | Boötes |
| Right ascension | 14h15m |
| Declination | 25°19′ |
| Magnitude | 12.1 |
| Size | 4.3′ by 1.3′ |
| Type | Spiral galaxy |

Find this galaxy 1° east of the magnitude 4.8 star 12 Boötis. This disk-shaped object looks best through 12-inch and larger telescopes, although, even through scopes twice that large, you won't see much detail.

What you may notice is a slightly brighter, oval-shaped central region surrounded by a tiny bit of haze. Only 2′ to the northwest, you'll find GSC 2010:1226, a magnitude 10.8 star.

**Object #357** NGC 5529 Bill and Sean Kelly/Adam Block/NOAO/AURA/NSF

| OBJECT #357 | NGC 5529 |
|---|---|
| Constellation | Boötes |
| Right ascension | 14h16m |
| Declination | 36°13′ |
| Magnitude | 11.9 |
| Size | 5.7′ by 0.7′ |
| Type | Spiral galaxy |

If you're like me and appreciate galaxies that look like lines drawn on the sky, don't miss NGC 5529. Look for it 3.9° west-southwest of magnitude 3.0 Seginus (Gamma [$\gamma$] Boötis).

Through an 8-inch telescope, you'll see a galaxy half a dozen times as long as it is wide. Only at the highest magnifications (and on the finest nights) will you detect the ever-so-slight central bulge.

Only 5′ off NGC 5529's south-southeastern tip lies magnitude 10.9 GSC 2552:903. The galaxy's long dimension points the way to this star.

| OBJECT #358 | Iota ($\iota$) Boötis |
|---|---|
| Constellation | Boötes |
| Right ascension | 14h16m |
| Declination | 51°22′ |
| Magnitudes | 4.9/7.5 |
| Separation | 38″ |
| Type | Double star |

Iota Boötis is an easy split through any telescope. The primary is yellow-white and the secondary glows white.

| OBJECT #359 | IC 4406 |
|---|---|
| Constellation | Lupus |
| Right ascension | 14h22m |
| Declination | −44°09′ |
| Magnitude | 10.2 |
| Size | 106″ |
| Type | Planetary nebula |
| Other name | The Retina Nebula |

As planetary nebulae go, the Retina Nebula is relatively bright although it measures more than 1.5′ across. What makes it obscure for northern observers is its location. It lies at a declination of −44°. My rule of thumb is that if you can see Omega Centauri (NGC 5139) from your observing site, you stand a good chance of picking up IC 4406, which lies 3° farther north.

IC 4406 sits a bit more than 3° southwest of magnitude 2.3 Eta ($\eta$) Centauri. Through an 8-inch telescope equipped with an OIII filter, first study the overall appearance of this object to see the "inner eye" that makes it the Retina Nebula. The "top" and "bottom" of this planetary orient east to west and appear remarkably straight. The northern edge is brighter than the southern one. At high powers, you'll see that the central region has indentations, lending the whole object the appearance of another dumbbell-shaped nebula.

# May

| OBJECT #360 | Alpha Centauri C |
|---|---|
| Constellation | Centaurus |
| Right ascension | 14h30m |
| Declination | −62°41′ |
| Magnitude | 11.0 |
| Type | Star |
| Other name | Proxima Centauri |

Our next target is the closest star to the Sun, Proxima Centauri (Alpha [α] Centauri C). As such, I think it easily makes our list of the sky's top 1,001 objects, despite its faintness. To find this star, first center its brilliant companion, magnitude −0.1 Rigil Kentaurus (Alpha Centauri A–B). Alpha Cen C lies a bit more than 2° south-southeast. At a distance of 4.22 light-years, C lies 0.17 light-year closer to us than its neighbor. However, it glows meekly at magnitude 11.05, so it's easy to miss.

"Proxima" comes from the Latin word for near. It's the same root that gives us our word "proximity."

| OBJECT #361 | NGC 5634 |
|---|---|
| Constellation | Virgo |
| Right ascension | 14h30m |
| Declination | −5°59′ |
| Magnitude | 9.5 |
| Size | 5.5′ |
| Type | Globular cluster |

Our next target is globular cluster NGC 5634 in Virgo. It lies halfway between magnitude 3.9 Mu (μ) Virginis and magnitude 4.1 Syrma (Iota (ι) Virginis).

M.E. Bakich, *1,001 Celestial Wonders to See Before You Die*, Patrick Moore's Practical Astronomy Series,
DOI 10.1007/978-1-4419-1777-5_5, 

Let's be honest. Virgo is known for its galaxies. The constellation contains some 200 deep-sky objects brighter than 13th magnitude. Only one — NGC 5634 — is a globular cluster.

Point a 4-inch telescope at this object, and you'll see lots of faint stars and one bright orange one — magnitude 8.0 SAO 139967, which sits a bit more than 1′ east-southeast of the cluster's center. The star isn't part of NGC 5634. It just happens to lie in the same direction from our viewpoint.

The cluster's stars are condensed, meaning you won't easily resolve them into individual points. But the back-and-forth visibility battle you'll encounter between the star and the cluster makes for a fascinating observation.

| OBJECT #362 | NGC 5643 |
|---|---|
| Constellation | Lupus |
| Right ascension | 14h33m |
| Declination | −44°10′ |
| Magnitude | 10.4 |
| Size | 5.1′ by 4.3′ |
| Type | Spiral galaxy |

To find our next object, look 2.1° south-southwest of magnitude 2.3 Eta ($\eta$) Centauri. This bright galaxy is visible through a 4-inch telescope as a round, evenly illuminated disk. Through a 12-inch scope at 300×, the northern half of the galaxy outshines the southern half. The bar, which runs east-west isn't faint, but it has to compete with the equally bright tightly wound spiral arms. The galaxy's eastern arm, which turns sharply to the north, is brighter than its western counterpart. Adding to the scene are several faint foreground stars superimposed on the galaxy's face.

| OBJECT #363 | NGC 5676 |
|---|---|
| Constellation | Boötes |
| Right ascension | 14h33m |
| Declination | 49°28′ |
| Magnitude | 11.2 |
| Size | 3.7′ by 1.6′ |
| Type | Spiral galaxy |

The next target sits 2.7° south-southeast of magnitude 4.0 Theta ($\theta$) Boötis. NGC 5676 is a high-surface-brightness galaxy that looks like a gray rectangle through a 6-inch telescope. If you double that aperture and increase the magnification past 200×, you'll see that NGC 5676 has an uneven distribution of brightness because of this galaxy's tilt and the way it presents its spiral arms to us. The northeastern half shines more brightly than the southwestern side.

| OBJECT #364 | NGC 5694 |
|---|---|
| Constellation | Hydra |
| Right ascension | 14h40m |
| Declination | −26°32′ |
| Magnitude | 9.2 |
| Size | 3.6′ |
| Type | Globular cluster |
| Other name | Caldwell 66 |

Our next target is globular cluster NGC 5694 in Hydra. English astronomer Sir William Herschel discovered it in 1784, but it wasn't until 1932 that astronomers identified it as a globular cluster.

To find NGC 5694, look about 2° west-southwest of a line of three 5th-magnitude stars. The stars carry the designations 55, 56, and 57 Hydrae and lie in the far eastern end of the sky's largest constellation. The roughly equal brightnesses and equal spacings between the stars makes me think of this grouping as a small, faint version of Orion's Belt.

NGC 5694 glows at magnitude 9.2, so even a 2.4-inch telescope will reveal it. The cluster's not huge, and its individual stars are faint, so it won't resolve into points of light well.

Most of NGC 5694's brightness comes from its compact core, which takes up more than 50% of this object's diameter. At magnifications above 150, you'll see several foreground stars superimposed on the cluster.

| OBJECT #365 | Pi ($\pi$) Boötis |
|---|---|
| Constellation | Boötes |
| Right ascension | 14h41m |
| Declination | 16°25′ |
| Magnitudes | 4.9/5.8 |
| Separation | 5.6″ |
| Type | Double star |

You'll find this nice binary 6.5° east-southeast of brilliant Arcturus (Alpha [$\alpha$] Boötis). The primary is white (or blue-white), and the secondary is yellow (or yellow-white).

| OBJECT #366 | NGC 5728 |
|---|---|
| Constellation | Libra |
| Right ascension | 14h42m |
| Declination | −17°15′ |
| Magnitude | 11.5 |
| Size | 3.7′ by 2.6′ |
| Type | Barred spiral galaxy |

You'll find NGC 5728 2.4° west-southwest of Zubenelgenubi (Alpha [$\alpha$] Librae). This barred spiral has an exceptionally bright nucleus. Such a feature classifies it as a Seyfert galaxy. Seyferts are a type of active galaxy that emit prodigious amounts of visible and infrared radiation from tiny regions in their core.

Through a 10-inch telescope, NGC 5728 appears as a faint halo with a much brighter, but tiny, nucleus. Don't confuse the core with a foreground star just 20″ to the northeast. Together, the star and the core give the impression of a double nucleus.

| OBJECT #367 | Epsilon ($\varepsilon$) Boötis |
|---|---|
| Constellation | Boötes |
| Right ascension | 14h45m |
| Declination | 27°04′ |
| Magnitudes | 2.7/5.1 |
| Separation | 2.8″ |
| Type | Double star |
| Common names | Izar; Pulcherrima |

This star's given proper name is Izar, which means "the girdle." The name "Pulcherrima" came to it much later. That's a Latin term meaning "most beautiful," and, indeed this is a gorgeous binary star. The two components do sit closely together, so you'll need a magnification above 150× to get a clean split. The primary is an orange giant star. Its companion resembles the Sun in evolution, but shines with a blue-white light.

| OBJECT #368 | NGC 5749 |
|---|---|
| Constellation | Lupus |
| Right ascension | 14h49m |

| (continued) | |
|---|---|
| Declination | −54°31′ |
| Magnitude | 8.8 |
| Size | 7′ |
| Type | Open cluster |

You'll find our next object along the southern border of Lupus, 4.2° southwest of magnitude 3.4 Zeta (ζ) Lupi. Through a 4-inch telescope, you'll spot a dozen stars, the brightest of which is magnitude 9.6 SAO 242013, which sits near the cluster's western edge. Larger apertures help make visible another level of faint background stars.

| OBJECT #369 | Alpha (α) Librae |
|---|---|
| Constellation | Libra |
| Right ascension | 14h51m |
| Declination | −16°02′ |
| Magnitudes | 2.8/5.2 |
| Separation | 231″ |
| Type | Double star |
| Other name | Zubenelgenubi |

This binary has such a wide separation that binoculars or a finder scope will split it. I think a magnification of about 50× works well. The primary is pale blue, and the secondary is orange or orange-white.

The name "Zubenelgenubi" is from the Arabic "Al Zuban al Janubiyyah," which means the "southern claw."

| OBJECT #370 | Xi (ξ) Boötis |
|---|---|
| Constellation | Boötes |
| Right ascension | 14h51m |
| Declination | 19°06′ |
| Magnitudes | 4.7/7.0 |
| Separation | 6.9″ |
| Type | Double star |
| Yellow and orange | |

You'll have no trouble locating this nicely colored binary. It sits 8.5° due east of Arcturus (Alpha [α] Boötis). The separation is reasonably close, so crank the power above 100×. The A component shines white, although many observers see it as yellow. Almost everyone sees the B star as orange, however.

**Object #371** NGC 5792 Brad Ehrhorn/Adam Block/NOAO/AURA/NSF

| OBJECT #371 | NGC 5792 |
|---|---|
| Constellation | Libra |
| Right ascension | 14h58m |
| Declination | –1°05′ |
| Magnitude | 11.2 |
| Size | 7.3′ by 1.9′ |
| Type | Barred spiral galaxy |

Our next target is a galaxy some observers think is the finest galaxy in Libra. NGC 5792 lies nearly edge-on to our line of sight, so it appears quite elongated. An 11-inch telescope shows a stellar nucleus with faint wings extending a total of nearly 5′. You won't actually see spiral structure, but the spiral arms do produce mottling near the core that's visible through larger scopes. Unfortunately, the magnitude 9.6 star GSC 4987:827 on the galaxy's northwestern edge hinders the view of faint detail.

| OBJECT #372 | IC 4499 |
|---|---|
| Constellation | Apus |
| Right ascension | 15h00m |
| Declination | –82°13′ |
| Magnitude | 9.4 |
| Size | 7.6′ |
| Type | Globular cluster |

Look for this cluster near the South Celestial Pole 0.8° north of magnitude 5.7 Pi$^2$ ($\pi^2$) Octantis. This cluster lies more than 60,000 light-years away, so its brightest stars glow feebly at 15th magnitude. Through an 8-inch telescope at 200×, the central region appears compact and condensed and the halo seems irregular. The magnitude 10.3 star GSC 9440:489 lies 2′ south of the cluster's center.

| OBJECT #373 | NGC 5812 |
|---|---|
| Constellation | Libra |
| Right ascension | 15h01m |
| Declination | −7°27′ |
| Magnitude | 11.2 |
| Size | 2.3′ by 1.9′ |
| Type | Elliptical galaxy |

Our next object lies 1° due north of Delta (δ) Librae. Through an 8-inch telescope, NGC 5812 looks perfectly round. Some observers liken its appearance to a planetary nebula. A 14-inch telescope will bring out a much fainter outer halo, and it may show the magnitude 14.9 galaxy IC 1084 about 5′ to the east.

| OBJECT #374 | 44 Boötis |
|---|---|
| Constellation | Boötes |
| Right ascension | 15h04m |
| Declination | 47°39′ |
| Magnitudes | 5.3/6.2 |
| Separation | 2.1″ |
| Type | Double star |

You'll find 44 Boö in the far northern part of the constellation 7.3° due north of Nekkar (Beta [β] Boötis). The primary and secondary are yellow-white and yellow-orange, respectively. The secondary is also a type of binary star astronomers call a contact binary. As the name implies, the two stars are so close that gravity distorts their surfaces, which touch. You won't be splitting that pair of the system. In fact, you'll need to crank the magnification above 150× just to split the A–B pair.

| OBJECT #375 | NGC 5822 |
|---|---|
| Constellation | Lupus |
| Right ascension | 15h04m |
| Declination | −54°25′ |
| Magnitude | 6.5 |
| Size | 35′ |
| Type | Open cluster |

Our next treat lies 2.6° south-southwest of magnitude 3.4 Zeta (ζ) Lupi. Easily visible to the naked eye from a dark site, this cluster explodes with detail through binoculars or any telescope. A 4-inch instrument at 100× shows 50 stars. Don't look for a center because the stars are evenly distributed across an area larger than the Full Moon. Larger apertures add to the star count.

| OBJECT #376 | NGC 5823 |
|---|---|
| Constellation | Circinus |
| Right ascension | 15h06m |
| Declination | −55°36′ |
| Magnitude | 7.9 |
| Size | 10′ |
| Type | Open cluster |
| Other name | Caldwell 88 |

You'll find this nice open cluster 3.6° west-northwest of magnitude 4.1 Beta (β) Circini, right on that constellation's northern border with Lupus. I'm not certain why this cluster is Caldwell 88 rather than NGC 5822 (Object #375), but Patrick Moore must have had his reasons. Through an 8-inch telescope at 150×, you'll see some three dozen mainly 10th- and 11th-magnitude stars. Many of them arrange in a long, winding, backward S-shape. The cluster's boundary is irregular.

| OBJECT #377 | NGC 5824 |
|---|---|
| Constellation | Lupus |
| Right ascension | 15h04m |
| Declination | −33°04′ |
| Magnitude | 9.1 |
| Size | 7.4′ |
| Type | Irregular galaxy |

To find our next target, look 4.9° northwest of magnitude 3.6 Phi$^1$ ($\phi^1$) Lupi. Through an 8-inch telescope, this globular has a dense center, a ragged edge, and few stars glowing meekly in the incredibly faint halo. One, magnitude 12.0 GSC 7315:514, lies 4′ north of the cluster's center.

| OBJECT #378 | M102 (NGC 5866) |
|---|---|
| Constellation | Draco |
| Right ascension | 15h07m |
| Declination | 55°46′ |
| Magnitude | 9.9 |
| Size | 6.6′ by 3.2′ |
| Type | Spiral galaxy |
| Other name | The Fool's Gold Galaxy |

Is M102 the galaxy NGC 5866? That depends on who you ask. Some astronomical historians argue that M102 is a duplicate observation of M101. Others say the evidence points to the galaxy NGC 5866.

Well, whether or not Messier meant NGC 5866 to be his 102nd entry, this lenticular shows up nicely through a 4-inch telescope as a bright streak with a brilliant center. On the best nights, a 10-inch scope reveals a thin dust lane extending almost as long as the galaxy.

To judge for yourself, point your scope 4.1° south-southwest of magnitude 3.3 Iota ($\iota$) Draconis.

*Astronomy* magazine Contributing Editor Stephen James O'Meara makes a good case that M102 is simply a more refined observation of M101 (Object #351). He therefore calls M102 the Fool's Gold Galaxy because, if you think you've found it when you're observing NGC 5866, the joke's on you.

| OBJECT #379 | UGC 9749 |
|---|---|
| Constellation | Ursa Minor |
| Right ascension | 15h09m |
| Declination | 67°12′ |
| Magnitude | 10.9 |
| Size | 41′ by 26′ |
| Type | Dwarf elliptical galaxy |
| Other name | The Ursa Minor Dwarf |

Our next object lies in the southernmost part of the northernmost constellation. The Ursa Minor dwarf is a dwarf elliptical galaxy whose name comes from its home constellation, Ursa Minor. It lies 4.7° south-southwest of magnitude 3.0 Pherkad (Gamma [$\gamma$] Ursae Minoris).

I suggest you use at least an 11-inch telescope, but don't crank the power up. In fact, you'll want to use the lowest-magnification, widest-field eyepiece you own.

That's because this galaxy covers one and a half times more area than the Full Moon. It measures 41′ by 26′.

It has a respectable magnitude, 10.9, but because that light is so spread out, the Ursa Minor Dwarf has a miserably low surface brightness.

The best approach for viewing this object is to head to the darkest observing site you can get to. Then, disengage your telescope's drive and slowly sweep the Dwarf's region of sky. What you're looking for is an ever-so-slight increase in the background glow of your eyepiece's field of view. Good luck.

| OBJECT #380 | NGC 5846 |
|---|---|
| Constellation | Virgo |
| Right ascension | 15h06m |
| Declination | 1°36′ |
| Magnitude | 10.1 |
| Size | 4′ by 3.7′ |
| Type | Elliptical galaxy |

Head to the easternmost edge of Virgo for our next object, which lies 1° east-southeast of the magnitude 4.4 star 110 Virginis. Through a 10-inch telescope, this object appears round with a broad bright central region and a wide halo. Crank up the power past 300×, and look for the magnitude 13.8 galaxy NGC 5846A buried in the southeast part of the halo.

NGC 5846 is the brightest in an east-west oriented slightly curved line of four galaxies. The westernmost, magnitude 12.7 NGC 5839, lies 15′ east of NGC 5846. Between those two sits magnitude 12.5 NGC 5845. Finally, the nice magnitude 10.8 spiral NGC 5850 lies 10′ east-southeast of NGC 5846.

| OBJECT #381 | NGC 5885 |
|---|---|
| Constellation | Libra |
| Right ascension | 15h15m |
| Declination | −10°05′ |
| Magnitude | 11.8 |
| Size | 3.2′ by 2.6′ |
| Type | Barred spiral galaxy |

NGC 5885 lies 0.8° south-southwest of Zubeneschamali (Beta [$\beta$] Librae). A 10-inch telescope shows this galaxy as a nearly uniform, faint haze. Unfortunately, not even the largest amateur telescope shows the elusive spiral structure that appears only after long-exposure imaging. Just 98″ northeast of the galaxy lies SAO 140412, a magnitude 10.1 foreground star.

| OBJECT #382 | NGC 5905 |
|---|---|
| Constellation | Draco |
| Right ascension | 15h15m |
| Declination | 55°31′ |
| Magnitude | 11.7 |
| Size | 4.3′ by 3.3′ |
| Type | Barred spiral galaxy |

Our next target sits 3.7° south-southwest of magnitude 3.3 Iota ($\iota$) Draconis. It lies less than 1° south of the more spectacular NGC 5907. Through a 10-inch telescope, NGC 5905 presents a circular halo that extends 3′ across. Although it may appear small and unspectacular to the eye, NGC 5905 measures 400,000 light-years across, making it one of the largest spiral galaxies known.

| OBJECT #383 | Delta ($\delta$) Boötis |
|---|---|
| Constellation | Boötes |
| Right ascension | 15h16m |
| Declination | 33°19′ |
| Magnitudes | 3.5/8.7 |

| (continued) | |
|---|---|
| Separation | 105″ |
| Type | Double star |

Although you can separate this wide binary through binoculars or a finder scope, use a telescope at 50× to bring out the stars' colors. The primary is yellow, while the fainter secondary shines white, or perhaps yellow-white.

**Object #384** NGC 5907 Adam Block/Mount Lemmon SkyCenter/University of Arizona

| OBJECT #384 | NGC 5907 |
|---|---|
| Constellation | Draco |
| Right ascension | 15h16m |
| Declination | 56°20′ |

| (continued) | |
|---|---|
| Magnitude | 10.3 |
| Size | 11.5′ by 1.7′ |
| Type | Spiral galaxy |

Our next target is another "needle" in the deep-sky sewing kit. You'll find NGC 5907 not quite 3° south-southwest of magnitude 3.3 Iota ($\iota$) Draconis. This wonderful object lies 35 million light-years away and travels through space with M102.

The plane of NGC 5907 tilts only 3.5° from our line of sight. Through a 4-inch telescope, you'll see the long, narrow center. Larger scopes won't reveal a lot more detail, but each increase in aperture will extend NGC 5907's apparent length.

| OBJECT #385 | Palomar 5 |
|---|---|
| Constellation | Serpens (Caput) |
| Right ascension | 15h16m |
| Declination | –0°07′ |
| Magnitude | 11.8 |
| Size | 6.9′ |
| Type | Globular cluster |

This faint object lies 9° west-northwest of magnitude 3.5 Mu ($\mu$) Serpentis. Alternatively, you can start at the bright globular cluster M5 (Object #389), and move 2.3° to the south-southwest.

Palomar 5 is the 5th entry on a list of only 15 mostly difficult-to-see globular clusters. German-born American astronomer Walter Baade discovered it in 1950, but it was American astronomer George Abell who gave them their "Palomar" designations. Astronomers discovered them on photographic plates taken during the first Palomar Observatory Sky Survey.

Seeing Palomar 5 isn't easy because the object has a low surface brightness. Through a 12-inch telescope use an eyepiece that magnifies 75×. Look for a subtle brightening of the sky brightness.

| OBJECT #386 | Beta Librae |
|---|---|
| Constellation | Libra |
| Right ascension | 15h17m |
| Declination | –9°23′ |
| Magnitude | 126 |
| Type | Colored star |
| Other name | Zubeneschamali |

Our next object is the brightest star in Libra, Zubeneschamali. The star's name comes from the Arabic for "the northern claw." That was long ago, when Libra didn't exist. Instead, its brightest stars represented the claws of Scorpius. Zubeneschamali was the northern claw, and Zubenelgenubi (Alpha [$\alpha$] Librae) was the southern claw. You'll find Zubenelgenubi about 9° southwest of Zubeneschamali.

Libra is a case where the normal Alpha, Beta, Gamma, etc., rule for denoting a constellation's brightest stars — in order — breaks down. Early celestial mapmaker Johannes Bayer rated Zubeneschamali as Libra's second-brightest star. We know today, however, that it's actually the brightest. It shines at magnitude 2.6, while Alpha's magnitude is 2.75, about 15% fainter.

Ok, you've now found two stars with funny names. But there's more. Go out some night, and locate Zubeneschamali. Don't use a telescope or even binoculars. Just look at it with your eyes. What color does it appear to you?

Since the mid-1970s, I've had an ongoing debate with some of my closest observing friends about the color of this star. I see it as green, and so have many others I've asked. But some amateur astronomers whose opinions I value and who I agree with about the majority of star colors think I'm crazy. Their take on Zubeneschamali is that it's either white or light-blue.

So please help me settle this debate. Cast your gaze on Zubeneschamali, and decide for yourself if it is indeed the only naked-eye star with a greenish tint. Then, e-mail me at mbakich@Astronomy.com with your conclusion.

| OBJECT #387 | NGC 5882 |
|---|---|
| Constellation | Lupus |
| Right ascension | 15h17m |
| Declination | −45°39′ |
| Magnitude | 9.4 |
| Size | 7″ |
| Type | Planetary nebula |

If you head 1.4° southwest from magnitude 3.4 Epsilon (ε) Lupi, you'll find a planetary nebula that is bright enough to exhibit impressive blue-green color. Through a 10-inch telescope at 250×, you'll see this object's shell, which appears circular. Larger scopes and higher magnifications show an inner tilted oval.

| OBJECT #388 | NGC 5897 |
|---|---|
| Constellation | Libra |
| Right ascension | 15h17m |
| Declination | −21°01′ |
| Magnitude | 8.6 |
| Size | 12.6′ |
| Type | Globular cluster |
| Other name | The Ghost Globular |

NGC 5897 is the best deep-sky object in Libra. To find it, travel exactly 8° southeast of Zubenelgenubi (Alpha [α] Librae). Because this object lies 40,000 light-years away, its brightest stars glow at only magnitude 13. Still, enough stars group together here that the cluster is easy to spot through 11×80 binoculars. The most striking aspect of NGC 5897 is how loosely its stars concentrate toward the center.

Through an 8-inch telescope at a dark site, you'll see only the brightest dozen or so stars against a faint, comet-like glow. Through a 13-inch scope, the star count rises to 50. If you're lucky enough to view NGC 5897 through a 20-inch scope crank the power past 200×. You'll see stars scatter widely across the cluster's center with little apparent concentration.

*Astronomy* magazine Contributing Editor Stephen James O'Meara called it the Ghost Globular because it resembles a ghost image of the globular M55 (Object #597).

**Object #389** M5 Sally and Curt King/Adam Block/NOAO/AURA/NSF

| OBJECT #389 | M5 (NGC 5904) |
|---|---|
| Constellation | Serpens (Caput) |
| Right ascension | 15h19m |
| Declination | 2°05′ |
| Magnitude | 5.7 |
| Size | 17.4′ |
| Type | Globular cluster |

Our next target is the great globular cluster M5 in Serpens — the brightest globular in the northern half of the sky. You can find it by starting at Zubeneschamali (Beta [$\beta$] Librae).

From that star, move 11.5° due north. If you have sharp eyes, and if your observing site is dark enough, you can spot M5 as a fuzzy magnitude 5.7 star without optical aid. Don't confuse it with 5 Serpentis, a magnitude 5.0 star 22′ to the southeast. To confirm your observation of M5, you should see both it and 5 Serpentis.

Through a 4-inch telescope, this cluster is full of detail. When you crank up the magnification to 150× or more, you'll see that M5 has a grainy structure. You'll spot several dozen stars around the core, which accounts for about one-quarter of the cluster's diameter.

Through an 11-inch scope, more than a hundred stars pop into view. Streamers of stars fill M5's outer regions. They provide a nice contrast to the relatively sparse background.

| OBJECT #390 | NGC 5899 |
|---|---|
| Constellation | Boötes |
| Right ascension | 15h15m |
| Declination | 42°03′ |
| Magnitude | 11.8 |
| Size | 3.3′ by 1.4′ |
| Type | Spiral galaxy |

You'll find this object, and three other galaxies, 3° northeast of magnitude 3.5 Nekkar (Beta [$\beta$] Boötis). The orange magnitude 6.1 star SAO 45445 lies 0.2° to the northwest. Through an 8-inch telescope, you'll see a disk-shaped object more than twice as long as it is wide oriented north-northeast to south-southwest. The galaxy's northern half glows a bit more brightly than the southern portion.

Through at least a 14-inch scope from a dark site, you can search for a trio of faint galaxies near NGC 5899. Look 9′ north for magnitude 14.1 NGC 5900. Magnitude 14.3 NGC 5895 lies 14′ west-southwest, and magnitude 13.2 NGC 5893 sits 4′ southwest of NGC 5895.

**Object #391** NGC 5921 Adam Block/NOAO/AURA/NSF

| OBJECT #391 | NGC 5921 |
|---|---|
| Constellation | Serpens (Caput) |
| Right ascension | 15h22m |
| Declination | 5°04′ |
| Magnitude | 10.8 |
| Size | 4.9′ by 4.2′ |
| Type | Barred spiral galaxy |

To find NGC 5921, move 5.7° west-southwest from magnitude 2.7 Unuk al Hai (Alpha [$\alpha$] Serpentis). Through a 10-inch telescope, the galaxy shows a bright center. You may spot its bar, probably with some difficulty, as a faint oval ring that hints at spiral structure.

NGC 5921 lies 75 million light-years from Earth, far beyond the magnitude 11.6 foreground star GSC 344:738, which appears on the southwestern edge of the galaxy's halo. UGC 9830, a magnitude 16.5 satellite galaxy to NGC 5921, lies 36′ to the south-southeast. Good luck seeing that one.

| OBJECT #392 | Abell 2065 |
|---|---|
| Constellation | Corona Borealis |
| Right ascension | 15h23m |

| (continued) | |
|---|---|
| Declination | 27°43′ |
| Size | 30.5′ |
| Type | Galaxy cluster |

Do you have access to a large telescope? Abell galaxy cluster 2065 is a fine challenge object for a 16-inch scope. Within the half-degree field that encompasses this cluster you'll see half a dozen galaxies.

I observed this object through a 30-inch scope under a pristine sky. Although I counted some three dozen galaxies, I judged the sighting moderately difficult because the individual members appear faint. You'll find this group 2° southwest of Nusukan (Beta [$\beta$] Coronae Borealis).

| OBJECT #393 | Mu ($\mu$) Boötis |
|---|---|
| Constellation | Boötes |
| Right ascension | 15h25m |
| Declination | 37°23′ |
| Magnitudes | 4.4/6.5 |
| Separation | 108″ and 2″ |
| Type | Double star |

This triple star system has a wide A–B separation. The two stars appear light-yellow and deep-yellow. When you have observed the pair at low magnification, crank up the power and target the fainter component. You'll discover it also is a close binary.

| OBJECT #394 | NGC 5925 |
|---|---|
| Constellation | Norma |
| Right ascension | 15h28m |
| Declination | −54°31′ |
| Magnitude | 8.4 |
| Size | 14′ |
| Type | Open cluster |

Norma, because of its proximity to the galactic plane, contains a large number of open clusters. This one lies 3.3° southeast of magnitude 3.4 Zeta ($\zeta$) Lupi. This cluster's individual stars scatter randomly across its face. The brighter tier of about three dozen stars ranges from 10th to 12th magnitude, and you can spot them through a 6-inch telescope. Larger scopes bring into view 50 additional stars that glow more faintly.

| OBJECT #395 | NGC 5927 |
|---|---|
| Constellation | Lupus |
| Right ascension | 15h28m |
| Declination | −50°40′ |
| Magnitude | 8.0 |
| Size | 12.0′ |
| Type | Globular cluster |

Our next target lies 2.9° east-northeast of magnitude 3.4 Zeta ($\zeta$) Lupi. An 8-inch telescope shows a dense core with a ragged outer edge. A 20-inch scope at 300× will resolve some 50 stars, but the core will appear just as dense as through the smaller instrument.

Look for a similarly difficult-to-resolve globular, magnitude 8.4 NGC 5946 (Object #398), 1.2° due east.

| OBJECT #396 | Delta ($\delta$) Serpentis |
|---|---|
| Constellation | Serpens |
| Right ascension | 15h35m |

| (continued) | |
|---|---|
| Declination | 10°32′ |
| Magnitudes | 4.2/5.2 |
| Separation | 4″ |
| Type | Double star |

This "star" actually is a pair of binary stars separated by 66″. The set you'll be observing is the A–B pair, which have a separation of 4″. The C-D pair is just a bit wider at 4.4″, but the stars glow dimly at magnitudes 14.7 and 15.2. You can try for them if you have a 16-inch telescope at your disposal.

| OBJECT #397 | NGC 5938 |
|---|---|
| Constellation | Triangulum Australe |
| Right ascension | 15h36m |
| Declination | –66°52′ |
| Magnitude | 11.7 |
| Size | 2.7′ by 2.4′ |
| Type | Barred spiral galaxy |

Look 0.5° south of magnitude 4.1 Epsilon (ε) Trianguli Australis for this distant galaxy. Indeed, it lies some 300 million light-years away. Through a 16-inch telescope at 300×, this face-on galaxy appears ragged with a small bright central region. A 12th magnitude star sits just south of the core.

| OBJECT #398 | NGC 5946 |
|---|---|
| Constellation | Norma |
| Right ascension | 15h35m |
| Declination | –50°40′ |
| Magnitude | 8.4 |
| Size | 3′ |
| Type | Globular cluster |

Our next target lies 3.9° east-northeast of magnitude 3.4 Zeta (ζ) Lupi. Through 8-inch and smaller telescopes, this cluster appears small with a central concentration and little resolution. A magnitude 11.8 star lies 30″ to the southwest of the core.

| OBJECT #399 | NGC 5962 |
|---|---|
| Constellation | Serpens (Caput) |
| Right ascension | 15h37m |
| Declination | 16°37′ |
| Magnitude | 11.3 |
| Size | 2.6′ by 1.8′ |
| Type | Spiral galaxy |

NGC 5962 lies in northern Serpens Caput, just west of the three stars (Beta [β], Gamma [γ], and Kappa [κ] Serpentis) that form the Serpent's head. This galaxy exhibits "flocculent" spiral structure, meaning its arms appear to be broken up into many small pieces in contrast to the "grand design" spirals with well-developed arms. At 150× and higher, you'll notice NGC 5962's broad central concentration. A sharp, bright nucleus some 15″ in diameter lies at its very center.

| OBJECT #400 | NGC 5965 |
|---|---|
| Constellation | Draco |
| Right ascension | 15h34m |
| Declination | 56°41′ |

| (continued) | |
|---|---|
| Magnitude | 11.9 |
| Size | 5.2′ by 0.7′ |
| Type | Spiral galaxy |

Our next object sits 2.6° south-southeast of magnitude 3.3 Iota ($\iota$) Draconis. Through a 12-inch telescope at 200×, this galaxy appears as a classic edge-on spiral, and it appears more than 4 times as long as it is wide oriented northeast to southwest. Smaller scopes don't show this length-to-width disparity as well because the spiral arms fade rapidly as they taper away from the core. The north-eastern arm outshines the other by quite a bit.

| OBJECT #401 | Struve 1962 |
|---|---|
| Constellation | Libra |
| Right ascension | 15h39m |
| Declination | −8°47′ |
| Magnitudes | 6.5/6.6 |
| Separation | 11.9″ |
| Type | Double star |

This pair of equally bright stars aligns along a rough north-south line, and both stars are white or blue-white. Struve 1962 lies 5.4° east of magnitude 2.6 Zubeneschamali (Beta [$\beta$] Librae).

| OBJECT #402 | Zeta ($\zeta$) Coronae Borealis |
|---|---|
| Constellation | Corona Borealis |
| Right ascension | 15h39m |
| Declination | 36°38′ |
| Magnitudes | 5.1/6.0 |
| Separation | 6.3″ |
| Type | Double star |

Zeta CrB lies in the northern part of the constellation, about a "crown's width" north of the Northern Crown asterism. Most observers peg the colors of these two stars as blue-white or white.

| OBJECT #403 | NGC 5985 |
|---|---|
| Constellation | Draco |
| Right ascension | 15h40m |
| Declination | 59°20′ |
| Magnitude | 11.1 |
| Size | 5.3′ by 2.9′ |
| Type | Spiral galaxy |

NGC 5985 is a nice spiral galaxy on its own, but it teams up with the magnitude 12.0 elliptical galaxy NGC 5982 and the magnitude 13.2 spiral galaxy NGC 5981 for one remarkable view. These three objects lie in an east-west line less than 14′ apart. You'll find this trio 1.8° east-northeast of magnitude 3.3 Iota ($\iota$) Draconis.

You'll need at least a 12-inch telescope to spot any detail in the spiral arms of NGC 5985. The other two galaxies won't reveal any details, although NGC 5981 is another of the universe's "needle" galaxies. To see the entire trio, use a magnification around 100×.

| OBJECT #404 | NGC 5986 |
|---|---|
| Constellation | Lupus |
| Right ascension | 15h46m |

| (continued) | |
|---|---|
| Declination | −37°47′ |
| Magnitude | 7.5 |
| Size | 9.8′ |
| Type | Globular cluster |

Our next target lies 2.8° west of magnitude 3.6 Eta (η) Lupi. Through a 6-inch telescope at 200×, this cluster appears irregular, mottled, and unresolved near its center except for one star 1′ northeast of the core. That's magnitude 11.2 GSC 7837:1334. Through a 12-inch scope, several dozen stars resolve into points. NGC 5986 lies 35,000 light-years from Earth and 15,000 light-years from the Milky Way's center.

| OBJECT #405 | R Coronae Borealis |
|---|---|
| Constellation | Corona Borealis |
| Right ascension | 15h49m |
| Declination | 28°09′ |
| Magnitude | 5.7 |
| Period | irregular |
| Type | Variable star |

Our next lies 3.4° east-northeast of magnitude 2.2 Alphecca (Alpha [α] Coronae Borealis). R Coronae Borealis also makes an isosceles triangle with magnitude 4.2 Gamma (γ) and magnitude 4.6 Delta (δ) Coronae Borealis.

But whether or not you see the star ... well, that's up to the star. Most of the time, R Coronae Borealis shines around magnitude 6. But at irregular intervals ranging from several months to many years, the star's brightness plunges.

How far its light dips also is irregular, but astronomers have observed R as faint as magnitude 14. That means at its maximum magnitude, the star is nearly 1,600 times as bright as when it's at minimum.

Although astronomers don't know when R Coronae Borealis' brightness will decrease, they have proposed two ideas as to why it drops. The first theory states that, intermittently, the star puffs out clouds of dust that cause it to dim. Because R also emits particles akin to the Sun's solar wind, the dust clouds disperse over time.

The second, less popular theory proposes that a huge cloud of dusty material orbits R Coronae Borealis. We observe R's variability as the cloud blocks out the star's light.

This, then, may be a recurring observation. Find R Coronae Borealis on the next clear night. Then return to the star to see if anything has changed. Odds are that you'll see the star disappear. I just can't say when.

**Object #406** NGC 6015 Paul and Dan Koblas/Adam Block/NOAO/AURA/NSF

| OBJECT #406 | NGC 6015 |
|---|---|
| Constellation | Draco |
| Right ascension | 15h51m |
| Declination | 62°19′ |
| Magnitude | 11.1 |
| Size | 6.4′ by 2.2′ |
| Type | Spiral galaxy |

You'll find our next target not quite 4° west-northwest of magnitude 2.7 Eta ($\eta$) Draconis. Through an 8-inch telescope you'll see a large, oval haze with central brightening that spans roughly 30% of the galaxy's diameter. Through larger aperture scopes, you may catch a glimpse of NGC 6015's structure, which appears peppered with faint knots.

| OBJECT #407 | NGC 6025 |
|---|---|
| Constellation | Triangulum Australe |
| Right ascension | 16h04m |
| Declination | –60°30′ |
| Magnitude | 5.1 |
| Size | 12′ |
| Type | Open cluster |
| Other name | Caldwell 95 |

Our next object lies at the northern edge of Triangulum Australe, right at that constellation's border with Norma. You can find it 3.1° north-northeast of magnitude 2.8 Beta ($\beta$) Trianguli Australis. Under a dark sky, most observers can spot NGC 6025 without optical aid. Through a 6-inch telescope, you'll count roughly 40 stars between magnitudes 7 and 11. Through a 14-inch or larger scope, crank up the magnification, and look 20′ south-southeast of the cluster's center for the magnitude 14.6 spiral galaxy PGC 56940.

| OBJECT #408 | NGC 6027 |
|---|---|
| Constellation | Serpens |
| Right ascension | 15h59m |
| Declination | 20°45′ |
| Magnitude | 14.0 |
| Size | 2′ by 1′ |
| Type | Galaxy group |
| Other name | Seyfert's Sextet |

Our next celestial treat is a group of six galaxies — three faint and three ultra-faint — that occupy a tiny area of sky. Seyfert's Sextet is made up of magnitude 13.8 NGC 6027, magnitude 13.9 NGC 6027a, magnitude 13.4 NGC 6027b, magnitude 16.5 NGC 6027c, magnitude 16.5 NGC 6027d, and magnitude 15.5 NGC 6027e.

French astronomers Édouard Jean-Marie Stephan discovered this object in 1882, but he didn't know what it was. American astronomer Carl Keenan Seyfert determined its true nature in 1951.

The Sextet also carries the designation Hickson 79 — the 79th entry in a catalog of compact groups of galaxies compiled in 1982 by Paul Hickson. Look for Seyfert's Sextet 1.9° east of magnitude 4.7 Rho ($\rho$) Serpentis.

To see Seyfert's Sextet you'll need at least a 16-inch telescope and terrific sky conditions. Even through such an instrument, at magnifications below 200× it's easy to mistake this group for a single object. On transparent and steady nights, you'll have the best luck spotting NGC 6027, NGC 6027a, and NGC 6027b. None of the galaxies shows any real structure, however, and at best appear barely non-stellar through any size scope. Merely detecting all six group members ranks as a major accomplishment.

| OBJECT #409 | Xi ($\xi$) Scorpii |
|---|---|
| Constellation | Scorpius |
| Right ascension | 16h04m |
| Declination | −11°22′ |
| Magnitudes | 4.8/7.3 |
| Separation | 7.6″ |
| Type | Double star |

To find Xi Sco. you'll have to search the extreme northern part of Scorpius. The star lies more than 16° west-northwest of Antares. A better star-hop originates at magnitude 2.5 Zeta ($\zeta$) Ophiuchi. Xi Sco lies only 8° west of that star. The primary is white, and the secondary is orange. Oh, and if you notice another binary in the same field of view as Xi, go on to the next entry.

| OBJECT #410 | Struve 1999 |
|---|---|
| Constellation | Scorpius |
| Right ascension | 16h04m |
| Declination | −11°27′ |
| Magnitudes | 7.4/8.1 |
| Separation | 11.6″ |
| Type | Double star |

Our previous target, Xi ($\xi$) Scorpii, has another designation — Struve 1998. Well, this binary, Struve 1999, lies less than 5′ south of it. The eyepiece field, therefore, will capture both doubles at once. This star is a slightly easier split than Xi. The most often-reported colors for the two components are pale yellow and orange.

| OBJECT #411 | NGC 6058 |
|---|---|
| Constellation | Hercules |
| Right ascension | 16h04m |

| (continued) | |
|---|---|
| Declination | 40°41′ |
| Magnitude | 12.9 |
| Size | 42″ |
| Type | Planetary nebula |

Our next target lies 2.8° southeast of magnitude 4.6 Chi (χ) Herculis. Through small telescopes, you'll see the triangle of stars that surrounds NGC 6058 before you see the planetary. Magnitude 9.0 SAO 45881 lies 6′ northeast. Magnitude 9.3 SAO 45874 sits 5′ northwest. And you'll see magnitude 10.7 GSC 3064:1181 less than 4′ south.

Through an 8-inch scope, the planetary appears faint and evenly illuminated. The tiny central region appears a bit brighter. A 14-inch instrument will reveal the 13th-magnitude central star surrounded by a small halo. A nebula filter (especially an Oxygen-III) helps a lot.

| OBJECT #412 | Beta (β) Scorpii |
|---|---|
| Constellation | Scorpius |
| Right ascension | 16h05m |
| Declination | −19°48′ |
| Magnitudes | 2.6/4.9 |
| Separation | 14″ |
| Type | Double star |
| Other name | Graffias |

Both components of Graffias are hot spectral type B stars. As such, they should appear quite blue. Perhaps because of the difference in brightness, amateur astronomers often describe the fainter component as yellowish or even orange! Spend some time observing this pair, and see what colors you detect.

Or, try this: Crank the magnification past 200×, and move the brighter component just outside your eyepiece's field of view. Do you detect a different color when you view the star by itself?

Richard Hinckley Allen in *Star-Names and Their Meanings* (1899) stated that the name "Graffias generally is said to be of unknown derivation." He goes on to say, however, "but since Γραψαιοσ signifies "Crab," it may be that here lies the origin of the title, for it is well known that the ideas and words for crab and scorpion were almost interchangeable in early days."

| OBJECT #413 | Abell 2151 |
|---|---|
| Constellation | Hercules |
| Right ascension | 16h05m |
| Declination | 17°45′ |
| Size | 68′ |
| Type | Galaxy cluster |
| Other name | The Hercules Galaxy Cluster |

All of the selections in this book qualify as "deep-sky" objects, but galaxy cluster Abell 2151 in Hercules takes that term to a whole new level. It lies at the astounding distance of 650 million light-years from Earth. Imagine that. The light that you glimpse from one of its galaxies started on its journey toward Earth several hundred million years before the first dinosaurs existed.

If your scope has a go-to drive, its database may not contain Abell galaxy clusters. That's not a problem here. Just target this cluster's brightest member, elliptical galaxy NGC 6041, which glows at magnitude 13.4. Without go-to, just find magnitude 5.0 Kappa (κ) Herculis. From that star, move 1° to the northwest, and your field of view will capture hundreds of galaxies, most of which glow too faintly to see. That still leaves several dozen within the range of moderate amateur instruments.

Success observing this galaxy cluster requires at least a 12-inch telescope and eyepieces that give powers in excess of 250×. High magnifications increase the contrast between extended objects like galaxies and the background sky. Abell 2151 spans more than 1°, so move your scope around a bit to see the maximum number of galaxies.

| OBJECT #414 | Kappa (κ) Herculis |
|---|---|
| Constellation | Hercules |
| Right ascension | 16h08m |
| Declination | 17°03′ |
| Magnitudes | 5.3/6.5 |
| Separation | 2.8″ |
| Type | Double star |

You'll find Kappa Her 3.9° west-southwest of magnitude 3.7 Gamma Herculis. The primary of this close binary appears yellow or yellow-white, and the secondary is orange. Use a magnification around 150×.

**Object #415** Nu Scorpii and IC 4592 Adam Block/NOAO/AURA/NSF

| OBJECT #415 | Nu (ν) Scorpii |
|---|---|
| Constellation | Scorpius |
| Right ascension | 16h12m |
| Declination | −19°28′ |
| Magnitudes | 4.3/6.4 |
| Separation | 41″ |
| Type | Double star |
| Other name | Jabbah |

Through 4-inch and larger telescopes, and with magnifications above 150×, you'll see Nu Scorpii (whose common name, "Jabbah," means "forehead") as a double-double star.

Separations are 1.3″ for the A–B pair and 2.4″ for the C-D pair. The brightest component glows with a yellow light, and the other three appear white. The A component sheds its light on the blue reflection nebula IC 4592, which, in turn, reflects that light in our direction. You can spot the nebula through a 4-inch scope, but be warned: It's huge. Use the lowest magnification and no filter.

| OBJECT #416 | IC 4593 |
|---|---|
| Constellation | Hercules |
| Right ascension | 16h12m |
| Declination | 12°04′ |
| Magnitude | 10.7 |
| Size | 42″ |
| Type | Planetary nebula |
| Other name | The White-Eyed Pea |

Our next treat lies 3.9° west-southwest of magnitude 4.6 Omega (ω) Herculis. Although this object has a common name, you'll need more than just a common telescope to see much detail here. Small instruments show only the magnitude 11.1 central star.

Through a 16-inch scope, the small halo that encases the central star appears, and it's blue. Magnifications above 350× show the halo has a slight oval shape, in a northwest to southeast orientation. For the best view through any scope, be sure to place magnitude 7.7 SAO 101998, which lies 11′ to the south-southeast, outside your eyepiece's field of view.

| OBJECT #417 | NGC 6067 |
|---|---|
| Constellation | Norma |
| Right ascension | 16h13m |
| Declination | –54°13′ |
| Magnitude | 5.6 |
| Size | 12′ |
| Type | Open cluster |

Our next treat — and it is a treat — lies 0.4° north of magnitude 5.0 Kappa (κ) Normae. This spectacular object is an observing prize through any size telescope. At its listed brightness, you might think you could see it without optical aid. Normally, that would be true, but NGC 6067 lies in such a rich star field that you'll spend valuable observing time trying to pick it out from the background.

Unless you're looking to find interesting patterns or double stars, use low-power eyepieces on this cluster. Even a 4-inch scope reveals more than 50 stars, while a 12-inch will show you more stars than you'll want to count. (I estimated more than 250.)

While you're in the area, check out two other open clusters. NGC 6031, which shines at magnitude 8.5, lies 0.8° to the west-northwest of NGC 6067. Then, 1.2° southeast of our first target, you'll find Collinder 299 (also known as Harvard 10), a cluster one-third of a degree in diameter with a listed magnitude of 6.9.

| OBJECT #418 | NGC 6072 |
|---|---|
| Constellation | Scorpius |
| Right ascension | 16h13m |
| Declination | –36°14′ |
| Magnitude | 11.7 |
| Size | 40″ |
| Type | Planetary nebula |

This bright planetary nebula sits 1.4° east-northeast of magnitude 4.2 Theta ($\theta$) Lupi. Through a 6-inch telescope, it appears as a featureless, circular disk. A 12-inch scope, an eyepiece that magnifies about 250×, and a nebula filter reveals a moderately dark rift splitting NGC 6072 into northern and southern halves. Note also how the edge appears irregularly bright.

| OBJECT #419 | Sigma ($\sigma$) Coronae Borealis |
|---|---|
| Constellation | Corona Borealis |
| Right ascension | 16h15m |
| Declination | 33°52′ |
| Magnitudes | 5.6/6.6 |
| Separation | 6.2″ |
| Type | Double star |

The quadruple star system Sigma CrB sits in the far-eastern part of this small constellation. Its two main components (A and B) are light-yellow and blue-white. A magnitude 13 C component lies 21″ to the east of the brighter pair. There's also a D component 87″ east of the A–B pair.

| OBJECT #420 | NGC 6087 |
|---|---|
| Constellation | Norma |
| Right ascension | 16h19m |
| Declination | –57°54′ |
| Magnitude | 5.4 |
| Size | 12.5′ |
| Type | Open cluster |
| Other names | The S Normae Cluster, Caldwell 89 |

Our next target lies in Norma 4.2° west-northwest of magnitude 3.8 Eta ($\eta$) Arae. Alternatively, you can find it 1.3° east of the much fainter star magnitude 5.6 Iota$^2$ ($\iota^2$) Normae. If you saw this cluster in a different part of the sky, it would be easy to see without optical aid. However, the richness of the star field coupled with the magnitude 5.6 star SAO 243509 just 0.4° to the west makes NGC 6087 a tough naked-eye catch.

Through a 4-inch telescope at 150×, you'll see some three dozen stars, the brightest of which shine at 8th magnitude. Note the cluster's rough triangular shape. A nice line of stars trending north-south sits at the southwest edge. You may see four, five, or six, depending on your scope's aperture.

Near the center, the brightest star in NGC 6087 is the Cepheid variable star S Normae, which was only recently proven to be a member. It glows with an orange hue and varies between magnitudes 6.12 and 6.77 over a period of 9.75 days. You'll easily identify this star by its color through binoculars or a finder scope.

**Object #421** M80 Gene Katz/Adam Block/NOAO/AURA/NSF

| OBJECT #421 | M80 (NGC 6093) |
|---|---|
| Constellation | Scorpius |
| Right ascension | 16h17m |
| Declination | –22°59′ |
| Magnitude | 7.3 |
| Size | 8.9′ |
| Type | Globular cluster |

Our next object is a fine small telescope target. M80 is easy to find. First, locate the 1st-magnitude luminary Antares (Alpha [$\alpha$] Scorpii). Then, move 4.5° northwest. M80 sits midway between Antares and magnitude 2.6 Graffias (Beta [$\beta$] Scorpii). Charles Messier discovered this object in January 1781. He later added it to his catalog as number 80.

At magnitude 7.3, you'll easily sweep up this globular through a 3-inch telescope. Its stars appear tightly packed, so a small scope won't let you resolve the ones near M80's bright core. When you observe this cluster, you'll notice the magnitude 8.5 star SAO 184288. It sits only 4′ to the northeast of M80's center. That star sits much closer to us than M80 and has nothing to do with the cluster.

In *Cycle of Celestial Objects*, Smyth writes about an interesting, but now outdated, concept: "This is a very important object when nebulae are considered in their relations to the surrounding spaces, which spaces, Sir William Herschel found, generally contain very few stars: so much so, that whenever it happened, after a short lapse of time, that no star came into the field of his instrument, he was accustomed to say to his assistant, 'Make ready to write, Nebulae are just approaching.'"

| OBJECT #422 | NGC 6101 |
|---|---|
| Constellation | Apus |
| Right ascension | 16h26m |
| Declination | –72°12′ |
| Magnitude | 9.2 |

| (continued) | |
|---|---|
| Size | 5′ |
| Type | Globular cluster |
| Other name | Caldwell 107 |

The next object on our list lies 3.7° south-southwest of magnitude 1.9 Atria (Alpha [α] Trianguli Australis). Through small telescopes, you'll see that this cluster appears small and faint, with a gradual concentration of light toward its center. Through a 14-inch scope at 300×, you'll begin to resolve a couple dozen of the cluster's brightest halo stars, which, unfortunately, aren't all that bright. Also through this aperture, you may notice the central region no longer appears uniform.

| OBJECT #423 | NGC 6118 |
|---|---|
| Constellation | Serpens (Caput) |
| Right ascension | 16h22m |
| Declination | –2°17′ |
| Magnitude | 11.7 |
| Size | 4.6′ by 1.9′ |
| Type | Spiral galaxy |

You'll need at least a medium size telescope to observe NGC 6118. Through a 10-inch scope you'll see only a faint, uniform haze. A larger telescope at a good observing site reveals a tiny, bright core and weak hints of spiral structure on the galaxy's eastern end. You'll also spot three 14th-magnitude stars embedded in the glow. To find NGC 6118, look 2.6° northeast of magnitude 2.7 Delta (δ) Ophiuchi.

**Object #424** M4 George Seitz/Adam Block/NOAO/AURA/NSF

| OBJECT #424 | M4 (NGC 6121) |
|---|---|
| Constellation | Scorpius |
| Right ascension | 16h24m |
| Declination | −26°32′ |
| Magnitude | 5.8 |
| Size | 26.3′ |
| Type | Globular cluster |

Globular cluster M4 ranks as one of the lucky objects on our list because it lies near a first-magnitude star that serves as a guide to it. Just center Antares (Alpha [$\alpha$] Scorpii) in your field of view, and M4 lies just to its west.

Swiss astronomer Jean-Philippe Loys de Cheseaux (1718–1751) discovered M4 in 1746. He described it as, "Close to Antares . . . It is white, round and smaller than the preceding ones. I do not think it has been found before." Messier, who added it to his catalog May 8, 1764, was the first to resolve it into a cluster of stars.

Astronomers classify M4 as a loose globular cluster of class IX. Globulars fall into classifications designated I to XII. A globular with a classification of I has the highest stellar density at its core. XII represents a homogenous globular with no increase in star concentration toward the center.

Point a 6-inch telescope toward M4, and you'll see dozens of stars scattered loosely across its diameter. A prominent chain of stars runs north-south through the cluster's center. A 12-inch or larger scope at 200× will reveal several hundred stars between 11th and 15th magnitude. At this magnification, many other star patterns mask the central chain of stars. Amateur scopes can resolve all but the central 10% of this cluster.

| OBJECT #425 | NGC 6124 |
|---|---|
| Constellation | Scorpius |
| Right ascension | 16h26m |
| Declination | −40°40′ |
| Magnitude | 5.8 |
| Size | 29′ |
| Type | Open cluster |
| Other name | Caldwell 75 |

Our next target forms the western tip of an isosceles triangle with two bright double stars, magnitude 3.0 Mu$^{1,2}$ ($\mu^{1,2}$) Scorpii and magnitude 3.6 Zeta$^{1,2}$ ($\zeta^{1,2}$) Scorpii, both of which lies in the Scorpion's tail.

At magnitude 5.8, sharp-eyed observers will pick out this cluster without optical aid from a dark site. You'll resolve a few of the object's outlying stars through binoculars or a finder scope.

Point a 4-inch telescope in NGC 6124's direction, and insert an eyepiece that gives a magnification around 75×, and 50 similarly bright stars will pop into view. Two dozen of them group near the center, with several nice double stars visible.

Now, back down the power to 35× or less, and take another look at the cluster. It has a distinct wedge shape pointing roughly to the southeast.

Although this cluster does lie within the nighttime band of our galaxy, its position places it in front of one of the many rifts composed of dark nebulae. Because of this, NGC 6124 sits in front of a sparse stellar background.

**Object #426** The Rho Ophiuchi region Jay Ballauer/Adam Block/NOAO/AURA/NSF

| OBJECT #426 | Rho Ophiuchi region |
|---|---|
| Constellation | Ophiuchus |
| Right ascension | 16h27m |
| Declination | –25°30′ |
| Size | 4° |
| Type | Emission and reflection nebulae |

Astronomers named the Rho Ophiuchi region for the magnitude 4.6 star Rho ($\rho$) Oph, but to begin searching this area, center on brilliant Antares (Alpha [$\alpha$] Scorpii). Stifle your natural urge to observe globular clusters M4 and NGC 6144 — you're here to hunt nebulae.

Start by scanning the area above (north) and to the left (east) of Antares through binoculars or the lowest-power, widest field of view your telescope will allow. For this initial hunt, look for the absence of stars caused by dark nebula Barnard 44 (B44). This dark, sharply defined lane starts at 22 Scorpii and runs eastward an incredible 6.5°, ending at 24 Ophiuchi.

Rho Oph is a double star, with yellow components of magnitudes 5.1 and 5.7 separated by 3″. Reflection nebula IC 4604 surrounds Rho. Look for the nebula's ribbed structure. Unfortunately, a nebula filter won't help with this type of object because such filters block the blue light from reflection nebulae. Some observers report more of the nebula visible through a light- or medium-blue filter.

Head back to 22 Sco, and examine the nebula IC 4605 around it. Because IC 4605 is an emission nebula, it responds well to a nebula filter. American astronomer Edward Emerson Barnard, who cataloged many of the sky's dark nebulae, described this nebulosity: "The star 22 Scorpii strikingly resembles a human eye, the lids being formed by two strips of nebulosity, one above it and one below."

| OBJECT #427 | NGC 6134 |
|---|---|
| Constellation | Norma |
| Right ascension | 16h28m |
| Declination | −49°09′ |
| Magnitude | 7.2 |
| Size | 8′ |
| Type | Open cluster |

You'll find our next target 1.6° northeast of magnitude 4.1 Gamma$^2$ ($\gamma^2$) Normae. Through an 8-inch telescope at 150×, you'll count 50 stars between magnitudes 11 and 14. The cluster shows no concentration but perhaps a general east-west trending of stars. NGC 6134's brightest star, magnitude 9.3 SAO 226781, sits at the southeastern edge.

| OBJECT #428 | NGC 6139 |
|---|---|
| Constellation | Scorpius |
| Right ascension | 16h28m |
| Declination | −38°51′ |
| Magnitude | 9.1 |
| Size | 8.2′ |
| Type | Globular cluster |

The next object sits 4.8° west-southwest of magnitude 3.0 Mu ($\mu$) Scorpii. A 4-inch telescope easily reveals NGC 6139 against a reasonably rich background. But simply seeing this cluster and seeing detail in it are two different things. You'll need at least a 14-inch telescope and a magnification above 300× to resolve even a few stars in this compact object. What you will see is a concentrated core surrounded by an unevenly lit halo.

| OBJECT #429 | NGC 6144 |
|---|---|
| Constellation | Scorpius |
| Right ascension | 16h27m |
| Declination | −26°02′ |
| Magnitude | 9.0 |
| Size | 9.3′ |
| Type | Globular cluster |

Our next object is incredibly easy to find. Just look 0.6° northwest of magnitude 1.1 Antares (Alpha [$\alpha$] Scorpii). In most low-power eyepieces, you'll capture both Antares and the cluster in the same field of view.

But that's definitely not how to observe it because Antares' glare will overwhelm everything else. So move Antares out of view to the southeast, crank up the power, and don't be diverted by the much brighter globular M4, which sits 1° west-southwest of NGC 6144.

An 8-inch telescope at 200× will begin to resolve the cluster's stars farthest from the center. Individual stars here are difficult to see because of the cluster's distance — it lies some 30,000 light-years from Earth. A better approach would be to view through a 16-inch scope, but I could say that about most celestial objects.

As with our small-telescope target NGC 6124, NGC 6144 sits at the edge of a vast array of dark nebulosity. Scan the region around it, especially northward. Pretty sparse, eh?

| OBJECT #430 | The Mini Coathanger |
|---|---|
| Constellation | Ursa Minor |
| Right ascension | 16h29m |
| Declination | 80°18′ |
| Type | Asterism |

Our next target looks best through small telescopes. It's the Mini Coathanger asterism way up north in Ursa Minor the Bear Cub. Amateur astronomer and *Astronomy* magazine Contributing Editor Phil Harrington named this asterism because of its resemblance to the more famous Coathanger, also known as Collinder 399, which lies in Vulpecula.

To find the Mini Coathanger, look 1.9° south-southwest of magnitude 4.2 Epsilon (ε) Ursae Minoris in the Little Dipper's handle. The Mini Coathanger is made up of 10 stars. They range in brightness from magnitude 9.2 SAO 2721 to magnitude 10.8 GSC 4574:802. From its hook to its base, the Mini Coathanger measures 9′. The base, at 17′, is nearly twice that length.

| OBJECT #431 | Alpha (α) Scorpii |
|---|---|
| Constellation | Scorpius |
| Right ascension | 16h29m |
| Declination | −26°26′ |
| Magnitudes | 1.1/5.4 |
| Separation | 2.6″ |
| Type | Double star |
| Other name | Antares |

Normally, a magnitude 5.4 star isn't hard to find. In this case, however, it sits only 2.6″ from magnitude 1.1 Antares A, so you'll need an 8-inch telescope and high power to split the two cleanly. When you do, the contrast effect will present a bright orange primary with an olive-green companion. I love this pair!

This brilliant star's name comes from the pairing of two Greek words, literally "anti" and the Greek god "Ares." Put together, they refer to a star that is the "rival of Mars" in color.

| OBJECT #432 | NGC 6152 |
|---|---|
| Constellation | Norma |
| Right ascension | 16h33m |
| Declination | –52°37′ |
| Magnitude | 8.1 |
| Size | 29′ |
| Type | Open cluster |

June's first object lies 3.2° southeast of magnitude 4.1 Gamma$^2$ ($\gamma^2$) Normae. NGC 6152 is visible through any size telescope, but you'll want to use a low-power eyepiece to distinguish this widely scattered open cluster from the incredibly rich Milky Way background. Through an 8-inch scope, you'll count more than 100 stars with ease. Note the two tiny clumps of brighter stars near the center.

| OBJECT #433 | NGC 6153 |
|---|---|
| Constellation | Scorpius |
| Right ascension | 16h32m |
| Declination | –40°15′ |
| Magnitude | 10.9 |
| Size | 25″ |
| Type | Planetary nebula |

M.E. Bakich, *1,001 Celestial Wonders to See Before You Die*, Patrick Moore's Practical Astronomy Series,
DOI 10.1007/978-1-4419-1777-5_6, 

Another of the many planetary nebulae in Scorpius lies 4.5° west-southwest of magnitude 3.0 Mu ($\mu$) Scorpii. Alternatively, point your telescope 1.2° east-northeast of NGC 6124, and you'll see this planetary nebula. Through an 8-inch or larger telescope, you'll need high magnification (above 200×) to reveal this object as anything other than a bloated "star." An Oxygen-III filter helps. Look for unevenness in the edge and an ever-so-slightly darker central region.

| OBJECT #434 | NGC 6167 |
|---|---|
| Constellation | Norma |
| Right ascension | 16h34m |
| Declination | −49°46′ |
| Magnitude | 6.7 |
| Size | 7′ |
| Type | Open cluster |

Our next target lies 2.4° east of magnitude 4.1 Gamma$^2$ ($\gamma^2$) Normae. In it, two dozen stars form a pattern resembling an H that orients east-west. The cluster's brightest star is magnitude 7.4 SAO 226901, which shines at the western end.

| OBJECT #435 | NGC 6169 |
|---|---|
| Constellation | Norma |
| Right ascension | 16h34m |
| Declination | −44°03′ |
| Magnitude | 6.6 |
| Size | 12′ |
| Type | Open cluster |
| Other name | The Mu Normae Cluster |

You want obscure? To see this object, most North American observers will have to scrape the southern horizon. This little-observed gem lies 4° west-southwest of magnitude 3.6 Zeta ($\zeta$) Scorpii. This open cluster lies in our line of sight with the magnitude 4.9 star Mu ($\mu$) Normae, thus the cluster's common name.

I take one of two approaches to observing this cluster through an 8-inch or larger telescope. Sometimes I use low power, around 50×, and just try to ignore the star Mu Normae. That's tough to do because Mu lies pretty much dead center.

My other approach is to use high magnification, say 250×, and observe each half of the cluster separately. Place Mu just outside the field of view twice, in opposite directions. That way, you can observe half of the 10th-magnitude and fainter cluster stars when Mu lies outside the field of view to the east and the other half when you've moved Mu out to the west.

**Object #436** M107 Bruce Hugo and Leslie Gaul/Adam Block/NOAO/AURA/NSF

| OBJECT #436 | M107 (NGC 6171) |
|---|---|
| Constellation | Ophiuchus |
| Right ascension | 16h33m |
| Declination | –13°03′ |
| Magnitude | 7.8 |
| Size | 10′ |
| Type | Globular cluster |

Our next target is a bright globular you'll easily locate through binoculars or a finder scope. To find M107, point your telescope 2.7° south-southwest of magnitude 2.6 Zeta (ζ) Ophiuchi. Through a 4-inch telescope at 100×, you'll see the cluster's bright core and faint halo.

Only 4′ west-northwest of the cluster's center lies the first individual star you'll notice. Unfortunately, it's a magnitude 10.3 star that sits in the foreground much closer than the 21,000-light-year distance to M107.

If your sky is exceptionally steady, crank the power up to 200× or beyond, and you'll begin to resolve several of the outer halo stars.

Take your time, and look carefully at the cluster. It's not round, is it? At high powers, M107 appears slightly flattened in an east-west orientation. Larger telescopes and even higher magnifications show that, although the outer cluster is slightly elliptical, the core is round.

| OBJECT #437 | NGC 6181 |
|---|---|
| Constellation | Hercules |
| Right ascension | 16h32m |
| Declination | 19°50′ |
| Magnitude | 11.9 |
| Size | 2.3′ by 0.9′ |
| Type | Spiral galaxy |

Look for the next object on our list 1.7° south-southeast of magnitude 2.8 Kornephoros (Beta [$\beta$] Herculis). NGC 6181 lies twice as far away as the galaxies in the Virgo Cluster, so don't expect to see much detail through small- or even medium-aperture telescopes.

Through a 10-inch scope, you'll see the oval halo oriented north-south. The core is broad and evenly lit. Through a 16-inch scope, you may glimpse the thin spiral arms that appear at the northern and southern ends. The north arm curves westward and the southern arm eastward.

| OBJECT #438 | NGC 6188 |
|---|---|
| Constellation | Ara |
| Right ascension | 16h41m |
| Declination | −48°47′ |
| Size | 20′ by 12′ |
| Type | Emission/reflection nebula |

Our next target lies 2.7° east-southeast of magnitude 4.8 Epsilon ($\varepsilon$) Normae. A 4-inch telescope at 150× under a dark sky reveals a faint haze that weaves through open cluster NGC 6193 (Object #439).

Through 10-inch and larger scopes, you'll see a lot more detail in the nebula. Note the dark region that separates the eastern section from the western one. Pay close attention to the nebula's edge, which appears irregularly bright and dark.

| OBJECT #439 | NGC 6193 |
|---|---|
| Constellation | Ara |
| Right ascension | 16h41m |
| Declination | −48°46′ |
| Magnitude | 5.2 |
| Size | 14′ |
| Type | Open cluster |
| Other name | Caldwell 82 |

Like our previous target, you'll find this object 2.7° east-southeast of magnitude 4.8 Epsilon ($\varepsilon$) Normae in the midst of the Ara OB1 association. And, if your sky is dark, you'll find it easily even without optical aid.

In the midst of the 1°-wide association, NGC 6193 is a loose, quarter-degree-wide wedge-shaped grouping made up of a couple bright stars and perhaps a dozen faint ones. The brightest is magnitude 5.7 SAO 227049. The double star DUN 206, two blue-white stars separated by only 10″, also lies within the cluster. Note how a dark nebula sharply cuts off the western edge of the cluster.

| OBJECT #440 | NGC 6192 |
|---|---|
| Constellation | Scorpius |
| Right ascension | 16h40m |
| Declination | −43°22′ |
| Magnitude | 8.5 |
| Size | 7′ |
| Type | Open cluster |

Look 2.8° west-southwest of magnitude 3.6 Zeta$^2$ ($\zeta^2$) Scorpii for our next object. Through a 6-inch telescope, you'll see a dozen stars surrounded by a rich Milky Way field. Despite the background, you'll identify NGC 6192 easily. A 10-inch scope reveals 25 stars whose brightness peaks at magnitude 11. Fainter stars sit at the limit of visibility as an unresolved haze surrounding the brighter members.

**Object #441** The Hercules Cluster (M13) Adam Block/Mount Lemmon SkyCenter/University of Arizona

| OBJECT #441 | M13 (NGC 6205) |
|---|---|
| Constellation | Hercules |
| Right ascension | 16h42m |
| Declination | 36°28′ |
| Magnitude | 5.7 |
| Size | 16.6′ |
| Type | Globular cluster |
| Other name | The Hercules Cluster |

Observers often refer to our next treat as the "Great" globular cluster. Of the sky's 10 brightest globulars, only two — M5 (seventh brightest) and M13 (eighth brightest) lie in the Northern Hemisphere. M13, however, sits nearly 35° higher than M5 for northern observers, so it's the brightest globular most amateur astronomers are familiar with.

English astronomer Edmond Halley discovered this object in 1714. Messier cataloged it June 1, 1764. The first line of his description seems almost comical now: "A nebula which I am sure contains no star."

Under a dark sky, you'll spot M13 easily as a fuzzy "star" two-thirds of the way from magnitude 3.0 Zeta ($\zeta$) to magnitude 3.5 Eta ($\eta$) Herculis. You'll resolve this globular through a 3-inch telescope.

Through 8-inch and larger scopes, you'll see hundreds of stars, some appearing as streamers emanating from the edge. Crank up your magnification to 200× or more and try to see a small Y-shaped region of three dark lanes near M13's center. Observers have dubbed this feature the "propeller."

The cluster's full extent fills the field of view of most telescope/eyepiece combinations. The core appears so bright that you may miss some detail because of it. Even around M13's edge, resolved stars appear against a background glow of thousands of unseen stars.

| OBJECT #442 | NGC 6207 |
|---|---|
| Constellation | Hercules |
| Right ascension | 16h43m |

| (continued) | |
|---|---|
| Declination | 36°50′ |
| Magnitude | 11.6 |
| Size | 3.0′ by 1.1′ |
| Type | Spiral galaxy |

Although our next target glows meekly at magnitude 11.6, it's really easy to locate. First, find the magnificent Hercules Cluster (M13), which lies one-third of the way from magnitude 3.5 Eta ($\eta$) Herculis to magnitude 3.0 Zeta ($\zeta$) Herculis. Those two stars form the western side of the famous Keystone asterism. Once you've found M13, look ever-so-slightly less than 0.5° northeast for NGC 6207. Many of your low-power eyepieces will show both objects in the same field of view.

But such views are deceiving. NGC 6207 lies some 1,600 times as far away as M13. In fact, if M13 were in orbit around the galaxy, no amateur telescope would reveal it. The globular would glow at magnitude 22, only slightly above the night sky's background brightness.

Through an 8-inch telescope, you'll see an evenly illuminated oval that measures a bit more than twice as long as it is wide. The galaxy orients north-northeast to south-southwest.

Step up to an 11-inch scope, and insert an eyepiece that gives a magnification of 250×. You'll notice the north-northeastern edge of this galaxy ends more abruptly than the other one, which appears much rounder. A 14-inch instrument at a power of 400× or above will reveal the ultra-thin outer halo that surrounds the bright central region.

| OBJECT #443 | NGC 6208 |
|---|---|
| Constellation | Ara |
| Right ascension | 16h50m |
| Declination | −53°49′ |
| Magnitude | 7.2 |
| Size | 12′ |
| Type | Open cluster |

This object sits 1.7° west-southwest of magnitude 4.1 Epsilon$^1$ ($\varepsilon^1$) Arae. Use a low-power eyepiece to view this cluster. Through a 6-inch telescope at 100×, you'll see the center of the cluster contains a nearly equilateral triangle of magnitude 10 stars. Other members, nearly as bright, trend in a general east-west orientation, giving NGC 6208 a rough rectangular shape. Many faint background stars add to the scene.

| OBJECT #444 | NGC 6210 |
|---|---|
| Constellation | Hercules |
| Right ascension | 16h45m |
| Declination | 23°49′ |
| Magnitude | 8.8 |
| Size | 14″ |
| Type | Planetary nebula |
| Other name | The Turtle Nebula |

Our next target lies 4° northeast of magnitude 2.8 Kornephoros (Beta [$\beta$] Herculis). Even through a small telescope, you can identify this planetary's light blue, turtle-shaped disk easily. Its high surface brightness allows you to really crank up the magnification but makes spotting the magnitude 12.5 central star a bit iffy. Through the largest amateur instruments at high magnification, you'll notice NGC 6210 is ever-so-slightly oval in an east-west orientation.

| OBJECT #445 | NGC 6217 |
|---|---|
| Constellation | Ursa Minor |
| Right ascension | 16h33m |

| (continued) | |
|---|---|
| Declination | 78°12′ |
| Magnitude | 11.2 |
| Size | 3.3′ |
| Type | Barred spiral galaxy |

Our next object rewards large-telescope owners. NGC 6217 is a nearly face-on, ringed spiral galaxy that hides its details from small scopes because it floats through space some 80 million light-years away.

It lies equidistant from magnitude 4.3 Zeta ($\zeta$) and magnitude 5.0 Eta ($\eta$) Ursae Minoris, 2.5° east-northeast of Zeta and 2.6° north-northeast of Eta. Through an 8-inch telescope, you'll see a faint oval twice as long as it is wide. That's not the galaxy's true shape, just the part you can spot through small instruments.

A 20-inch or larger scope begins to fill in the gaps. At magnifications above 300×, you'll spot the bar and the beginnings of both spiral arms. The northernmost (and definitely brighter) arm curves eastward, and the southernmost one curves toward the west. The arms actually continue and form a complete ring around the central region, but that lies outside the realm of most amateur telescopes.

**Object #446** M12 Michael Gariepy/Adam Block/NOAO/AURA/NSF

| OBJECT #446 | M12 (NGC 6218) |
|---|---|
| Constellation | Ophiuchus |
| Right ascension | 16h47m |
| Declination | −1°57′ |
| Magnitude | 6.1 |
| Size | 14.5′ |
| Type | Globular cluster |

Our next gorgeous treat lies 7.7° east-northeast of magnitude 3.2 Yed Posterior (Epsilon [$\varepsilon$] Ophiuchi). If M12 sat in Virgo instead of Ophiuchus, sharp-eyed observers could spot it without optical aid. Its position within such a rich star field, however, probably eliminates that possibility.

Through a 4-inch telescope, you'll see a bright, compact core surrounded by a faint halo. Four 10th-magnitude foreground stars sit in front of the cluster. Through a 10-inch scope M12 resolves into hundreds of stars. At high magnifications, the stars evenly distribute, the center shows little concentration, and the cluster's edge appears irregular.

| OBJECT #447 | 16/17 Draconis |
|---|---|
| Constellation | Draco |
| Right ascension | 16h36m |
| Declination | 52°55′ |
| Magnitudes | 5.4/6.4/5.5 |
| Separations | 3.4″/90″ |
| Type | Double star |

The stars 16 and 17 Draconis form a widely spaced double star. But look more closely, and you'll see that 16 Draconis also is a much closer binary star. The primary of 16 Dra shines with a bluish light, and the secondary is white. You'll find all of these stars 8.2° west of magnitude 2.8 Rastaban (Beta [$\beta$] Draconis).

| OBJECT #448 | NGC 6221 |
|---|---|
| Constellation | Ara |
| Right ascension | 16h53m |
| Declination | −59°13′ |
| Magnitude | 10.1 |
| Size | 4.9′ by 3.2′ |
| Type | Spiral galaxy |

Our next target lies 25′ east-southeast of magnitude 3.8 Eta ($\eta$) Arae, so when you view it, be sure to move the bright star out of the field of view. This barred spiral galaxy lies 70 million light-years away and interacts with NGC 6225, which you'll find only 10′ to the northwest.

A 6-inch telescope shows an oval glow oriented north-south. Through a 12-inch scope, NGC 6221 appears well defined with a central bar oriented east-west. You may detect traces of the spiral arms emanating from the ends of the bar, but you won't see them through the majority of amateur instruments.

**Object #449** The False Comet Adam Block/NOAO/AURA/NSF

| OBJECT #449 | The False Comet |
|---|---|
| Constellation | Scorpius |
| Right ascension | 16h54m |
| Declination | –42° |
| Size | 4° |
| Type | Asterism |

This naked-eye object is an asterism known as the False Comet, which lies at the southernmost reaches of Scorpius. British astronomer Sir John Herschel christened this region the False Comet during his stay in South Africa in the 1830s. And although I haven't confirmed it by reading Herschel's words yet, I'm certain he named his "comet" after False Bay, the place where his ship first touched land in South Africa.

Likewise, he called the wider region of the sky here the Table of Scorpius. That seems to me to be a reference to Table Mountain, which he could see every night as he gazed at the stars from his observatory.

The head of the False Comet is the spectacular NGC 6231, the sky's sixth-brightest open cluster, and an object sometimes called the Northern Jewel Box. I've included a lot about NGC 6231 as a separate entry. It's object #450.

The False Comet covers a 4°-long region from the double star Zeta (ζ) Scorpii north to the double star Mu (μ) Scorpii. Clusters other than NGC 6231 in the region of the False Comet are the huge magnitude 3.4 Collinder 316, magnitude 8.6 Trumpler 24, and magnitude 6.4 NGC 6242.

The False Comet is a naked-eye object, and an easy one from a dark observing location. It also looks great through binoculars. However, if you can point a small scope in its direction, you'll thank me.

| OBJECT #450 | NGC 6229 |
|---|---|
| Constellation | Hercules |
| Right ascension | 16h47m |
| Declination | 47°32′ |
| Magnitude | 9.4 |
| Size | 4.5′ |
| Type | Globular cluster |

This object sits 4.8° east-northeast of magnitude 3.9 Tau (τ) Herculis. At low to medium magnifications, you'll see a nice triangle formed by the cluster, magnitude 8.0 SAO 46278 only 6′ to the west, and magnitude 8.4 SAO 46280, which sits 7′ to the southwest.

NGC 6229 lies 90,000 light-years away, so its stars appear faint. The brightest glow at only magnitude 15.5. Through an 8-inch telescope at 200×, you'll see an unresolved glow with an irregular outline. Through a 14-inch scope, you'll pick up about half a dozen individual stars, and the cluster's face will appear quite mottled.

| OBJECT #451 | NGC 6231 |
|---|---|
| Constellation | Scorpius |
| Right ascension | 16h54m |
| Declination | −41°48′ |
| Magnitude | 2.6 |
| Size | 14′ |
| Type | Open cluster |
| Other names | The Northern Jewel Box, Caldwell 76 |

At the southernmost reaches of Scorpius, a stellar comet shines brightly. The head of this asterism, known as the False Comet, is the spectacular open cluster NGC 6231.

Australian chemist Ernst Johannes Hartung wrote about NGC 6231, "This cluster produces the impression of a handful of glittering diamonds displayed on black velvet." Hartung was the author of the popular *Astronomical Objects for Southern Telescopes*, which first appeared in 1968.

Hodierna discovered NGC 6231 and included it on his list of 40 nebulous objects, printed in 1654. This object shines so brightly, however, that skywatchers earlier than Hodierna certainly noticed it.

The region from Zeta (ζ) to Mu (μ) Scorpii contains a group of hot, young stars called the Scorpius OB1 association. Twenty member stars shine brighter than 9th magnitude, but their real brightnesses are astounding — the brightest stars of NGC 6231 outshine the Sun by 60,000 times.

Through a 4-inch telescope, you'll see more than 100 stars. Particularly striking is the knot of half a dozen bright stars at the cluster's center. The cluster is so bright that it stands out even from the surrounding Milky Way star field. It has sharply defined eastern and southwestern edges.

| OBJECT #452 | NGC 6235 |
|---|---|
| Constellation | Ophiuchus |
| Right ascension | 16h53m |
| Declination | −22°11′ |
| Magnitude | 8.9 |
| Size | 5′ |
| Type | Globular cluster |

The next target sits 5° east of magnitude 4.5 Omega ($\omega$) Ophiuchi. This small globular demands high magnification before it will reveal details, and even then they're scant. Through a 10-inch telescope at 250×, the cluster appears irregular with several moderately dark indentations crossing its border. The most prominent of these triangular depressions lies at the eastern edge. The magnitude 10.8 foreground star GSC 6230:1844 lies 2′ north-northwest of the cluster's center. Step up to a 16-inch scope, and you'll resolve about a dozen faint member stars.

| OBJECT #453 | NGC 6250 |
|---|---|
| Constellation | Ara |
| Right ascension | 16h58m |
| Declination | −45°57′ |
| Magnitude | 5.9 |
| Size | 7′ |
| Type | Open cluster |

This modest cluster lies 3.6° southwest of magnitude 3.3 Eta ($\eta$) Scorpii. A 4-inch telescope will show about a dozen stars brighter than magnitude 12. An 8-inch scope shows the same. Through an instrument with a 12-inch aperture, however, dozens of background stars appear, giving NGC 6250 a multi-layered appearance.

The five brightest stars in the grouping form a rough, wide "M" shape. The two luminaries of the M are magnitude 7.6 SAO 227508 and magnitude 7.9 SAO 227500, both of which sit at the eastern edge.

**Object #454** M10 Michael and Michael McGuiggan/Adam Block/NOAO/AURA/NSF

| OBJECT #454 | M10 (NGC 6254) |
|---|---|
| Constellation | Ophiuchus |
| Right ascension | 16h57m |
| Declination | –4°06′ |
| Magnitude | 6.6 |
| Size | 15.1′ |
| Type | Globular cluster |

Our next target, M10 in Ophiuchus, is a great one for small telescopes. Ophiuchus contains many globulars. In fact, all seven of Charles Messier's finds within this star group were that type of object.

You'll find the 10th entry on Messier's list 8.1° northeast of magnitude 2.6 Zeta (ζ) Ophiuchi. M10 glows at magnitude 6.6 and measures 15′ across, half the Full Moon's diameter. Its brightness and size are such that sharp-eyed observers may spot it from a dark site without optical aid. M10 is easy to see through binoculars or a finder scope.

The cluster lies about 14,000 light-years away. Because of its distance and its location near the Milky Way's plane, dust between us and M10 dims its stars by nearly a magnitude.

Through a 4-inch telescope at low power, you'll see a bright central region surrounded by a faint halo. Crank the magnification up to 200×, however, and you'll resolve a swarm of just-visible stars hovering around the compact core. Through larger telescopes, note how incredibly rich the halo is, and how its brightness slowly decreases with distance from the core. The magnitude 9.9 star HIP 82905 lies 10′ west-southwest of M10's center.

| OBJECT #455 | The Keystone |
|---|---|
| Constellation | Hercules |
| Right ascension | 17 h |
| Declination | 34° |
| Size | 7.5° by 5.5° |
| Type | Asterism |

Our next target is a naked-eye object called the Keystone of Hercules. To find it, draw a line from magnitude –0.04 Arcturus (Alpha [α] Boötis) to magnitude 0.03 Vega (Alpha Lyrae). Arcturus outshines Vega by only seven hundredths of a magnitude — that's a brightness difference even the most seasoned observer would find difficult to distinguish.

Anyway, two-thirds of the way from Arcturus to Vega, you'll encounter four medium-bright stars that make up the Keystone, a figure defined as the wedge-shaped piece at the top of an arch that locks the other pieces in place.

The brightest of the four, magnitude 2.8 Zeta (ζ) Herculis, sits at the southwestern corner. A bit more than 7° north lies magnitude 3.5 Eta (η) Herculis. Move to the east-southeast 6.5° for magnitude 3.2 Pi (π) Herculis. Finally, magnitude 3.9 Epsilon (ε) Herculis lies 6.5° south-southwest of Pi.

The space outlined by these stars isn't empty. From a dark site, most people can spot half a dozen stars without optical aid. And through large telescopes, hundreds of faint galaxies come into view.

| OBJECT #456 | NGC 6259 |
|---|---|
| Constellation | Scorpius |
| Right ascension | 17h01m |
| Declination | –44°40′ |
| Magnitude | 8.0 |
| Size | 8′ |
| Type | Open cluster |

Our next target sits 2.5° southwest of magnitude 3.3 Eta (η) Scorpii. Through a 4-inch telescope, you'll see an evenly distributed glow. A 10-inch scope at 200× reveals nearly a hundred stars fainter than 12th magnitude Several regions exhibit clumping. Among them, one just east of the core and one at the western edge stand out.

After you've had your fill of NGC 6259, look 0.6° west-southwest for open cluster NGC 6249. This magnitude 8.4 object spans 6′ and has an unusual figure.

| OBJECT #457 | M62 (NGC 6266) |
|---|---|
| Constellation | Ophiuchus |
| Right ascension | 17h01m |
| Declination | −30°07′ |
| Magnitude | 6.7 |
| Size | 14.1′ |
| Type | Globular cluster |

You'll find our next treat as far south in Ophiuchus as you can go, 5.8° east-southeast of magnitude 2.8 Tau ($\tau$) Scorpii. Although astronomers classify this as a "globular" cluster, you'll immediately notice that it's not spherical. The most radical departure from a round shape lies in a fan of stars that extends from the northwestern edge. Through an 8-inch telescope at 200×, the entire border appears irregular, and the southeast side looks flat. M12's core is bright and concentrated.

Admiral Smyth, writing in the *Cycle of Celestial Objects* (1844), quotes a famous astronomer and then goes on to chasten a different observer for making a mistake that Messier instituted his catalog for: "Sir William Herschel, who first resolved it, pronounced it a miniature of Messier's No. 3, and adds, 'By the 20-feet telescope, which at the time of these observations was of the Newtonian construction, the profundity of this cluster is of the 734th order.' To my annoyance, it was started (sic) as a comet a few years ago, by a gentleman who ought to have known better."

| OBJECT #458 | M19 (NGC 6273) |
|---|---|
| Constellation | Ophiuchus |
| Right ascension | 17h03m |
| Declination | −26°16′ |
| Magnitude | 6.7 |
| Size | 13.5′ |
| Type | Globular cluster |

The next object lies 4.5° west-southwest of magnitude 3.3 Theta ($\theta$) Ophiuchi. Seen on photographs or at high enough magnification, all globular clusters exhibit some degree of oblateness (flattening) because they rotate. M19, however, appears more flattened than any other Milky Way globular. Astronomers initially thought its ellipticity was real, but more recent investigations have shown that this is an illusion created by foreground absorption of starlight on its eastern and western edges.

Whatever the reason, you'll spot M19's oval form through any telescope. An 8-inch scope at 250× will resolve less than 20 of its stars. Through a 14-inch telescope, you'll see 50 stars of nearly equal brightness. The brightest star, magnitude 11.2 GSC 6819:282, sits 1′ northwest of the cluster's center.

| OBJECT #459 | NGC 6284 |
|---|---|
| Constellation | Ophiuchus |
| Right ascension | 17h04m |
| Declination | −24°46′ |
| Magnitude | 8.9 |
| Size | 6.2′ |
| Type | Globular cluster |

Our next target lies 4° west of magnitude 3.3 Theta ($\theta$) Ophiuchi, or 1° east-northeast of the magnitude 5.7 star 26 Ophiuchi. Although NGC 6284 is visible through all telescopes, I suggest at least a 14-inch instrument if you want to see any details. At 300×, the cluster has a bright core with an irregular border. This is most apparent on the eastern side, which appears to contain a small dark gap. If sky conditions warrant, increase the magnification to 450×, and look for half a dozen resolved stars.

| OBJECT #460 | NGC 6287 |
|---|---|
| Constellation | Ophiuchus |
| Right ascension | 17h05m |
| Declination | −22°42′ |
| Magnitude | 9.3 |
| Size | 4.8′ |
| Type | Globular cluster |

You'll find our next object 2° north of our previous target, NGC 6284. Alternatively, it lies 3.3° west-northwest of magnitude 5.1 Omicron ($o$) Ophiuchi. You may have trouble convincing yourself that this object is a globular cluster, and that, in itself, makes it worth a look.

Through a 6-inch telescope at 200×, NGC 6287 appears small, faint, irregular, and remarkably unconcentrated. Step up to a 14-inch scope, crank the power past 350×, and you'll resolve about a dozen stars against an ultra-faint background haze that you won't see if your sky contains scattered light.

| OBJECT #461 | Mu ($\mu$) Draconis |
|---|---|
| Constellation | Draco |
| Right ascension | 17h05m |
| Declination | 54°28′ |
| Magnitudes | 5.7/5.7 |
| Separation | 1.9″ |
| Type | Double star |
| Other name | Arrakis |

Our next target will test a small telescope. A 4-inch scope, however, should have no problem splitting this binary. Use high power, above 150×, and you'll see an equally bright pair of yellow-white suns.

This star's common name comes from the astronomical catalog of Ulug Beg (c. 1393–1449), who called it "Al Rakis." This designation, according to Allen, could mean "the dancer" or, more probably, "the trotting camel." Hmm. Not a very nice comparison if you're a dancer.

| OBJECT #462 | NGC 6293 |
|---|---|
| Constellation | Ophiuchus |
| Right ascension | 17h10m |
| Declination | −26°35′ |
| Magnitude | 8.2 |
| Size | 7.9′ |
| Type | Globular cluster |

This object lies 1.2° west of the magnitude 4.3 star 36 Ophiuchi. Through a 10-inch telescope at 200×, you'll notice this globular's bright, condensed core and a much fainter irregular halo. Crank the magnification to 350×, and NGC 6293 subdivides into clumps of unresolved stars. Several foreground stars lie within the field of view. The brightest is magnitude 8.4 SAO 185111, which sits 12′ northeast of the cluster's center.

| OBJECT #463 | NGC 6300 |
|---|---|
| Constellation | Ara |
| Right ascension | 17h17m |

| (continued) | |
|---|---|
| Declination | –62°49′ |
| Magnitude | 10.1 |
| Size | 5.2′ by 3.3′ |
| Type | Barred spiral galaxy |

Look 2.7° southwest of magnitude 3.6 Delta (δ) Arae to see our next object. This galaxy appears oval, stretched twice as long as it is wide, with a west-northwest to east-southeast orientation. Through a 12-inch telescope, the core appears much brighter than the surrounding halo. No less than four 13th-magnitude stars superimpose on this galaxy's face, two over the core, and one each on the northern and southern end. Through a 20-inch scope, I assumed the presence of faint spiral arms by the dark areas near NGC 6300's the central region.

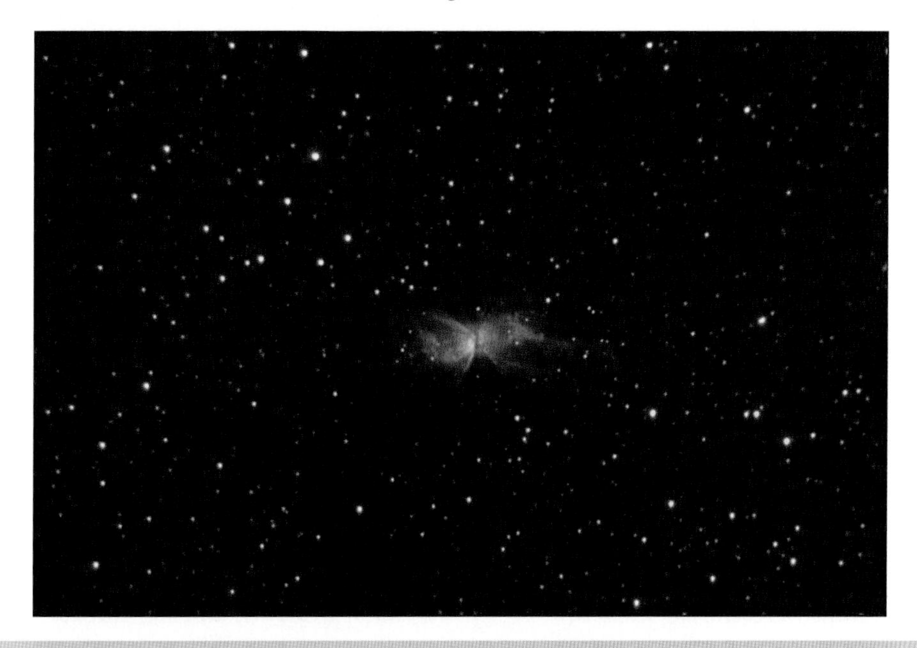

**Object #464** The Bug Nebula (NGC 6302) Adam Block/NOAO/AURA/NSF

| OBJECT #464 | NGC 6302 |
|---|---|
| Constellation | Scorpius |
| Right ascension | 17h14m |
| Declination | –37°06′ |
| Magnitude | 9.6 |
| Size | 50″ |
| Type | Planetary nebula |
| Other names | The Bug Nebula, Caldwell 69 |

Look for our next target 3.9° west of magnitude 1.6 Shaula (Lambda [λ] Scorpii). The Bug Nebula, named for its insect-like shape, is one of the brightest and most massive planetary nebulae known.

Through a 6-inch telescope at low power, NGC 6302 appears like a bright galaxy four times as long as it is wide, oriented east to west. At magnifications above 150×, the bipolar nature of this object is apparent. Look for a prominent lobe with a tapered end on the western side. Then, try to spot the faint "arm" emerging from the eastern side. Don't look for this object's central star. Intervening dust dims it by as much as five magnitudes.

| OBJECT #465 | NGC 6304 |
|---|---|
| Constellation | Ophiuchus |
| Right ascension | 17h15m |
| Declination | −29°28′ |
| Magnitude | 8.3 |
| Size | 8′ |
| Type | Globular cluster |

You'll find our next treat 4.8° south-southwest of magnitude 3.3 Theta (θ) Ophiuchi. This globular appears quite condensed with a faint outer halo. Through a 14-inch telescope, it appeared mottled with an irregular border whose southern side is flat.

| OBJECT #466 | IC 4633 |
|---|---|
| Constellation | Apus |
| Right ascension | 17h14m |
| Declination | −77°32′ |
| Magnitude | 11.8 |
| Size | 4.0′ by 3.0′ |
| Type | Spiral galaxy |

Our next target lies 1.7° east of magnitude 4.2 Beta (β) Apodis. This galaxy lies at the western extreme of a vast tract of nebulosity. Through an 8-inch telescope you'll first spot the magnitude 8.9 star GSC 9447:1844, which lies just east of the galaxy's central region. That area appears as an oval 50% longer than it is wide oriented northwest to southeast.

Look just less than 7′ to the east-northeast of IC 4633 for the much fainter galaxy IC 4635. This spiral glows weakly at magnitude 14.0.

**Object #467** The Box Nebula (NGC 6309) Adam Block/NOAO/AURA/NSF

| OBJECT #467 | NGC 6309 |
|---|---|
| Constellation | Ophiuchus |
| Right ascension | 17h14m |
| Declination | −12°55′ |
| Magnitude | 11.5 |
| Size | 18″ |
| Type | Planetary nebula |
| Other name | The Box Nebula |

Our next target isn't big, bright, or famous. Sound like fun? Well, then, look for the Box Nebula (NGC 6309) in Ophiuchus 1.6° west of magnitude 4.3 Nu (ν) Serpentis.

At least the Box Nebula doesn't spread its light across a large area. NGC 6309 is tiny, and that's a plus in this case because the small size keeps the planetary's surface brightness high.

Through an 8-inch telescope, crank the magnification past 250× to identify the box's shape. A nebula filter such as an Oxygen-III (OIII) will help a lot. Some observers use lower powers and combine NGC 6309 with the 9th-magnitude star 1′ to its northwest to form the Exclamation Point Nebula. Try it yourself.

If you're lucky enough to view the Box Nebula through a telescope with 16′ of aperture or more, you'll see tiny but distinct details at magnifications approaching 500×. For example, the northwest half appears slightly brighter than the southwestern part. Faint nebulous tendrils roughly one-quarter the Box's length extend outward from the northwestern edge. And the central star will pose no problem. That magnitude 14 point of light lies at the nebula's center.

Oh, by the way, this is one of two celestial wonders amateur astronomers call the Box Nebula. The other is NGC 6445 (Object #504).

| OBJECT #468 | 36 Ophiuchi |
|---|---|
| Constellation | Ophiuchus |
| Right ascension | 17h15m |
| Declination | −26°36′ |
| Magnitudes | 5.1/5.1 |
| Separation | 4.4″ |
| Type | Double star |

Here's an odd double: Both stars appear equally bright, and both have the same yellowish-white color. You'll get good results through any size telescope, but be sure to use a magnification of 100× or more.

| OBJECT #469 | NGC 6316 |
|---|---|
| Constellation | Ophiuchus |
| Right ascension | 17h17m |
| Declination | −28°08′ |
| Magnitude | 8.1 |
| Size | 5.4′ |
| Type | Globular cluster |

This target sits 1.4° north-northeast of Object #464 (NGC 6304). Alternatively, you'll find it 1.6° south of our previous object, the magnitude 4.3 star 36 Ophiuchi.

Through a 10-inch telescope, this globular has a bright, broad core with a small faint halo. You won't resolve many cluster members, but you will see a magnitude 11.0 foreground star 1′ to the southeast of the core.

| OBJECT #470 | Alpha (α) Herculis |
|---|---|
| Constellation | Hercules |
| Right ascension | 17h15m |

| (continued) | |
|---|---|
| Declination | 14°23′ |
| Magnitudes | 3.5/5.4 |
| Separation | 4.6″ |
| Type | Double star |
| Other name | Ras Algethi |

I love observing this star. Because of a wonderful contrast effect, the secondary appears olive-green. The brighter primary is yellow, with a trace of orange.

"Ras Algethi" means the Kneeler's head, and it refers to the way the mythological figure the constellation Hercules represents kneels in the sky. Smyth calls this pair orange and either emerald or bluish green, later saying this star is a "lovely object, one of the finest in the heavens."

| OBJECT #471 | Delta (δ) Herculis |
|---|---|
| Constellation | Hercules |
| Right ascension | 17h15m |
| Declination | 24°50′ |
| Magnitudes | 3.1/8.2 |
| Separation | 8.9″ |
| Type | Double star |
| Other name | Sarin |

Our next target is an unequally bright pair located 10.5° north of 3rd-magnitude Ras Algethi (Alpha [α] Herculis). The primary shines with a pale-blue light. The secondary is a colorless white, but look closely. Its brightness is only 1% that of the primary. Interestingly, Smyth describes the colors of these stars as greenish-white and grape-red.

The name commonly associated with this star, Sarin, may be a twentieth-century addition. In the great reference for stellar nomenclature, *Star-Names and Their Meanings*, Allen doesn't mention Delta Herculis at all.

| OBJECT #472 | Omicron (ο) Ophiuchi |
|---|---|
| Constellation | Ophiuchus |
| Right ascension | 17h18m |
| Declination | −24°17′ |
| Magnitudes | 5.4/6.9 |
| Separation | 10.3″ |
| Type | Double star |

You can split this pair even through a 2.4-inch telescope. The yellow primary shines 4 times brighter than the blue secondary. To find Omicron Oph, look 1.2° northwest of magnitude 3.3 Theta (θ) Ophiuchi.

| OBJECT #473 | M9 (NGC 6333) |
|---|---|
| Constellation | Ophiuchus |
| Right ascension | 17h19m |
| Declination | −18°31′ |
| Magnitude | 7.6 |
| Size | 9.3′ |
| Type | Globular cluster |

The ninth entry on Charles Messier's fabled list lies 3.5° southeast of magnitude 2.4 Eta (η) Ophiuchi and just east of dark nebula Barnard 64. The nearness of this object probably reduces M9's light by a full magnitude.

Through an 8-inch telescope at 200×, you'll see a bright, broad, irregularly illuminated central region. The edge looks irregular even at lower magnifications. The core appears slightly oval with a north-south orientation.

| OBJECT #474 | NGC 6334 |
|---|---|
| Constellation | Scorpius |
| Right ascension | 17h20m |
| Declination | −35°51′ |
| Size | 35′ by 20′ |
| Type | Emission nebula |
| Other name | The Cat's Paw Nebula |

Some named deep-sky objects look exactly like their namesakes. The Dumbbell Nebula and the North America Nebula come to mind. Add one more — the Cat's Paw Nebula.

Sir John Herschel discovered NGC 6334 June 7, 1837, from the Cape of Good Hope in South Africa. He labeled it "h 3678" in his 1847 catalog. More recently, Australian astronomer Colin S. Gum gave this nebula six entries (Gum 61, 62, 63, 64a, 64b, and 64c) in his 1955 catalog, the first major survey of southern-sky HII regions.

The Cat's Paw Nebula ranks among the Milky Way's largest star-forming regions. It comprises five individual nebulous patches in a circular area. The brightest, which measures 6′ across and contains a 9th-magnitude star, lies on the southeastern end of the complex.

Because this object is larger than the Full Moon, you'll need a wide-field telescope/eyepiece combination to view it all. Another approach is to crank up the power a bit, add a nebula filter, and view each of the five areas separately.

To find NGC 6334, look 3° west-northwest of Shaula (Lambda [λ] Scorpii).

| OBJECT #475 | NGC 6337 |
|---|---|
| Constellation | Scorpius |
| Right ascension | 17h22m |
| Declination | −38°29′ |
| Magnitude | 12.3 |
| Size | 48″ |
| Type | Planetary nebula |
| Other name | The Cheerio Nebula |

Our next target, named for the breakfast cereal it resembles, lies inside the arc of stars that forms the Scorpion's stinger. Specifically, you can find it 2° southwest of magnitude 2.7 Upsilon (υ) Scorpii. Through a 12-inch telescope at 300×, you'll see a thin ring with a superposed star on both the northeastern and southwestern edges. A nebula filter like an Oxygen-III really helps. Although this object has a low magnitude, its surface brightness is high.

| OBJECT #476 | M92 (NGC 6341) |
|---|---|
| Constellation | Hercules |
| Right ascension | 17h17m |
| Declination | 43°08′ |
| Magnitude | 6.5 |
| Size | 11.2′ |
| Type | Globular cluster |

This treat sits 6.3° north of the Keystone (Object #454) star Pi (π) Herculis. If you're a beginning observer who has observed M13 (Object #441), take the time to seek out this constellation's other Messier object. With stars nearly as bright as those in M13, M92 easily resolves through small telescopes. A 6-inch instrument at 100× will reveal the slight oval nature of this object. Can you see it? The orientation is north-south.

Through an 8-inch scope, the core appears concentrated and huge, surrounded by an outer halo of myriad faint stars. High-power views through a 14-inch scope will let you count stars until you tire of the process.

| OBJECT #477 | NGC 6342 |
|---|---|
| Constellation | Ophiuchus |
| Right ascension | 17h21m |
| Declination | −19°35′ |
| Magnitude | 9.5 |
| Size | 4.4′ |
| Type | Globular cluster |

This object lies 1.5° north of magnitude 4.4 Xi (ξ) Ophiuchi. Through an 8-inch telescope, you'll see the moderately bright (but tiny) central region oriented northwest to southeast. Through a 14-inch telescope at 300×, you'll resolve no more than half a dozen stars that lie at the limit of visibility.

| OBJECT #478 | UGC 10822 |
|---|---|
| Constellation | Draco |
| Right ascension | 17h20m |
| Declination | 57°55′ |
| Magnitude | 9.9 |
| Size | 51.0′ by 31.0′ |
| Type | Elliptical galaxy |
| Other name | The Draco Dwarf |

Our next target is for large telescopes and is the galaxy UGC 10822. The "UGC" stands for the *Uppsala Observatory General Catalog*, which astronomers compiled in 1971 from photographic plates taken during the Palomar Observatory Sky Survey between 1950 and 1957. The UGC contains 13,073 mainly Northern Hemisphere galaxies as faint as magnitude 14.5.

But UGC 10822 has another name: the Draco Dwarf. Astronomers classify this object as a dwarf spheroidal galaxy, a low-luminosity companion of the Milky Way. The Draco Dwarf is also a member of our Local Group of galaxies. It lies approximately 275,000 light-years away.

Now here's a great test for observers with large telescopes. The Draco Dwarf glows at 12th magnitude and spans 37′ by 24′. Its size means that it covers an area 27% larger than the Full Moon.

So the Draco Dwarf is huge and it has ultra-low surface brightness. You'll need at least a 12-inch telescope at a dark site. Insert your lowest-magnification, widest-field eyepiece and start scanning.

Begin your search 4.6° west-northwest of magnitude 3.7 Grumium (Xi [ξ] Draconis). Seeing this galaxy is tough. Look for an ever-so-slight increase in the sky's background glow that's bigger than the Full Moon.

When you locate the Draco Dwarf, you'll notice that about 10 foreground stars brighter than magnitude 11 lie strewn across its face. The brightest is magnitude 8.8 SAO 30348, which has a bit of a yellowish hue.

Faint as it is, the Draco Dwarf is a scientific treasure trove. American astronomer Albert George Wilson discovered it in 1954 when he was working at Lowell Observatory in Flagstaff, Arizona. Recently, studies of large internal velocity differences among the stars of this galaxy point to the existence of vast quantities of dark matter.

**Object #479** The Snake Nebula (B72) Tom McQuillan/Adam Block/NOAO/AURA/NSF

| OBJECT #479 | Barnard 72 |
|---|---|
| Constellation | Ophiuchus |
| Right ascension | 17h24m |
| Declination | −23°38′ |

| (continued) | |
|---|---|
| Size | 4′ |
| Type | Dark nebula |
| Other name | The Snake Nebula |

The Milky Way, which contains hundreds of billions of stars, also has areas of great darkness. Pre-twentieth century astronomers believed these starless voids were literal holes in the sky. In Volume 38 of *The Astrophysical Journal* (1913), Edward Emerson Barnard published "Dark Regions in the Sky Suggesting an Obscuration of Light."

He wrote, "The so-called 'black holes' in the Milky Way are of very great interest. Some of them are so definite that, possibly, they suggest not vacancies, but rather some kind of obscuring body lying in the Milky Way, or between us and it, which cuts out the light from the stars."

Barnard cataloged 349 dark nebulae. The Snake Nebula was the 74th entry. It carries the designation B72, however, because Barnard added two "extra" numbers to his list before B72: B44a and B67a.

Cold molecular gas and dust, primarily carbon, comprise dark nebulae. Such clouds absorb light from nearby stars but do not reradiate it as visible light. Rather, molecular clouds such as the Snake Nebula give off infrared light.

You'll get your best views of the Snake Nebula through a telescope/eyepiece combination that yields a field of view around 0.5°. Filters don't help because they dim the surrounding stars, reducing the contrast between them and the dark nebula. To find this object, look 1.4° north-northeast of magnitude 3.3 Theta ($\theta$) Ophiuchi.

| OBJECT #480 | NGC 6352 |
|---|---|
| Constellation | Ara |
| Right ascension | 17h26m |
| Declination | −48°25′ |
| Magnitude | 8.1 |
| Size | 7.1′ |
| Type | Globular cluster |
| Other name | Caldwell 81 |

You'll find this target 1.8° northwest of magnitude 3.0 Alpha ($\alpha$) Arae. Before you turn to your telescope, peer at this object through binoculars. You'll see NGC 6352 easily, but what I want you to notice is the incredibly bright star field around it.

Through a 4-inch telescope, the cluster appears round and quite hazy with a slightly brighter center. A 12-inch scope resolves approximately 25 stars. The magnitude 11.0 star GSC 8345:1542 lies 1′ to the southwest of the globular's core.

| OBJECT #481 | NGC 6356 |
|---|---|
| Constellation | Ophiuchus |
| Right ascension | 17h24m |
| Declination | −17°49′ |
| Magnitude | 8.2 |
| Size | 7.2′ |
| Type | Globular cluster |

This object sits 3.8° east-southeast of magnitude 2.4 Eta ($\eta$) Ophiuchi. Although you'll spot this globular easily through a finder scope, its great distance makes it a challenge to resolve individual stars. Through an 8-inch telescope at 200×, you'll see the broad central region surrounded by an irregular border. I saw no halo region even through a 14-inch scope.

| OBJECT #482 | Rho ($\rho$) Herculis |
|---|---|
| Constellation | Hercules |
| Right ascension | 17h24m |
| Declination | 37°09′ |
| Magnitudes | 4.6/5.6 |
| Separation | 4.1″ |
| Type | Double star |

This pair lies 1.8° east-northeast of magnitude 3.2 Pi ($\pi$) Herculis. Descriptions of the stars' colors vary. What do you see? Are both yellow? Both white? Do either have a bluish tinge? Remember, human color receptors vary, so there's no wrong answer.

| OBJECT #483 | NGC 6357 |
|---|---|
| Constellation | Scorpius |
| Right ascension | 17h25m |
| Declination | –34°12′ |
| Size | 25′ |
| Type | Emission nebula |
| Common names | The Lobster Nebula; The War and Peace Nebula |

You'll find this great nebula 3.3° north-northwest of magnitude 1.6 Shaula (Lambda ($\lambda$) Scorpii.

Sir John Herschel discovered and cataloged the Lobster Nebula in 1837, during his 4-year stay at the Cape of Good Hope.

The nebula attained its colloquial name because of its appearance in mid-infrared wavelengths as seen by the Midcourse Space Experiment (MSX) scientists. They worked for the Ballistic Missile Defense Organization, of which the MSX satellite was an experiment. They said the bright portion of the nebula, to the northwest, looked like a dove, while the mid-infrared filaments to the east seemed to trace out a human skull. You can still see the "dove" in the 2MASS near-infrared image, but the "skull" is not nearly as apparent as in the MSX image.

NGC 6357 surrounds the magnitude 9.6 open cluster Pismis 24. Some of this cluster's bright blue stars are among the most massive ever discovered. For example, astronomers using the Hubble Space Telescope discovered a double star weighing 100 solar masses.

Although the apparent size of the Lobster Nebula exceeds that of the Full Moon, through amateur telescopes you'll see only one-tenth its extent. The nebula's brightest region lies to the west of center. Use a nebula filter to improve the view. Note the four 7th-magnitude stars in a line south of the nebula. The northernmost, magnitude 7.1 SAO 208790, lies atop NGC 6357's center.

| OBJECT #484 | IC 4651 |
|---|---|
| Constellation | Ara |
| Right ascension | 17h25m |
| Declination | –49°57′ |
| Magnitude | 6.9 |
| Size | 12′ |
| Type | Open cluster |

Our next target lies 1.1° west of magnitude 3.0 Alpha ($\alpha$) Arae. This beautiful cluster has a rough arrowhead shape that points west-northwest. A 3-inch telescope at 75× resolves 50 stars standing out nicely from the surrounding field. The brightest star, HIP 85245 in the cluster's northeast corner, glows reddish and shines at magnitude 8.9.

| OBJECT #485 | NGC 6362 |
| --- | --- |
| Constellation | Ara |
| Right ascension | 17h32m |
| Declination | –67°03′ |
| Magnitude | 7.5 |
| Size | 10.7′ |
| Type | Globular cluster |

To find this nice object set in a rich field of faint stars, look in southern Ara 1.2° northeast of magnitude 4.7 Zeta (ζ) Apodis. Through a 4-inch telescope at 150×, you'll see a slight central concentration surrounded by a grainy halo composed of unresolved stars. A 12-inch scope at 250× will reveal 25 individual stars. The two brightest are foreground stars shining at 10th magnitude. Use high magnifications if your seeing will allow them for the most rewarding views of this cluster.

| OBJECT #486 | NGC 6366 |
| --- | --- |
| Constellation | Ophiuchus |
| Right ascension | 17h28m |
| Declination | –5°05′ |
| Magnitude | 8.9 |
| Size | 8.3′ |
| Type | Globular cluster |

Our next target lies 3.9° northwest of magnitude 4.6 Mu (μ) Ophiuchi. Although this object shines several magnitudes fainter than some better-known globulars, it has a sparse condensation, so its stars resolve more easily. In fact, through large telescopes at high magnifications, you might think you're looking at an open cluster. Through a 12-inch scope, you'll see about a dozen faint stars floating above a hazy background. The brightest, magnitude 10.5 GSC 5075:701, lies less than 5′ west of NGC 6366's center. Also, be sure to move magnitude 4.5 SAO 141665 out of the field of view. That star sits only 16′ to the west of the cluster.

**Object #487** The Little Ghost (NGC 6369) Jay Gabany/Adam Block/NOAO/AURA/NSF

| OBJECT #487 | NGC 6369 |
|---|---|
| Constellation | Ophiuchus |
| Right ascension | 17h29m |
| Declination | –23°46′ |
| Magnitude | 11.4 |
| Size | >30″ |
| Type | Planetary nebula |
| Other name | The Little Ghost |

Our next target is another planetary nebula worth lengthy observation. It lies 0.5° west-northwest of the magnitude 4.8 star 51 Ophiuchi.

The Little Ghost measures about 30″ across and glows at magnitude 11.4. Through an 8-inch telescope at 200×, you'll see a circular ring with a slightly brighter northern half. Although the central star shows up well in astroimages, you'll need at least 14 inches of aperture to spot this magnitude 16 stinker. Through even larger instruments, the bright northern rim turns into a vivid streak buried in the ring's nebulosity.

| OBJECT #488 | Nu (ν) Draconis |
|---|---|
| Constellation | Draco |
| Right ascension | 17h32m |
| Declination | 55°11′ |
| Magnitudes | 4.9/4.9 |

| (continued) | |
|---|---|
| Separation | 61.9″ |
| Type | Double star |
| Other name | Kuma |

You can split this wide binary through steadily held binoculars or a finder scope. The two stars are equally bright, and both are white.

This star's name ties into the Arabic notion that the Head of Draco represents four mother camels protecting a baby camel from the attack of two hyenas. Kuma is one of the mother camels.

| OBJECT #489 | NGC 6380 |
|---|---|
| Constellation | Scorpius |
| Right ascension | 17h35m |
| Declination | –39°04′ |
| Magnitude | 11.1 |
| Size | 3.9′ |
| Type | Globular cluster |

This challenging object lies 1.5° west of magnitude 2.4 Kappa ($\kappa$) Scorpii. We view this globular through 35,000 light-years of interstellar material, so just spotting it is prize enough. Through a 12-inch telescope, NGC 6380 appears as a faint, broadly concentrated haze.

NGC 6380 also carries the designation Ton 1, derived from Tonantzintla Observatory, which lies near Puebla, Mexico. From that observatory, Turkish astronomer Paris Pismis (1911–1999) discovered this globular cluster and another, even fainter one, Ton 2, in 1959. To see magnitude 12.2 Ton 2, look 25′ east of NGC 6380.

| OBJECT #490 | NGC 6383 |
|---|---|
| Constellation | Scorpius |
| Right ascension | 17h35m |
| Declination | –32°35′ |
| Magnitude | 5.5 |
| Size | 20′ |
| Type | Open cluster |

To find this object, look 4.5° north of magnitude 1.6 Shaula (Lambda [$\lambda$] Scorpii). Alternatively, find M6 (Object #496), and look 1.2° to the west-southwest. The first star that you'll see will be magnitude 5.7 SAO 208977. That's the luminary supplying most of this cluster's brightness, but the faint stars around it contribute to the nice overall display. Through a 4-inch scope at 100×, the central part of NGC 6383 reminded me a little of the constellation Grus. One crooked line of stars winds northwest to southeast, and a small bar crosses that line at the south end. SAO 208977 lies at the intersection.

Two lesser-known clusters lie near NGC 6383. Slightly less than 0.5° east lies magnitude 7.7 Trumpler 28. It measures 8′ across. After you locate Tr 28, look 1° west for another, extremely sparse, open cluster, magnitude 8.8 Antalova 2.

| OBJECT #491 | NGC 6384 |
|---|---|
| Constellation | Ophiuchus |
| Right ascension | 17h32m |
| Declination | 7°04′ |
| Magnitude | 10.4 |
| Size | 6.2′ by 4.1′ |
| Type | Spiral galaxy |

Now here's an odd duck: a reasonably bright galaxy in Ophiuchus. You'll find this object lies 3.7° northwest of magnitude 2.8 Cebelrai (Beta [$\beta$] Ophiuchi). Through an 8-inch telescope at 150×, this galaxy appears nearly rectangular. It has a broad, bright central region that orients from the north-northeast to the south-southwest. Larger-aperture scopes and higher magnifications reveal a faint halo that hugs the core.

Done with NGC 6384, and ready for a bit of a challenge? Look for magnitude 14.0 NGC 6378, a small spiral galaxy, 0.9° to the south-southwest.

| OBJECT #492 | NGC 6388 |
|---|---|
| Constellation | Scorpius |
| Right ascension | 17h36m |
| Declination | −44°44′ |
| Magnitude | 6.8 |
| Size | 10.4′ |
| Type | Globular cluster |

Our next target sits 1.7° south of magnitude 1.9 Theta ($\theta$) Scorpii. This is a bright globular cluster that might frustrate you because you won't resolve any of its stars through telescopes with apertures smaller than 20′. What you will see is a highly concentrated core surrounded by a thin halo. A pair of 10th-magnitude stars lie nearby. One sits 1′ north of the core, and the other is almost 2′ to the west-southwest.

| OBJECT #493 | NGC 6397 |
|---|---|
| Constellation | Ara |
| Right ascension | 17h41m |
| Declination | −53°40′ |
| Magnitude | 5.3 |
| Size | 25.7′ |
| Type | Globular cluster |
| Other name | Caldwell 86 |

Our next fine target lies 2.9° northeast of magnitude 2.9 Beta ($\beta$) Arae, or 0.9° north-northeast of magnitude 5.3 Pi ($\pi$) Arae. As the sky's fourth-brightest globular cluster (tied for that honor with NGC 6752 in Pavo [Object #582]), NGC 6397 certainly would have earned a Messier designation if it were in the northern sky.

If you observe from a location that puts this cluster more than even 10° up in the sky, you'll find it easily visible with your naked eyes as a fuzzy star. Through an 8-inch telescope at 200×, you'll resolve some 50 stars slightly concentrated toward the brighter middle.

Through a 14-inch scope, the view is remarkable. Stars of all brightnesses form patterns, groups, and distinct layers. The entire cluster resolves, and some observers have likened its appearance at high magnifications to an open cluster.

At a distance of 7,200 light-years, NGC 6397 is either the closest or the second-closest globular cluster. M4 (Object #424) lies at the same distance.

| OBJECT #494 | NGC 6401 |
|---|---|
| Constellation | Ophiuchus |
| Right ascension | 17h39m |
| Declination | −23°55′ |
| Magnitude | 9.5 |
| Size | 4.8′ |
| Type | Globular cluster |

Unlike our previous entry, this object won't dazzle you with hundreds of stars. But the contrast between it and NGC 6397 couldn't be greater, so that makes it worth observing. It lies 3.9° east-northeast of magnitude 3.3 Theta (θ) Ophiuchi, or 1.6° east of the magnitude 4.8 star 51 Ophiuchi. Through a 6-inch telescope, it appears tiny and round. A 12-inch scope at 250× makes it less round with little concentration of stars, but it's still small.

| OBJECT #495 | M14 (NGC 6402) |
|---|---|
| Constellation | Ophiuchus |
| Right ascension | 17h38m |
| Declination | -3°15′ |
| Magnitude | 7.6 |
| Size | 11.7′ |
| Type | Globular cluster |

Our next target is number 14 on Charles Messier's list. He discovered this cluster June 1, 1764.

M14 lies in a lonely area of sky far from any bright star. The best way to find it is to start at magnitude 3.8 Gamma (γ) Ophiuchi, and move 6.5° south-southwest.

Through a 4-inch telescope at a magnification of 100×, you'll see a tightly concentrated ball of stars that's not easy to resolve. The reason is that M14 lies some 30,000 light-years away, so its brightest stars glow at magnitude 14.

Another thing you'll notice about this cluster, especially at higher magnifications, is that it's not round. It appears stretched slightly in an east-west orientation. The outer halo fades out smoothly.

| OBJECT #496 | M6 (NGC 6405) |
|---|---|
| Constellation | Scorpius |
| Right ascension | 17h40m |
| Declination | -32°13′ |
| Magnitude | 4.2 |
| Size | 33′ |
| Type | Open cluster |
| Other name | The Butterfly Cluster |

Our next object is the Butterfly Cluster, also known as M6. From a dark site, you'll pick out this open cluster easily without optical aid. It lies 5° north-northeast of magnitude 1.6 Shaula (Lambda [λ] Scorpii), Scorpius' 2nd-brightest star.

Because Scorpius' tail lies near the southern horizon for Northern Hemisphere observers, M6 often appears hazy, like a nebula. If you see it high in the sky, however, you'll have no doubt as to its starry nature.

Although M6 is a naked-eye or binocular object, it looks best through a telescope. Start with your eyepiece that gives the lowest magnification, and increase the power as you try to pick out the butterfly's wings, one to the north and the other to the south.

A 4-inch telescope will reveal 50 stars, and through an 11-inch scope, you'll count 200 stars looking more like an angry beehive than a butterfly. M6's brightest star is the orange, magnitude 6.0 star SAO 209132, which lies on the cluster's eastern edge.

| OBJECT #497 | The Lozenge |
|---|---|
| Constellation | Draco |
| Right ascension | 17h43m |
| Declination | 53°53′ |
| Type | Asterism |

This easy-to-see asterism called the Lozenge. Amateur astronomers also know it as the Head of Draco the Dragon. Draco is the 8th-largest constellation. It winds between the Big and Little Dippers halfway around the North Celestial Pole.

To find the Lozenge, draw a line from brilliant Vega (Alpha [$\alpha$] Lyrae) to Kochab (Beta [$\beta$] Ursae Minoris). Kochab is the second-brightest of the Little Dipper's stars (next to Polaris), and the brightest in that asterism's bowl. The Lozenge lies near the halfway point of this line, but a bit closer to Vega.

The four stars that form the Lozenge are Rastaban, Eltanin, Grumium, and Al Rakis. That's quite a mouthful. By Greek letter, they are Beta ($\beta$), Gamma ($\gamma$), Xi ($\xi$) and Nu ($\nu$) Draconis, respectively. Gamma's the brightest, coming in at magnitude 2.2, but the double star Nu's the best target.

| OBJECT #498 | NGC 6416 |
|---|---|
| Constellation | Scorpius |
| Right ascension | 17h44m |
| Declination | −32°22′ |
| Magnitude | 5.7 |
| Size | 15′ |
| Type | Open cluster |

Our next treat lies in an incredibly rich region of the Milky Way 5.2° north-northeast of magnitude 1.6 Shaula (Lambda [$\lambda$] Scorpii). Another way to find this cluster is to look a bit less than 1° east of M6.

Through a 4-inch telescope at 75×, you may have trouble picking this cluster out of the background. Look for a slightly denser region of irregularly spaced stars in the shape of an isosceles triangle whose apex points toward the southeast. If you observe NGC 6416 though larger instruments, I'd suggest you keep the power low unless you're trying to split the few double stars within the cluster.

| OBJECT #499 | Psi ($\psi$) Draconis |
|---|---|
| Constellation | Draco |
| Right ascension | 17h42m |
| Declination | 72°09′ |
| Magnitudes | 4.9/6.1 |
| Separation | 30.3″ |
| Type | Double star |
| Other name | Dziban |

You'll have no trouble splitting this wide binary. The primary is white, and the secondary is nearly so with just a trace of yellow. Psi Dra tops an isosceles triangle with Chi ($\chi$) and Phi ($\Phi$) Draconis forming the base. Psi lies roughly 3° from each of these stars.

Remember the story of the four mother camels protecting a baby camel from two hyenas (Object #488)? Well, $Psi^1$ and $Psi^2$ Draconis may represent the two hyenas. The other possibility is that the beasts are the stars Zeta and Eta Draconis.

| OBJECT #500 | IC 4665 |
|---|---|
| Constellation | Ophiuchus |
| Right ascension | 17h46m |
| Declination | 5°43′ |
| Magnitude | 4.2 |
| Size | 70′ |
| Type | Open cluster |
| Other names | The Black Swallowtail Butterfly Cluster, the Little Beehive |

It's ok if you've never observed (or even heard of) IC 4665. Although it's a relatively bright open cluster, it won't look that bright to you because its light spreads out over a diameter of 70′. That's 5.5 times the area covered by a Full Moon.

IC 4665 lies in northern Ophiuchus, 1.3° north-northeast of magnitude 2.8 Beta ($\beta$) Ophiuchi. To observe the cluster, use a low-power eyepiece that will give you at least a 1° field of view. You'll see several dozen 7th- to 9th-magnitude stars and another 20 stars around magnitude 10 set against a rich background haze of fainter points.

My best view of IC 4665 came through an 8-inch f/4.5 Newtonian reflector equipped with a binoviewer. Using eyepieces that gave a magnification of around 45×, the cluster took on a layered 3-D appearance.

*Astronomy* magazine Contributing Editor Stephen James O'Meara gave this object the common name I list. He likens IC 4665 to a Black Swallowtail butterfly, which has a black body and wings lined with bright white spots, like the stars in this cluster.

| OBJECT #501 | IC 4662 |
|---|---|
| Constellation | Pavo |
| Right ascension | 17h47m |
| Declination | −64°38′ |
| Magnitude | 11.3 |
| Size | 3.7′ by 2.5′ |
| Type | Irregular galaxy |

Point at least a 10-inch telescope at magnitude 3.6 Eta ($\eta$) Pavonis and you'll find our next target just 10′ to the star's northeast. Because the star's brightness overwhelms that of IC 4662, move Eta out of the field of view before you begin looking for details in the galaxy.

A large stellar association that contains two emission nebulae dominates the appearance of this dwarf galaxy. In the eyepiece, the association reveals itself on the northeast side as a bright spot in an otherwise featureless haze.

| OBJECT #502 | NGC 6440 |
|---|---|
| Constellation | Sagittarius |
| Right ascension | 17h49m |
| Declination | −20°22′ |
| Magnitude | 9.3 |
| Size | 4.4 |
| Type | Globular cluster |

Our next target lies 5.9° west of magnitude 3.8 Mu ($\mu$) Sagittarii. Admittedly, this object is small and not all that bright. Its main claim to fame is that it sits less than 0.4° south-southwest of NGC 6445 (Object #504). Insert your highest-power eyepiece that gives at least a half-degree field of view. The globular itself shows a wide, concentrated core surrounded by a thin halo of unresolved stars.

| OBJECT #503 | NGC 6441 |
|---|---|
| Constellation | Scorpius |
| Right ascension | 17h50m |
| Declination | −37°03′ |
| Magnitude | 7.2 |
| Size | 7.8′ |
| Type | Globular cluster |
| Other name | The Silver Nugget Cluster |

This object sits 3.3° due east of magnitude 1.6 Shaula (Lambda [$\lambda$] Scorpii), right next to the 3rd-magnitude star G Scorpii. Although this orange star just 4′ west of the cluster's center provides a nice contrast, move it out of the field of view when you're ready to study the cluster. Through an 8-inch telescope at 150×, the core appears bright and concentrated with a thin but easily visible, slightly irregular, halo surrounding it.

The magnitude 10.0 star GSC 7389:2031 lies a bit more than 1′ southwest of the globular's core. NGC 6441 is one of only four globular clusters known to contain a planetary nebula, but you'll need a 25-inch or larger scope, a finder chart, ultra-high magnification, and superb seeing to spot it.

*Astronomy* magazine Contributing Editor Stephen James O'Meara gave this cluster the common name I list here. He thought that, under moderate magnification, the stars of NGC 6441 and the single star G Scorpii looked like silver and gold nuggets lying face up in the sand.

| OBJECT #504 | NGC 6445 |
|---|---|
| Constellation | Sagittarius |
| Right ascension | 17h49m |
| Declination | −20°01′ |
| Magnitude | 11.2 |
| Size | 34″ |
| Type | Planetary nebula |
| Other name | The Box Nebula |

Our next target lies 5.8° west of magnitude 3.8 Mu ($\mu$) Sagittarii, and less than 0.4° north-northeast from NGC 6440 (Object #502). NGC 6445 is the second "Box Nebula" I've described in this book. The other is NGC 6309 (Object #467).

Astronomers classify NGC 6445 as a bipolar planetary nebula, and, as its name implies, it does have a distinct rectangular shape. Through a 12-inch telescope at 250×, you'll see the two relatively bright lobes, one to the northwest and the other to the southeast, separated by a dark central region. Larger telescopes show this planetary as a thin rectangle with a large, dark central void.

**Object #505** Ptolemy's Cluster (M7) Allan Cook/Adam Block/NOAO/AURA/NSF

| OBJECT #505 | M7 (NGC 6475) |
|---|---|
| Constellation | Scorpius |
| Right ascension | 17h54m |
| Declination | –34°49′ |
| Magnitude | 3.3 |
| Size | 80′ |
| Type | Open cluster |
| Other name | Ptolemy's Cluster |

Scorpius offers several deep-sky treats visible with the unaided eye. One of its standout deep-sky objects — open cluster M7 — lies 4.7° east-northeast of the two bright stars that mark the scorpion's stinger, Lambda ($\lambda$) and Upsilon ($\varepsilon$) Scorpii. You'll spot M7 easily from any reasonably dark site.

Greek philosopher Ptolemy mentioned M7 around 130, calling it a "Nebula following the sting of Scorpius." Messier added it to his catalog in 1764, writing, "Star cluster more considerable than the preceding [M6]; this cluster appears to the naked eye like a nebulosity." M7 is the southernmost object in Messier's catalog.

M7 is large enough that four Full Moons could fit within its borders. Because it is so big, you'll need to use a low-magnification eyepiece to observe the entire cluster. When you do, notice how the background richness of the Milky Way enhances the view.

Alternately, you can crank up the power and look for double stars, patterns the cluster's stars form, or gaps between lines of stars noted by many observers. Through 10×50 binoculars, you'll count 50 stars in M7. Double the aperture (to a 4-inch telescope), and you'll double the number of stars you see.

| OBJECT #506 | M23 (NGC 6494) |
|---|---|
| Constellation | Sagittarius |
| Right ascension | 17h57m |
| Declination | –19°01′ |
| Magnitude | 5.5 |
| Size | 27′ |
| Type | Open cluster |

M23 is a small telescope target and one of Charles Messier's least-observed objects. That's a shame, because it's gorgeous. To locate this fine sight, point your telescope, or even binoculars, 4.4° west-northwest of magnitude 3.8 Mu ($\mu$) Sagittarii, toward the thickest part of the Milky Way.

This open cluster shines at magnitude 5.5 and measures 27′ across, a size equal to the Full Moon. It shows up clearly in a finder scope, and you might even catch its faint glow without optical aid from a dark site.

Through a 4-inch telescope at 100×, you'll see a gorgeous cluster with 50 stars spreading out into several curving rows of stars. (I can make five distinct chains.) Although the surrounding star field is rich, you'll have no trouble identifying where M23 ends. The brightest non-cluster star in the area is magnitude 6.5 SAO 160909, which lies 20′ northwest of M23's center. If you have access to larger telescopes, you'll see more stars up to a limit of about 100 cluster members.

| OBJECT #507 | NGC 6496 |
|---|---|
| Constellation | Scorpius |
| Right ascension | 17h59m |
| Declination | –44°16′ |
| Magnitude | 8.6 |
| Size | 5.6 |
| Type | Globular cluster |

Our next target sits 4.1° east-southeast of magnitude 1.9 Theta ($\theta$) Scorpii, right on that constellation's border with Corona Australis. When you first train your telescope on this object, you'll spot the orange magnitude 4.8 star SAO 228562, which lies 0.4° to the west-southwest. Admittedly, it makes the scene more attractive overall, but move the star out of the field of view when you observe the globular.

Although this object is a globular cluster, it's one of the loosest in the sky. You won't even imagine that it has a central condensation, so look for the half dozen or so magnitude 11 and 12 stars that mark its location.

**Object #508** The Lost in Space Galaxy (NGC 6503) Adam Block/NOAO/AURA/NSF

| OBJECT #508 | NGC 6503 |
|---|---|
| Constellation | Draco |
| Right ascension | 17h49m |
| Declination | 70°09′ |
| Magnitude | 10.3 |
| Size | 7.3′ by 2.4′ |
| Type | Spiral galaxy |
| Other name | The Lost in Space Galaxy |

Our next target is a bright spiral galaxy you can spot easily through a 4-inch telescope. It displays a broadly concentrated region offset to the north of center.

The surface mottling becomes apparent through a 16-inch scope. Measurements of the brightest blue stars place NGC 6503 at 17 million light-years. Find this galaxy by starting at magnitude 4.9 Omega ($\omega$) Draconis, and moving 1.8° northeast.

When I first heard *Astronomy* magazine Contributing Editor Stephen James O'Meara's common name for this object, I thought he was referring to the television series *Lost in Space*, which ran from 1965 to 1968. As a big fan of science fiction, I loved the perceived tie-in. But no. O'Meara dubbed it the Lost in Space Galaxy simply because of its isolation and neglect (by observers).

**Object #509** The Trifid Nebula (M20) Adam Block/Mount Lemmon SkyCenter/University of Arizona

| OBJECT #509 | M20 (NGC 6514) |
|---|---|
| Constellation | Sagittarius |
| Right ascension | 18h02m |
| Declination | −23°02′ |
| Size | 20′ by 20′ |
| Type | Emission nebula |
| Other name | The Trifid Nebula |

Our next treat is the magnificent Trifid Nebula, an object Messier discovered June 5, 1764. This object owes its common name to three dust lanes that converge in front of the emission nebulosity. Look for a nice triple-star system just west of center. These three stars, and two more, provide the ultraviolet radiation that causes M20 to glow. On the Trifid's northern edge sits a reflection nebula, shown as blue in photos. This region reflects the light of the magnitude 9 star that sits at its center. You'll want to set a lot of time aside to observe this object.

You'll find M20 3.3° southwest of magnitude 3.8 Mu ($\mu$) Sagittarii.

| OBJECT #510 | Barnard's Star |
|---|---|
| Constellation | Ophiuchus |
| Right ascension | 17h58m |
| Declination | 4°42′ |
| Magnitude | 9.5 |
| Type | Star |

You'll find the star that displays the greatest proper motion of any in the sky 3.6° east of magnitude 2.8 Cebelrai (Beta [$\beta$] Ophiuchi). Proper motion is a measure of how much a star (or any celestial object) changes its position relative to objects that lie farther away. Astronomers use arcseconds per year (″/year) to quantify a star's proper motion. Barnard's Star has a proper motion of 10.4″/year.

Once you've pointed your telescope toward the region of Barnard's Star, you'll recognize it by its color — red — rather than by its brightness. An 11th-magnitude white star lies 1′ to its east-southeast. Barnard's Star lies 6 light-years from Earth.

**Object #511** The Castaway Cluster (NGC 6520) Fred Calvert/Adam Block/NOAO/AURA/NSF

| OBJECT #511 | NGC 6520 |
|---|---|
| Constellation | Sagittarius |
| Right ascension | 18h03m |
| Declination | –27°54′ |
| Magnitude | 7.6 |
| Size | 6′ |
| Type | Open cluster |
| Other name | The Castaway Cluster |

Our next target lies 2.6° north-northwest of magnitude 3.0 Al Nasl (Gamma$^2$ [$\gamma^2$] Sagittarii). Through a small telescope, you'll see a well-defined cluster against a bright background of distant stars. Through an 8-inch telescope, 30 stars of magnitudes 10 and 11 sit in the central one-third of the cluster. At about 150×, scan the region to the west and south, and you'll see far fewer stars. What you're looking at is the dark nebula Barnard 86 (Object #527).

*Astronomy* magazine Contributing Editor Stephen James O'Meara calls this object the Castaway Cluster in honor of Robinson Crusoe because, to him, the view around NGC 6520 is so grand that the cluster looks like some lonely, uncharted isle in a tempestuous sea.

**Object #512** NGC 6522 (left) Adam Block/Mount Lemmon SkyCenter/University of Arizona

| OBJECT #512 | NGC 6522 |
|---|---|
| Constellation | Sagittarius |
| Right ascension | 18h04m |
| Declination | –30°02′ |
| Magnitude | 8.4 |
| Size | 5.6′ |
| Type | Globular cluster |

For our next treat, head 0.6° northwest of magnitude 3.0 Al Nasl (Gamma$^2$ [$\gamma^2$] Sagittarii). A region of low obscuration known as Baade's Window, named for German astronomer Walter Baade, enables us to see this cluster, which lies 90% of the way to our galaxy's center.

A 12-inch telescope will resolve several dozen stars around a strongly concentrated core you won't resolve. You'll also spot a magnitude 11.4 foreground star less than 2′ from the cluster's eastern edge.

Finally, look 16′ to the east of NGC 6522 for another globular cluster, magnitude 9.6 NGC 6528.

**Object #513** The Lagoon Nebula (M8) Adam Block/NOAO/AURA/NSF

| OBJECT #513 | M8 (NGC 6523) |
|---|---|
| Constellation | Sagittarius |
| Right ascension | 18h04m |
| Declination | −24°23′ |
| Size | 45′ by 30′ |
| Type | Emission nebula |
| Other name | The Lagoon Nebula |

Look 5.5° west of magnitude Lambda ($\lambda$) Sagittarii for the magnificent Lagoon Nebula. Around 1680, English astronomer John Flamsteed discovered what became the eighth object on Messier's list.

You'll spot this object with your naked eyes from a dark site. It measures 3 times as wide as the Full Moon, and you'll be able to follow most of the nebulosity through your telescope. A dark lane (the lagoon) cuts the object in half. On the eastern side of the rift, you'll see three dozen stars of open cluster NGC 6530 embedded in the gas.

The brightest star west of the lane is the magnitude 5.9 star 9 Sagittarii, which is responsible for the nebula's glow. A bit more to the west is M8's core, a region of intense brightness. Look for the Hourglass Nebula, a star-forming region with lots of young stars.

**Objects #514** NGC 6528 (right) Adam Block/Mount Lemmon SkyCenter/University of Arizona

| OBJECT #514 | NGC 6528 |
|---|---|
| Constellation | Sagittarius |
| Right ascension | 18h05m |
| Declination | −30°03′ |
| Magnitude | 9.6 |
| Size | 5′ |
| Type | Globular cluster |

To find this object, first locate NGC 6522 (Object #512). NGC 6528 lies 16′ to the east. The stars in this small cluster are even more difficult to resolve than those of its partner. Through a 10-inch telescope at 200×, the cluster appears tiny and uniformly bright. Increasing the magnification to 350× (if your sky conditions allow it) will reveal an irregular, ultra-thin halo.

| OBJECT #515 | M21 (NGC 6531) |
|---|---|
| Constellation | Sagittarius |
| Right ascension | 18h05m |
| Declination | −22°30′ |
| Magnitude | 5.9 |
| Size | 13′ |
| Type | Open cluster |

You'll find this relatively unobserved Messier object 2.6° southwest of magnitude 3.8 Mu ($\mu$) Sagittarii. I'm pretty certain you'll sweep up the Trifid Nebula (M20) first, so when you do, move 42′ northeast for M21.

Through a 6-inch telescope, you'll see two dozen stars brighter than 12th magnitude. The magnitude 7.2 star SAO 186215 shines at the cluster's center.

| OBJECT #516 | NGC 6535 |
|---|---|
| Constellation | Serpens (Cauda) |
| Right ascension | 18h04m |
| Declination | –0°18′ |
| Magnitude | 9.3 |
| Size | 3.4 |
| Type | Globular cluster |

You'll find our next object 5.1° west-northwest of magnitude 3.2 Eta ($\eta$) Serpentis. Because this globular is so small, you'll need a pretty large telescope to pull any details out of it. Through a 10-inch scope at 200×, you'll see a mottled, granular appearance. Crank the magnification up to 300×, and you'll see the cluster's outer edge as irregular. A 20-inch instrument shows this irregularity as resolved stars.

| OBJECT #517 | NGC 6537 |
|---|---|
| Constellation | Sagittarius |
| Right ascension | 18h05m |
| Declination | –19°51′ |
| Magnitude | 13.0 |
| Size | 1.5′ |
| Type | Planetary nebula |
| Other name | The Red Spider Nebula |

Our next target sits 2.3° west-northwest of magnitude 3.8 Mu ($\mu$) Sagittarii. Pull out a 12-inch or larger telescope to view this object. At any magnification below 250×, the Red Spider Nebula appears starlike. Above that limit, you'll see a round bluish disk. An Oxygen-III filter helps bring out this nebula, but you won't see the object's blue color through it.

This object's common name comes from its appearance on long-exposure images taken through large telescopes.

| OBJECT #518 | NGC 6539 |
|---|---|
| Constellation | Serpens (Cauda) |
| Right ascension | 18h05m |
| Declination | –7°35′ |
| Magnitude | 9.8 |
| Size | 6.9′ |
| Type | Globular cluster |

You'll find this object 0.7° northeast of magnitude 5.2 Tau ($\tau$) Ophiuchi. This cluster is so small that you'll find it difficult to resolve individual stars through any size telescope. Its brightest stars glow at 16th magnitude, so through a 12-inch telescope at 300×, you'll see a granular appearance.

| OBJECT #519 | NGC 6541 |
|---|---|
| Constellation | Corona Australis |
| Right ascension | 18h08m |
| Declination | –43°42′ |
| Magnitude | 6.1 |
| Size | 13.1′ |
| Type | Globular cluster |
| Other name | Caldwell 78 |

You'll find this nice cluster 4° west-northwest of magnitude 3.5 Alpha ($\alpha$) Telescopii. Through a 6-inch telescope, you'll see a circular glow of unresolved suns. Double the aperture, and move to 200×, and more than 100 stars pop into view, although stars in the core remain tough to separate. NGC 6541 has an irregular outer boundary where thin dark lanes impinge on the outer halo. Just 21′ to the northwest of the cluster lies magnitude 4.9 SAO 228708.

**Object #520** The Cat's Eye Nebula (NGC 6543) Adam Block/Mount Lemmon SkyCenter/University of Arizona

| OBJECT #520 | NGC 6543 |
|---|---|
| Constellation | Draco |
| Right ascension | 17h59m |
| Declination | 66°38′ |
| Magnitude | 8.1 |
| Size | >18″ |
| Type | Planetary nebula |
| Other names | The Cat's Eye Nebula, Caldwell 6 |

Our next object is a wonderful planetary nebula you'll find just a bit more than 5° east-northeast of magnitude 3.2 Zeta (ζ) Draconis. Through telescopes as small as 4 inches in aperture, this colorful object looks blue, blue-green, greenish-blue, or green, depending on your eyes' color sensitivity.

Use a magnification of 200× in an 8-inch telescope, and you'll see some hazy spiral structure around the bright central star. A faint outer shell 5′ across surrounds NGC 6543. This halo contains more mass than the core. Past observers have misidentified a bright part of the halo as a galaxy. It even carries its own designation — IC 4677.

| OBJECT #521 | NGC 6544 |
|---|---|
| Constellation | Sagittarius |
| Right ascension | 18h07m |
| Declination | –25°00′ |

| (continued) | |
|---|---|
| Magnitude | 7.5 |
| Size | 9.2′ |
| Type | Globular cluster |
| Other name | The Starfish Cluster |

Our next object lies 1° southeast of the magnificent Lagoon Nebula (M8). Any size telescope will show you this cluster, but few amateur scopes, no matter their size, will resolve the stars within it. A magnitude 10.7 star sits 28″ to the east-northeast of the small core.

*Astronomy* magazine Contributing Editor Stephen James O'Meara calls this the Starfish Cluster because at high magnification its brightest stars form a starfish-shaped pattern.

| OBJECT #522 | 40/41 Draconis |
|---|---|
| Constellation | Draco |
| Right ascension | 18h00m |
| Declination | 80°00′ |
| Magnitudes | 5.7/6.1 |
| Separation | 19.3″ |
| type | Double star Yellow-white |

The easiest way to find this pair of stars is to look 3.5° east-southeast of magnitude 4.4 Epsilon (ε) Ursae Minoris. Most observers see the primary as yellow and the secondary as white. You'll also find "both yellow" and "both white" as color descriptors.

| OBJECT #523 | NGC 6553 |
|---|---|
| Constellation | Sagittarius |
| Right ascension | 18h09m |
| Declination | −25°54′ |
| Magnitude | 8.3 |
| Size | 9.2′ |
| Type | Globular cluster |

You'll find our next object 4.2° west of magnitude 2.8 Kaus Borealis (Lambda [λ] Sagittarii). Through an 8-inch telescope at 150× from a dark site, you'll see this cluster's triangular appearance. I think of it as an arrowhead pointing toward the southeast. The arrowhead's northeast tip is the brightest. Look for two stars of about magnitude 11.6, each less than 1′ from the cluster's center. One lies to the northeast and the other toward the west-southwest.

| OBJECT #524 | NGC 6558 |
|---|---|
| Constellation | Sagittarius |
| Right ascension | 18h10m |
| Declination | −31°46′ |
| Magnitude | 8.6 |
| Size | 4.2′ |
| Type | Globular cluster |

Our next target sits 1.6° southeast of magnitude 3.0 Al Nasl (Gamma [γ] Sagittarii). Through a 12-inch telescope at 150×, I noticed four 12th-magnitude stars in the shape of the constellation Corvus that frame this cluster. The cluster itself appears faint, small, and unresolvable. Several curving lines of stars, especially toward the south, add to the overall appeal of the view.

**Object #525** NGC 6559 John and Christie Connor/Adam Block/NOAO/AURA/NSF

| OBJECT #525 | NGC 6559 |
|---|---|
| Constellation | Sagittarius |
| Right ascension | 18h10m |
| Declination | −24°07′ |
| Size | 8′ |
| Type | Emission nebula |

An often overlooked object in Sagittarius is NGC 6559. You'll find it 1.4° east of the Lagoon Nebula (M8). This spectacular star-forming region would get a lot more ink if it lay elsewhere. But most amateurs focus on nearby M8 and M20.

For imagers, few regions of nebulosity show the variety of color and detail that NGC 6559 does. Wispy dark clouds hang delicately in the foreground of diffuse, glowing hydrogen gas. Follow the clouds toward a bright red arc of gas that shields luminous stars beneath it.

Although these stars try to push away the gas and dust near them, thick portions remain and continue to scatter light. This colors the area with hints of blue and purple. Use a nebula filter to see the emission nebulosity, but remove the filter if you're looking for blue reflection nebulosity.

| OBJECT #526 | 95 Herculis |
|---|---|
| Constellation | Hercules |
| Right ascension | 18h02m |
| Declination | 21°36′ |
| Magnitudes | 5.0/5.1 |
| Separation | 6.3″ |
| Type | Double star |

Our next stellar pair lies in another of those "no-man's lands" where few bright stars abound. One way to locate it is to find magnitude 2.8 Ras Algethi (Alpha [α] Herculis). 95 Her lies a bit more than 13° northeast of this star. The (barely) brighter of these two stars glows yellow, and its companion is white.

| OBJECT #527 | Barnard 86 |
|---|---|
| Constellation | Sagittarius |
| Right ascension | 18h03m |
| Declination | –27°53′ |
| Size | 5′ by 5′ |
| Type | Dark nebula |
| Other name | The Ink Spot |

Our next object is a dark nebula called the Ink Spot, also known as Barnard 86 (B86).

If you're uncertain how to target B86, just aim for NGC 6520 (Object #511), a nearby star cluster. Together, these objects present a wonderful contrast. Through an 8-inch telescope, you'll see about 30 stellar members of NGC 6520 against a bright background of distant stars.

No such background exists for B86. Its starless, irregular form stands out against myriad faint stars. Look for the orange magnitude 6.7 double star WDS HDS2541 on B86's western edge — a nice complement to the scene.

| OBJECT #528 | 70 Ophiuchi |
|---|---|
| Constellation | Ophiuchus |
| Right ascension | 18h06m |
| Declination | 2°30′ |
| Magnitudes | 4.2/6.0 |
| Separation | 4.1″ |
| Type | Double star |

This close binary lies 4.4° east of magnitude 3.8 Gamma ($\gamma$) Ophiuchi. The two components exhibit a nice color contrast. The primary is a deep-yellow, and the secondary is red.

**Object #529** NGC 6563 Adam Block/NOAO/AURA/NSF

| OBJECT #529 | NGC 6563 |
|---|---|
| Constellation | Sagittarius |
| Right ascension | 18h12m |
| Declination | −33°52′ |
| Magnitude | 11.0 |
| Size | 48″ |
| Type | Planetary nebula |

To find our next object, look 2.5° west of magnitude 1.8 Kaus Australis (Epsilon [ε] Sagittarii). Through an 8-inch telescope at 200×, this nebula shows a pale disk. A 14-inch scope reveals that the outer edge is ever-so-slightly brighter. A nebula filter (such as an Oxygen-III) really helps. Rather than round, this planetary shows a slight oval shape inclined northeast to southwest.

| OBJECT #530 | NGC 6569 |
|---|---|
| Constellation | Sagittarius |
| Right ascension | 18h14m |
| Declination | −31°50′ |
| Magnitude | 8.4 |
| Size | 6.4′ |
| Type | Globular cluster |

Our next target sits 2.2° southeast of magnitude 3.0 Al Nasl (Gamma [γ] Sagittarii) in a rich and appealing star field. Through a 10-inch telescope at 200×, this globular appears stretched ever so slightly with a general northeast to southwest orientation. The tiny core is a bit brighter than the small halo, and a faint mottling suggests the presence of individual stars, but I didn't resolve any.

A rough "W" shape of stars lies just to NGC 6569's south. Continue almost 9′ south of the cluster, and you'll arrive at magnitude 6.8 SAO 209873.

**Object #531** The Emerald Nebula (NGC 6572) Bruce Bodner/Adam Block/NOAO/AURA/NSF

| OBJECT #531 | NGC 6572 |
|---|---|
| Constellation | Ophiuchus |
| Right ascension | 18h12m |
| Declination | 6°51′ |
| Magnitude | 8.1 |
| Size | 18″ |
| Type | Planetary nebula |
| Other name | The Emerald Nebula |

This treat sits 2.2° south-southeast of the magnitude 4.6 star 71 Ophiuchi. Even a small telescope will reveal the Emerald Nebula. This planetary nebula is small, only 18″ across, but it has a high surface brightness, and it's colorful.

Through an 8-inch telescope, you'll see the Emerald Nebula's oval shape. A small but bright central region also appears. The observing strategy I suggest to pull out NGC 6572's color is to keep the magnification low, but I've had good results from 100× to 400× through a 12-inch scope. In June 2009, I viewed this object through a 30-inch Newtonian reflector. The color held strong and steady at magnifications through 450×.

| OBJECT #532 | NGC 6584 |
|---|---|
| Constellation | Telescopium |
| Right ascension | 18h19m |
| Declination | −52°13′ |
| Magnitude | 7.9 |
| Size | 6.6′ |
| Type | Globular cluster |

Our next object lies 3.5° south-southwest of magnitude 4.1 Zeta (ζ) Telescopii. First, score a point if you can locate Telescopium. This nondescript constellation lies directly south of Corona Australis. The cluster lurks within a loose arrangement of four 12th-magnitude stars, the farthest of which lies less than 3′ away. Widening out the view, two magnitude 7.5 stars lie within a quarter-degree north and northwest.

This cluster is one that really benefits from large-aperture telescopes. Through a 14-inch scope at 125×, the central region appeared broad and bright, with a slight east-west orientation. At 300×, many faint stars lie at the limit of resolution in an irregular halo. At least three faint streamers of (probably) foreground stars arc toward the northwest.

**Object #533** M24 Fred Calvert/Adam Block/NOAO/AURA/NSF

| OBJECT #533 | M24 (NGC 6603) |
|---|---|
| Constellation | Sagittarius |
| Right ascension | 18h17m |
| Declination | –18°50′ |

| (continued) | |
|---|---|
| Magnitude | 4.6 |
| Size | 95′ by 35′ |
| Type | Asterism |
| Other names | The Small Sagittarius Star Cloud, Delle Caustiche |

With everything going on in this incredibly rich region, it's tough to explain why Charles Messier placed M24, an obvious star cloud, on his list. It spans nearly 2° and measures 0.6° wide. I suggest a first look at M24 though 15× binoculars. You'll find it 3° north of magnitude 3.8 Mu ($\mu$) Sagittarii.

Toward the northwest end of the star cloud is the object actually designated NGC 6603, a magnificent open cluster of several dozen stars. Only 4′ south of the cluster, you'll see the magnitude 7.4 foreground star SAO 161294.

Less than 1° west-northwest of NGC 6603 lies Barnard 92, a prominent, small, dark nebula shaped like a fingerprint. The only break in its darkness is the magnitude 11.4 star GSC 6268:517.

Father Pietro Angelo Secchi (1818–1878) called the Sagittarius Star Cloud "Delle Caustiche," as he said, "from the peculiar arrangement of its stars in rays, arches, caustic curves, and intertwined spirals."

| OBJECT #534 | NGC 6604 |
|---|---|
| Constellation | Serpens (Cauda) |
| Right ascension | 18h18m |
| Declination | –12°14′ |
| Magnitude | 6.5 |
| Size | 4′ |
| Type | Open cluster |

Our next target lies 3.6° northwest of magnitude 4.7 Gamma ($\gamma$) Scuti. The first thing you'll notice is the magnitude 7.4 star SAO 161292, which sits just to the east of the cluster's center.

Through a 10-inch telescope, you'll see two dozen stars. An emission nebula surrounds the cluster. It shows up well as a faint mist. Use a nebula filter to dim the cluster's stars.

| OBJECT #535 | NGC 6605 |
|---|---|
| Constellation | Serpens Cauda |
| Right ascension | 18h17m |
| Declination | –15°01′ |
| Magnitude | 6.0 |
| Size | 7′ |
| Type | Open cluster |

You'll find this target 3.2° west of magnitude 4.7 Gamma ($\gamma$) Scuti. This object may or may not be a cluster. There's some debate among astronomers. To me, that makes it interesting, so have a look. A 4-inch telescope will show only the barest brightening of the background star field. Through a 12-inch scope, you'll count 75 stars, but are they a cluster? You decide.

**Object #536** The Eagle Nebula (M16) Adam Block/Mount Lemmon SkyCenter/University of Arizona

| OBJECT #536 | M16 (NGC 6611) |
|---|---|
| Constellation | Serpens Cauda |
| Right ascension | 18h19m |
| Declination | −13°47′ |
| Magnitude | 6.0 |
| Size | 21′ |
| Type | Open cluster |
| Common names | The Eagle Nebula, the Star-Queen Nebula |

Our next target lies just over Sagittarius' border into Serpens. The Eagle Nebula is a combination object: The open cluster is NGC 6611, and the nebula is IC 4703.

Through a 6-inch scope, the cluster comprises about a dozen stars brighter than magnitude 10 and several dozen fainter components, which gives it a nice three-dimensional appearance. The nebula engulfs the cluster and continues to the south. The dark nebula that helps make up the eagle's body lies well to the southeast.

You'll find M16 2.6° west-northwest of magnitude 4.7 Gamma (γ) Scuti.

In *Burnham's Celestial Handbook*, Robert Burnham Jr. explains the not-so-common proper name for M16: "The present author has introduced the name The Star-Queen Nebula for this exceptional wonder of deep-space; the name 'Eagle Nebula' has also appeared in modern observing guides, but seems perhaps a little too prosaic for a vista of such cosmic splendor, aside from the fact that the eagle is already honored by two first magnitude stars, Vega and Altair."

| OBJECT #537 | M18 (NGC 6613) |
|---|---|
| Constellation | Sagittarius |
| Right ascension | 18h20m |
| Declination | −17°08′ |

| (continued) | |
|---|---|
| Magnitude | 6.9 |
| Size | 10′ |
| Type | Open cluster |

You'll find another of Charles Messier's lesser-known objects 4.2° north-northeast of magnitude 3.8 Mu ($\mu$) Sagittarii. A 4-inch telescope reveals about a dozen stars. Larger telescopes don't show many more, and higher magnifications spread the cluster out so much that, in effect, it gets lost among the background stars.

**Object #538** The Swan Nebula (M17) Adam Block/NOAO/AURA/NSF

| OBJECT #538 | M17 (NGC 6618) |
|---|---|
| Constellation | Sagittarius |
| Right ascension | 18h21m |
| Declination | −16°11′ |
| Size | 20′ by 15′ |
| Type | Emission nebula |
| Common names | The Omega Nebula, the Swan Nebula, the Checkmark Nebula, the Horseshoe Nebula |

The Omega Nebula lies 2.6° southwest of magnitude 4.7 Gamma ($\gamma$) Scuti. Swiss astronomer Jean-Phillippe Loys de Chéseaux (1718–1751) discovered M17 (and M16) about 1746. Through a 6-inch telescope, it appears as a bright bar 7′ long with a short extension from the west end to the south. Increase the magnification past 150×, and the extension reveals itself as a hook-shaped feature with dark material obscuring light from its central region. Through a 12-inch or larger scope, you'll see much more nebulosity — and the brightest regions will show fine striations.

| | |
|---|---|
| OBJECT #539 | NGC 6624 |
| Constellation | Sagittarius |
| Right ascension | 18h24m |
| Declination | −30°22′ |
| Magnitude | 7.6 |
| Size | 8.8′ |
| Type | Globular cluster |

This object lies 0.8° southeast of magnitude 2.7 Kaus Media (Delta [δ] Sagittarii). Through a 4-inch telescope, you'll spot this cluster easily as round and evenly bright. A 10-inch scope at 250× reveals the outer edge as irregular and the tiny center brighter than the halo.

| | |
|---|---|
| OBJECT #540 | M28 (NGC 6626) |
| Constellation | Sagittarius |
| Right ascension | 18h25m |
| Declination | −24°52′ |
| Magnitude | 6.8 |
| Size | 11.2′ |
| Type | Globular cluster |

This target lies 1° northwest of magnitude 2.8 Lambda (λ) Sagittarii. Through an 8-inch telescope at 150×, you'll resolve several dozen stars in the wide halo. Through a 14-inch scope, your star count will climb past 150. Viewing the core at 250× or above gives a three-dimensional effect. A relatively bright chain of stars extends to the north. Another fainter one heads to the north-northwest.

| | |
|---|---|
| OBJECT #541 | NGC 6633 |
| Constellation | Ophiuchus |
| Right ascension | 18h28m |
| Declination | 6°34′ |
| Magnitude | 4.6 |
| Size | 27′ |
| Type | Open cluster |
| Other names | The Captain Hook Cluster, the Wasp-Waist Cluster |

Our next object sits in northern Ophiuchus, but the easiest way to find it is to look 7.6° west-northwest of magnitude 4.6 Alya (Theta [θ] Serpentis). Because NGC 6633 lies in an incredibly rich star field, you'll have to be both sharp-eyed and patient to spot it without optical aid, but it can be done.

The cluster lies only 1,000 light-years away, so it's huge (as big as the Full Moon), and many of its stars are bright. Through a 4-inch telescope, more than a dozen outshine 10th magnitude. That group appears to sit in front of an additional 50 fainter stars. If you use larger-aperture instruments, keep the magnification around 100× so you don't disperse the stars so much you're looking "through" the cluster.

The brightest nearby star is SAO 123516, which lies 0.4° south-southeast of NGC 6633. Note the tight clump of magnitude 8.5–10.5 stars at the cluster's eastern edge.

*Astronomy* magazine Contributing Editor Stephen James O'Meara gave this cluster both of the common names I've listed here. He called it the Captain Hook Cluster because, to him, the main form of the cluster appears hook-shaped. If, however, you see the hook's thin shaft as the waist of a wasp whose wings are the perpendicular row of stars at the cluster's northeast end, then, to him, it becomes the Wasp-Waist Cluster.

| | |
|---|---|
| OBJECT #542 | M69 (NGC 6637) |
| Constellation | Sagittarius |
| Right ascension | 18h31m |

| (continued) | |
|---|---|
| Declination | −32°21′ |
| Magnitude | 7.6 |
| Size | 7.1′ |
| Type | Globular cluster |

To find our next target, which Charles Messier put in the middle of his list, look 2.5° northeast of magnitude 1.8 Kaus Australis (Epsilon [ε] Sagittarii). M69 sits near the southwest corner of the Teapot asterism.

M69 lies 30,000 light-years from us and only 6,000 light-years from the center of the Milky Way. Astronomers classify it as one of the most metal-rich globulars, which means its stars have relatively high concentrations of elements heavier than helium. That amount is much less than that of our Sun, however, dating M69's stars to be much older than the one we orbit.

Through an 8-inch telescope, you may have a bit of trouble resolving this globular's stars. Its core appears broad but concentrated, and a thin halo surrounds it. The cluster lies within a rich star field, so crank up the magnification — it can take it.

Through a 14-inch scope, you'll be able to resolve about a dozen of M69's stars, but even with that size instrument, it won't be easy. Look for three clumps of faint stars that form a triangle close to and surrounding the cluster's center, one to the northwest, one southwest, and the third one to the east. Oh, and the magnitude 8.0 star that sits a bit more than 4′ northwest of M69's center is SAO 210259.

| OBJECT #543 | NGC 6638 |
|---|---|
| Constellation | Sagittarius |
| Right ascension | 18h31m |
| Declination | −25°30′ |
| Magnitude | 9.2 |
| Size | 7.3′ |
| Type | Globular cluster |

Our next target lies 0.7° east of magnitude 2.8 Kaus Borealis (Lambda [λ] Sagittarii). Through an 8-inch telescope at 150×, this cluster appears round with a small, condensed core and a thin halo. Magnitude 9.9 SAO 186904 lies a bit more than 3′ to the south-southwest of the cluster's center.

| OBJECT #544 | NGC 6642 |
|---|---|
| Constellation | Sagittarius |
| Right ascension | 18h32m |
| Declination | −23°29′ |
| Magnitude | 8.9 |
| Size | 5.8′ |
| Type | Globular cluster |

This object sits 2.1° north-northeast of magnitude 2.8 Kaus Borealis (Lambda [λ] Sagittarii). Through a 10-inch telescope at 250×, you might notice a slight widening of the central area in a northwest to southeast orientation. That core region remains unresolvable through much larger scopes, although it somewhat resembles a dumbbell. A wide double star lies 2′ north-northwest. The components shine at magnitudes 10.7 and 12.4 and lie 30″ apart. Another, with orange magnitude 7.7 SAO 186912 as the primary and a magnitude 10.9 secondary, lies 12′ northwest. That pair's separation also measures 30″.

| OBJECT #545 | M25 (IC 4725) |
|---|---|
| Constellation | Sagittarius |
| Right ascension | 18h32m |

| (continued) | |
|---|---|
| Declination | −19°15′ |
| Magnitude | 4.6 |
| Size | 32′ |
| Type | Open cluster |

Our next target lies 4.4° east-northeast of magnitude 3.8 Mu ($\mu$) Sagittarii. It's visible to the naked-eye from a dark site, but not easily because of the richness of the Milky Way background.

A 6-inch telescope at 125× reveals 50 member stars. Two chains of stars stretching east to west lie near the center. A starless gap divides them. Many of M25's stars exceed 11th magnitude, making this a nice sight through small telescopes. Yellow magnitude 6.8 SAO 161557 sits at the northwest edge.

| OBJECT #546 | NGC 6645 |
|---|---|
| Constellation | Sagittarius |
| Right ascension | 18h33m |
| Declination | −16°54′ |
| Magnitude | 8.5 |
| Size | 10′ |
| Type | Open cluster |

Move your telescope 2.4° north of our previous treat, M25, to discover NGC 6645, another fine open cluster. Through an 8-inch telescope, you'll count 50 stars. The feature you want to see lies near the center of the cluster. It's a circlet of 15 stars whose center is empty. A line of half a dozen stars protrudes from the eastern side and heads toward the east-northeast for nearly 10′.

| OBJECT #547 | NGC 6649 |
|---|---|
| Constellation | Scutum |
| Right ascension | 18h34m |
| Declination | −10°24′ |
| Magnitude | 8.9 |
| Size | 6.6′ |
| Type | Open cluster |

NGC 6649 lies 2.2° south of magnitude 3.9 Alpha ($\alpha$) Scuti. It's the only Scutum star cluster located in front of a large dusty area. This compact cluster contains some 50 stars 10th magnitude and fainter. A 4-inch telescope shows it as nebulous. Through a 6-inch scope, however, you'll resolve dozens of stars.

| OBJECT #548 | NGC 6652 |
|---|---|
| Constellation | Sagittarius |
| Right ascension | 18h36m |
| Declination | –32°59′ |
| Magnitude | 8.5 |
| Size | 6′ |
| Type | Globular cluster |

July's first target sits 2.8° east-northeast of magnitude 1.8 Kaus Australis (Epsilon [ε] Sagittarii). Through a 6-inch telescope at 125×, this object has a compact core with an irregular halo around it. A 14-inch scope at 300× didn't resolve any of its stars, but it did expand the core enough to make it appear wedge-shaped, pointed toward the east-southeast. Magnitude 6.9 SAO 210344 lies 7′ northwest of the cluster's center.

M.E. Bakich, *1,001 Celestial Wonders to See Before You Die*, Patrick Moore's Practical Astronomy Series,
DOI 10.1007/978-1-4419-1777-5_7, 

**Object #549** M22 Doug Matthews/Adam Block/NOAO/AURA/NSF

| OBJECT #549 | M22 (NGC 6656) |
|---|---|
| Constellation | Sagittarius |
| Right ascension | 18h36m |
| Declination | –23°54′ |
| Magnitude | 5.1 |
| Size | 24′ |
| Type | Globular cluster |

One of the most spectacular objects in Sagittarius is globular cluster M22. This easy naked-eye object ranks as the sky's third-brightest globular, exceeded only by Omega Centauri (NGC 5139) and 47 Tucanae (NGC 104).

In the *Cycle of Celestial Objects*, Admiral Smyth describes M22 as "A fine globular cluster, outlying that astral stream, the Via Lactea, in the space between the Archer's head and bow, not far from the point of the winter solstice, and midway between $\mu$ and $\sigma$ Sagittarii. It consists of very minute and thickly condensed particles of light, with a group of small stars preceding by 3′, somewhat in a crucial form. Halley ascribes the discovery of this in 1665, to Abraham Ihle, the German; but it has been thought this name should have been Abraham Hill, who was one of the first council of the Royal Society, and was wont to dabble with astronomy. Hevelius, however, appears to have noticed it previous to 1665, so that neither Ihle nor Hill can be supported."

When you observe M22, altitude — how high it appears in the sky — is everything. Many amateur astronomers in Northern Europe, Canada, and the northern United States are unaware of this cluster's magnificence because, for them, it hugs the southern horizon. See it high in the sky, however, and you'll understand the hype.

Through even a 4-inch telescope, you'll see several dozen stars. Pay attention to the bright starry background, which adds character to the scene. Try to discern exactly where the outer boundary of M22 ends and the background begins.

Move up to a 10-inch scope, and you'll be overwhelmed by hundreds of member stars. You can try counting them by dividing M22 into eight wedge-shaped pieces. Count the stars in just one wedge, then multiply by 8.

M22 lies 2.4° northeast of magnitude 2.8 Kaus Borealis (Lambda [$\lambda$] Sagittarii).

Turning to Smyth once more, we learn a bit about the state of astronomy in 1844: "This object is a fine specimen of the compression on which the nebula-theory is built. The globular systems of stars appear thicker in the middle than they would do if these stars were all at equal distances from each other; they must, therefore, be condensed towards the centre. That the stars should be thus accidentally disposed is too improbable a supposition to be admitted; whence Sir William Herschel supposes that they are thus brought together by their mutual attractions, and that the gradual condensation towards the centre must be received as proof of a central power of such a kind."

| OBJECT #550 | NGC 6664 |
|---|---|
| Constellation | Scutum |
| Right ascension | 18h37m |
| Declination | –8°13′ |
| Magnitude | 7.8 |
| Size | 12′ |
| Type | Open cluster |

Our next target lies only 0.3° east of Alpha ($\alpha$) Scuti. You'll sweep both up with a low-power eyepiece, so move Alpha Sct out of the field of view for a better look at the cluster. Even through a 2.4-inch telescope, you'll see NGC 6664's brightest stars. Through larger scopes, about 50 stars appear. Look for the "M" or "U" pattern formed by the brightest stars, which some observers have reported.

| OBJECT #551 | IC 4756 |
|---|---|
| Constellation | Serpens (Cauda) |
| Right ascension | 18h39m |
| Declination | 5°27′ |
| Magnitude | 4.6 |
| Size | 52′ |
| Type | Open cluster |
| Other name | Graff's Cluster |

At the northern edge of Serpens Cauda sits a pleasantly surprising open cluster and a rare example of a naked-eye object with an IC (*Index Catalogue*) designation. To find it, look 4.5° west-northwest of magnitude 4.6 Alya (Theta [$\theta$] Serpentis).

IC 4756 is a widely scattered open cluster that appears at dark observing sites as a small bright haze near the edge of the Milky Way. Through a 4-inch telescope, you'll see a huge, beautiful collection of 50 magnitude 9 and 10 stars. The cluster's brightest star, magnitude 6.4 SAO 123778, sits at IC 4756's southeast edge.

This cluster's common name comes from German astronomer Kasimir Romuald Graff (1878–1950) who, in 1922, independently discovered it. Solon Bailey had found it earlier on photographic plates taken at Harvard College's Arequipa station in Peru.

| OBJECT #552 | M70 (NGC 6681) |
|---|---|
| Constellation | Sagittarius |
| Right ascension | 18h43m |
| Declination | –32°18′ |
| Magnitude | 8.0 |
| Size | 7.8′ |
| Type | Globular cluster |

This southerly Messier object lies midway between magnitude 2.6 Ascella (Zeta [ζ] Sagittarii) and magnitude 1.8 Kaus Australis (Epsilon [ε] Sagittarii). M70 (NGC 6681) is similar in brightness to M69 (Object #542), but has a more sharply concentrated center. Through an 8-inch telescope at 200×, you'll see a bright core and a thin halo in which you can resolve half a dozen stars. A short line of relatively bright stars shoots northward from the cluster's eastern side.

M70 has an association with Comet C/1995 O1 (Hale-Bopp). Neither discoverers Alan Hale nor Thomas Bopp was actively comet-hunting when they made their discoveries July 23, 1995. They merely pointed telescopes at M70 and spotted a foreign object in the field of view.

**Object #553** Epsilon Lyrae Adam Block/NOAO/AURA/NSF

| OBJECT #553 | Epsilon (ε) Lyrae |
| --- | --- |
| Constellation | Lyra |
| Right ascension | 18h44m |
| Declination | 39°39′ |
| Magnitudes | 5.0/6.1;5.5/5.5 |
| Separations | 2.6″/2.3″ |
| Type | Double star |
| Other name | The Double Double |

Our next object will surprise you through a small telescope. It's the star Epsilon ($\varepsilon$) Lyrae, but most observers call it the Double Double. As the name indicates, Epsilon is a pair of double stars that lie close together.

You'll find Epsilon easily. It sits 1.7° east-northeast of Vega. Through binoculars, you'll see the initial double nature of this object, and you might think, "Ah, a double star." You're half right.

Point a telescope toward it, and you'll see both "stars" resolve into pairs. To assure success, use a magnification above 75×. I've split both pairs many times through a 2.4-inch (60-millimeter) telescope, so you shouldn't have a problem.

| OBJECT #554 | NGC 6684 |
|---|---|
| Constellation | Pavo |
| Right ascension | 18h49m |
| Declination | −65°11′ |
| Magnitude | 10.4 |
| Size | 4.5′ by 3.3′ |
| Type | Barred spiral galaxy |

Center magnitude 5.7 Theta ($\theta$) Pavonis in your eyepiece, and you'll be only 6′ north-northwest of this object. Through an 8-inch telescope at 200×, you'll see a bright core surrounded by a circular halo 2′ across. Increasing the aperture brightens the galaxy but fails to reveal additional detail.

For an extra bit of challenge, point your telescope 0.5° northeast of NGC 6684, and try to spot NGC 6684A. This magnitude 14.3 irregular galaxy measures 2.6′ by 1.3′.

| OBJECT #555 | M26 (NGC 6694) |
|---|---|
| Constellation | Scutum |
| Right ascension | 18h45m |
| Declination | −9°24′ |
| Magnitude | 8.0 |
| Size | 14′ |
| Type | Open cluster |

Our next target, M26, lies in a rich star field in a Milky Way region called the Scutum Star Cloud. Through a 4-inch scope, you'll see two dozen stars. Increase the aperture to 12′, and you'll have a good chance to resolve another 70 cluster members.

To find this object, draw a line from magnitude 3.9 Alpha ($\alpha$) through magnitude 4.7 Delta ($\delta$) Scuti. Then extend the line in the same direction another half the distance between Alpha and Delta.

| OBJECT #556 | The Scutum Star Cloud |
|---|---|
| Constellation | Scutum |
| Right ascension (approx.) | 18h37m |
| Declination (approx.) | −10° |
| Size (approx.) | 4.2° by 2.4° |

You won't need any optical aid to find this sky region, but you will need to find the location of the small, faint constellation Scutum the Shield. First, locate the three bright stars of the Summer Triangle (Vega, Altair, and Deneb) as a starting point. Altair's associated constellation is Aquila the Eagle. Scutum lies just off Aquila's tail.

Under a dark, clear sky, you'll notice this region of the summer Milky Way looks like it contains more stars than nearby areas. We call such spots star clouds, and, indeed, lots of star formation is happening here.

Once you can find Scutum easily, view it through binoculars. Now that's a lot of stars!

| OBJECT #557 | Zeta ($\zeta$) Lyrae |
|---|---|
| Constellation | Lyra |
| Right ascension | 18h45m |
| Declination | 37°36′ |
| Magnitudes | 4.3/5.9 |
| Separation | 44″ |
| Type | Double star |

This pretty binary combines a light-blue primary with a pale-yellow or yellowish-white secondary. Any telescope will split this bright, wide pair.

| OBJECT #558 | Beta ($\beta$) Lyrae |
|---|---|
| Constellation | Lyra |
| Right ascension | 18h50m |
| Declination | 33°22′ |
| Magnitude | 3.3–4.3 |
| Period | 12.936 days |
| Type | Variable star |
| Other name | Sheliak |

Sheliak is the prototype for a class of variable stars. Beta Lyrae variables are eclipsing binary stars. Here, the larger star totally eclipses its smaller companion to produce its main minimum. Half the period later, the smaller star occults part of the larger one. This generates a secondary minimum of about magnitude 3.8. Beta Lyrae also is a double star with an magnitude 8.6 secondary located 46″ away.

This star's proper name is easy enough to explain. According to Richard Hinckley Allen in *Star-Names and Their Meanings*, it comes from Al Shilyak, one of the Arabian names for Lyra.

**Object #559** The Wild Duck Cluster (M11) Adam Block/NOAO/AURA/NSF

| OBJECT #559 | M11 (NGC 6705) |
|---|---|
| Constellation | Scutum |
| Right ascension | 18h51m |
| Declination | –6°16′ |
| Magnitude | 5.8 |
| Size | 13′ |
| Type | Open cluster |
| Other name | The Wild Duck Cluster |

The Northern Hemisphere's summer is — by far — the best time to view our galaxy's rich star fields. And it's during summer that the spectacular Wild Duck Cluster flies through the Milky Way.

German astronomer Gottfried Kirch discovered the Wild Duck Cluster in 1681. Messier made it his 11th catalog entry May 30, 1764.

The common name originated in *A Cycle of Celestial Objects*, written by Admiral William H. Smyth in 1844. Of M11, Smyth wrote, "A splendid cluster of stars on the dexter chief [upper-right corner] of Sobieski's shield. This object, which somewhat resembles a flight of wild ducks in shape, is a gathering of minute stars with a prominent 8 mag. star in the middle and two following: these are decidedly between us and the cluster."

From a dark site, sharp-eyed observers will spot M11 with their unaided eyes by following a curved line of stars of decreasing brightness. Start with magnitude 3.4 Lambda (λ) Aquilae; move to 4th-magnitude 12 Aql; finally, proceed to magnitude 4.8 Eta (η) Scuti, which will lead you to M11.

An ideal telescope size and magnification for this object are 8 inches and 75×. More aperture and higher powers will help you resolve additional stars, but through this combo you'll see more than a hundred stars. M11's core, which resembles a poor globular cluster, is tightly packed, and streamers of stars and dark lanes emanate from the central region in all directions.

| OBJECT #560 | NGC 6709 |
|---|---|
| Constellation | Aquila |
| Right ascension | 18h52m |
| Declination | 10°21′ |
| Magnitude | 6.7 |
| Size | 13′ |
| Type | Open cluster |

This rich cluster sits 4.9° southwest of magnitude 3.0 Zeta (ζ) Aquilae. It lies some 4,000 light-years away from Earth. You won't spot it with your naked eyes, but small binoculars resolve the cluster well against a background crowded with stars. A 6-inch telescope shows approximately 50 stars of mostly magnitudes 9, 10, and 11. A 12-inch scope more than doubles the number of stars you'll see and provides an illusory layered effect where the fainter stars appear to lie in the background.

| OBJECT #561 | NGC 6712 |
|---|---|
| Constellation | Scutum |
| Right ascension | 18h53m |
| Declination | –8°42′ |
| Magnitude | 8.2 |
| Size | 7.2′ |
| Type | Globular cluster |

NGC 6712 is Scutum's only globular cluster. It's bright enough that you can see it through binoculars at a dark site. To get a good look at the individual stars in this object, use an aperture of 10 inches. You'll find this cluster 2.4° east of magnitude 4.9 Epsilon (ε) Scuti.

| OBJECT #562 | Barnard 114 through Barnard 118 |
|---|---|
| Constellation | Scutum |
| Right ascension | 18h53m |
| Declination | −6°58′ |
| Size | 6′ |
| Type | Dark nebulae |

This complex of dark nebulae lies 1.8° south-southwest of magnitude 4.8 Eta ($\eta$) Scuti. Alternatively, you can find its northern edge 0.5° southeast of the Wild Duck Cluster (Object #559). From there, the darkness drifts southward some two Full Moon widths. Best results come through telescope/eyepiece combinations that yield approximately 75×.

| OBJECT #563 | IC 1296 |
|---|---|
| Constellation | Lyra |
| Right ascension | 18h53m |
| Declination | 33°03′ |
| Magnitude | 14.8 |
| Size | 0.9′ by 0.5′ |
| Type | Barred spiral galaxy |

Before you attempt to observe IC 1296, please note both its extremely faint magnitude and its small size. If you're still interested then you must love challenge objects. You also must have at least a 16-inch telescope under a dark sky.

First, find the Ring Nebula (M57). Then gently nudge your scope 4′ to the northwest. Finally, crank up the magnification past 250×, and look for the tiny, round galaxy. Don't expect to see even a hint of its bar or arms.

| OBJECT #564 | M54 (NGC 6715) |
|---|---|
| Constellation | Sagittarius |
| Right ascension | 18h55m |
| Declination | −30°29′ |
| Magnitude | 7.6 |
| Size | 9.1′ |
| Type | Globular cluster |

Our next treat lies 1.7° west-southwest of magnitude 2.6 Ascella (Zeta [$\zeta$] Sagittarii). Although its total magnitude is bright, M54's stars only resolve through the largest amateur telescopes. That's because it lies 87,000 light-years away, making it the most distant globular cluster in Charles Messier's catalog. Its brightest stars shine at only magnitude 15.5.

Through a 12-inch telescope, M54's center appears broad and bright. A razor-thin halo surrounds it, but you'll need to crank up the magnification past 350× to see it.

| OBJECT #565 | NGC 6716 |
|---|---|
| Constellation | Sagittarius |
| Right ascension | 18h55m |
| Declination | −19°54′ |
| Magnitude | 7.5 |
| Size | 10′ |
| Type | Open cluster |

This object sits 1° northwest of magnitude 5.0 Xi$^1$ ($\xi^1$) Sagittarii, and it's one I put into the class I call the "great equalizer" clusters. That means NGC 6716 looks as good through a 4-inch telescope as it does through a 12-inch one. Sure, you may see more ultra-faint stars when you use a larger aperture, but the essence of the cluster will be the same.

So, through a 4-inch or larger scope, you'll see two distinct groupings of stars with a gap between them. A curved arc of half a dozen 9th- and 10th-magnitude stars sits to the north, and a second group whose shape reminds me of the Coathanger (Object #592) lies to the southwest. Adding a splash of color is the orange magnitude 7.0 star SAO 161947, which lies 12′ to the west-northwest of NGC 6716's center.

| OBJECT #566 | NGC 6717 |
|---|---|
| Constellation | Sagittarius |
| Right ascension | 18h55m |
| Declination | −22°42′ |
| Magnitude | 8.4 |
| Size | 5.4′ |
| Type | Globular cluster |

Our next object lies 1.7° south-southwest of magnitude 3.5 Xi$^2$ ($\xi^2$) Sagittarii. When you arrive at that location, however, you'll encounter a problem: Magnitude 5.0 Nu$^2$ ($\nu^2$) Sagittarii lies less than 2′ north of our target. Through small telescopes, the globular appears as a hazy patch near the star. Be sure to crank the power up to at least 200× to put some separation between the two objects.

At that magnification, or through a larger scope like the 12-inch with which I last observed NGC 6717, you'll first spot several foreground stars, the brightest of which shines at magnitude 11.7. Those stars seem superimposed on a nebulous background glow, and that's what you're trying to see. I think of Bruce Lee's instruction to his student in the 1973 movie *Enter the Dragon*: "It's like a finger pointing the way to the Moon. Don't concentrate on the finger, or you'll miss all that heavenly glory."

NGC 6717 carries another astronomical designation — Palomar 9. In fact, it's the brightest (by more than a magnitude) of any from that relatively obscure list of 15 deep-sky targets. For more difficult quarry, try Palomar 12 (Object #688).

| OBJECT #567 | M57 (NGC 6720) |
|---|---|
| Constellation | Lyra |
| Right ascension | 18h54m |
| Declination | 33°02′ |
| Magnitude | 9.7 |
| Size | 71″ |
| Type | Planetary nebula |
| Other name | The Ring Nebula |

Our next object, the Ring Nebula — also known as M57 — is a great target for small telescopes.

Through a small telescope (one with a mirror or lens 4 inches [100 mm] across or smaller), you'll see the Ring Nebula as a pale grey ball. If you use a magnification greater than 100×, you'll notice the outer part of the ball looks thicker than the central region. This gives M57 its distinctive "ring" appearance.

The Ring Nebula lies in the direction of the constellation Lyra the Harp, which we see best during the Northern Hemisphere's summer and fall. The main part of Lyra is one brilliant star — Vega — and a crooked box of four fainter stars nearby. On a star chart (and then in the sky), locate Beta ($\beta$) Lyrae and Gamma ($\gamma$) Lyrae. These two stars make the end of the box that lies farthest from Vega. Roughly midway between them, you'll find the Ring Nebula.

Spotting M57's central star ranks as a difficult observing challenge that will test you, your telescope, and the quality of your observing site. Through a 16-inch or larger telescope on a night of excellent seeing, use an eyepiece that yields between 300× and 400×. Keep in mind you're searching for a 15th-magnitude star against a background that's not completely dark. if the central star doesn't show itself immediately, slightly tap on the telescope's tube. At such high magnification, tapping with one finger should do. Because the human eye is sensitive to motion, you may see the white dwarf at this point. German astronomer Friedrich von Hahn discovered the star in 1800.

| OBJECT #568 | Theta ($\theta$) Serpentis |
|---|---|
| Constellation | Serpens |
| Right ascension | 18h56m |
| Declination | 4°12′ |
| Magnitudes | 4.5/5.4 |
| Separation | 22.3″ |
| Type | Double star |
| Other name | Alya |

You'll find Theta Ser 7.4° west of magnitude 3.4 Delta ($\delta$) Aquilae. The brighter component is blue, and its companion is a pale yellow.

According to Allen in *Star-Names and Their Meanings*, the common name Alya is from the same source as Unukalhai (Alpha [$\alpha$] Serpentis), which also carried the names Alioth, Alyah, and Alyat. Allen says that each of these is a term "for the broad and fat tail of the Eastern sheep that may have been at some early day figured here in the Orientals' sky." He goes on to say, however, that they most likely derive from "Al Hayyah," which means "the snake."

| OBJECT #569 | NGC 6723 |
|---|---|
| Constellation | Sagittarius |
| Right ascension | 19h00m |
| Declination | –36°38′ |
| Magnitude | 7.9 |
| Size | 11.0′ |
| Type | Globular cluster |
| Other name | The Chandelier Cluster |

Our next target sits 0.5° north-northeast of magnitude 4.8 Epsilon ($\varepsilon$) Coronae Australis, within a region rich in reflection nebulosity. In fact, Objects #571 and #572 sit 0.5° and 0.6° east-southeast of this globular.

Through an 8-inch telescope at 200×, you'll see a large, concentrated core that exhibits mottling (alternating bright and dark regions). Two dark gaps lie at the cluster's southwestern edge.

Through a 14-inch scope, crank the magnification up to 300×, and you'll see NGC 6723's irregular shape plus many more individual stars. Do you see the draping arcs of stars for which this globular got its popular name the Chandelier Cluster? A 16-inch scope will reveal more than 100 faint points of light. The brightest star in view glows at magnitude 10.4 and lies 3′ from the cluster's center.

| OBJECT #570 | IC 1295 |
|---|---|
| Constellation | Scutum |
| Right ascension | 18h55m |
| Declination | –8°50′ |
| Magnitude | ~12 |
| Size | 86″ |
| Type | Planetary nebula |

Our next planetary nebula is brighter than its listed magnitude would lead you to believe. The reason is IC 1295's relatively high surface brightness. At only 1.5′ across, it doesn't spread out its light far. Telescopes as small as 6′ in diameter will show its disk shape. With a larger scope, the nebula's irregular nature becomes more apparent.

You'll find this object 4.8° east of magnitude 3.9 Alpha ($\alpha$) Scuti and only 0.4° east-southeast of NGC 6712 (Object #561).

| OBJECT #571 | NGC 6726 |
|---|---|
| Constellation | Corona Australis |
| Right ascension | 19h02m |
| Declination | −36°53′ |
| Size | 9′ by 7′ |
| Type | Reflection nebula |

Our next target lies in the small constellation Corona Australis the Southern Crown. You'll find it 1° west of magnitude 4.2 Gamma ($\gamma$) Coronae Australis.

NGC 6726 shines by reflecting the light of TY Coronae Australis, a variable star that ranges between magnitude 8.8 and magnitude 12.6. Because reflected starlight is made up of all wavelengths of light, don't use any filter when observing this nebula.

In fact, this entire region abounds with bright and dark nebulosity. While NGC 6726 is the brightest area, don't ignore our next object, NGC 6729.

Finally, a much brighter treat awaits you only 0.5° to the west-northwest of NGC 6729. There, you'll find the bright globular cluster NGC 6723 (Object #569), which sits over the border into Sagittarius. This object shines at magnitude 7.9 and measures 11′ across.

| OBJECT #572 | NGC 6729 |
|---|---|
| Constellation | Corona Australis |
| Right ascension | 19h02m |
| Declination | −36°57′ |
| Size | 1′ by 1′ |
| Type | Emission nebula |
| Other name | Caldwell 68 |

NGC 6729 sits only 5′ to the southeast of our previous target, NGC 6726. This nebula glows more faintly than NGC 6726, but it's at least as interesting (if not more so) because it contains another variable star — R Coronae Australis. This star's brightness varies from magnitude 9.7 to about magnitude 12.

Many observers have noted NGC 6729's distinct, cometary shape. See if you don't agree. Through a 12-inch telescope at 200×, you'll see the reflection nebula extending away from R Coronae Australis. NGC 6729 appears five times as long as it is wide.

| OBJECT #573 | NGC 6738 |
|---|---|
| Constellation | Aquila |
| Right ascension | 19h01m |
| Declination | 11°36′ |
| Magnitude | 8.3 |
| Size | 15′ |
| Type | Open cluster |

You'll find our next target 2.5° south-southwest of magnitude 3.0 Zeta ($\zeta$) Aquilae. A 4-inch telescope will pick up perhaps two dozen stars in this area, but many fainter points await revealing through larger apertures. The two brightest stars, magnitude 9.0 SAO 104365 and magnitude 9.2 SAO 104371, lie at the cluster's center and align east-west.

Now here's the kicker: NGC 6738 is not an actual open cluster. In 2003, a team of five researchers concluded that "NGC 6738 is definitely not a physical stellar ensemble: photometry does not show a defined mean sequence, proper motions and radial velocities are randomly distributed, spectrophotometric parallaxes range between 10 and 1600 pc, and the apparent luminosity function is identical to that of the surrounding field. NGC 6738 therefore appears to be an apparent concentration of a few bright stars projected on patchy background absorption." — The Smithsonian/NASA Astrophysics Data System.

I hope that such a conclusion, rather than causing you to cross this object off the list, actually makes you want to go out and observe it. Go see the "cluster" that fooled astronomers for so long. How does it appear to you? Even knowing the end result, it's hard for most observers to classify NGC 6738 as anything but another of the sky's intriguing open clusters.

| OBJECT #574 | NGC 6741 |
|---|---|
| Constellation | Aquila |
| Right ascension | 19h03m |
| Declination | −0°27′ |
| Magnitude | 11.4 |
| Size | 6″ |
| Type | Planetary nebula |
| Other name | The Phantom Streak |

Here's a tough catch for those with telescopes with apertures less than 11 inches across. You'll find it 4.5° north-northwest of magnitude 3.4 Lambda ($\lambda$) Aquilae. It shows up at low magnifications as a dim "star" that jumps into view when a nebula filter is placed in front of the eyepiece. Actually, it's all the other stars dimming because of the filter, but the effect is the same.

For the Phantom Streak, however, high magnification is the way to go. Through a 12-inch telescope at 300×, the tiny disk becomes apparent when you compare it to the half dozen or so other similarly bright stars in the field of view. And here's another thing you can look for through this size scope: color. I can pick up an ever-so-faint robin's-egg blue, although it's easier to spot when I drop the magnification to 200×. Through larger telescopes, like the 30-inch reflector I used in June 2009 at Rancho Hidalgo in Animas, New Mexico, NGC 6741's color isn't in doubt.

It's easy to see why the Phantom Streak Nebula has its common name. It appears faint, and, through a small telescope, the only way you might see it is to use averted vision, thereby only occasionally viewing the "streak." Unfortunately, I have been unable to find out who bestowed this object's moniker on it.

| OBJECT #575 | NGC 6742 |
|---|---|
| Constellation | Draco |
| Right ascension | 18h59m |
| Declination | 48°28′ |
| Magnitude | 13.4 |
| Size | 30″ |
| Type | Planetary nebula |
| Other name | Abell 50 |

Our next target is the neglected planetary nebula NGC 6742, which lies near Draco's southern borders with Cygnus and Lyra. Although its magnitude implies that it's quite faint, it has a surface brightness high enough for you to easily view it through a 10-inch telescope. It appears as a sharp-edged, round disk. Increase the magnification, and try to spot the ring structure in this object.

This object doesn't lie close to any bright star. To find NGC 6742, first find the magnitude 5.0 star 16 Lyrae. Then look 1.5° north-northwest.

| OBJECT #576 | IC 4808 |
|---|---|
| Constellation | Corona Australis |
| Right ascension | 19h01m |
| Declination | −45°19′ |
| Magnitude | 12.9 |
| Size | 1.9′ by 0.8′ |
| Type | Spiral galaxy |

You'll find this galaxy right on Corona Australis' southern border with Telescopium. The easiest way to find it is to look 3.9° west-southwest of magnitude 4.9 Beta[1] ($\beta^1$) Sagittarii. IC 4808 represents one of the faintest targets in the 1,001 Objects, so this one's best suited for large-scope users. Still, catching it through a 6-inch telescope under a dark, steady sky makes a noteworthy observation.

Through a 14-inch scope at 200×, you'll see an oval glow twice as long as it is wide, oriented northeast to southwest, with a slightly brighter, evenly illuminated central area. Crank the power up to 350×, and the outer region's even brightness distribution begins to break down, especially at the southwestern end. That's the telltale sign of spiral structure, although you won't see this galaxy's arms through amateur instruments.

| OBJECT #577 | NGC 6744 |
|---|---|
| Constellation | Pavo |
| Right ascension | 19h10m |
| Declination | −63°51′ |
| Magnitude | 8.6 |
| Size | 15.5′ by 10.0′ |
| Type | Spiral galaxy |
| Other name | Caldwell 101 |

For most readers, seeing this wonder will require a trip south of the border. The constellation Pavo the Peacock sits so far south it butts up against Octans, the star figure that surrounds the South Celestial Pole. From Miami, NGC 6744 climbs less than 0.5° above the horizon.

Scottish-born Australian astronomer James Dunlop discovered NGC 6744 between 1823 and 1827, and made it number 262 in *A Catalogue of Nebulae and Clusters of Stars in the Southern Hemisphere observed in New South Wales* (*Philosophical Transactions of the Royal Society*, volume 118, 1828). Dunlop's catalog contained 629 deep-sky objects.

Although NGC 6744 exceeds 9th magnitude, its surface brightness is low, as in the case of M101. In fact, to see NGC 6744's spiral arms, you'll need at least a 10-inch telescope. Through it, you'll see the galaxy's oval shape; the halo is about 50-percent longer than wide. The outer reaches (the spiral arms) appear clumpy due to large, massive star clusters. Find a 20-inch scope to peer through, perhaps at a star party, and this galaxy's structure will amaze you.

To find NGC 6744, look 2.6° southeast from magnitude 4.2 Lambda ($\lambda$) Pavonis.

| OBJECT #578 | NGC 6745 |
|---|---|
| Constellation | Lyra |
| Right ascension | 19h02m |
| Declination | 40°45′ |
| Magnitude | 12.3 |
| Size | 1.4′ by 0.7′ |
| Type | Irregular galaxy |

Our next target isn't hard to spot through medium-size telescopes because of its high surface brightness, but, admittedly, you won't see a lot of detail. The disk is uniformly bright and no halo or spiral arms are visible. Can you spot the magnitude 14 star that sits in front of the galaxy's northern tip? NGC 6745 lies 2.8° northwest of magnitude 4.4 Eta ($\eta$) Lyrae.

Owners of large telescopes would do well to spend some time observing this galaxy at high power. The galaxy orients north-south, but its western edge is straight. Compare it to the eastern edge, which puffs out like a normal galaxy. This odd shape is the result of a collision between NGC 6745 — a former spiral — and another, smaller galaxy. Deep exposures through large telescopes show regions of star formation triggered by this collision.

| OBJECT #579 | Bernes 157 |
|---|---|
| Constellation | Corona Australis |
| Right ascension | 19h03m |
| Declination | –37°08′ |
| Size | 55′ by 18′ |
| Type | Dark nebula |

Look midway between magnitude 4.8 Epsilon ($\varepsilon$) Coronae Australis and magnitude 4.2 Gamma ($\gamma$) Coronae Australis to find the dark nebula Bernes 157. This ultra-dark spot sits at the southeastern edge of the region inhabited by NGC 6726 (Object #571) and NGC 6729 (Object #572). Also, less than 1° to the northwest, you'll find NGC 6723 (Object #569). Magnitude 4.2 Gamma ($\gamma$) Coronae Australis lies at the nebula's northern edge.

Visual absorption of starlight in this region is as high as 8 magnitudes, so you'll need a large aperture to capture the few 13th-magnitude stars strewn across Bernes 157. I find it fascinating that dark nebulae blot out starlight because, in say 10 million years, this star-forming region will be alive with new stars.

The best views of Bernes 157 are through either large-aperture binoculars that magnify at least 15 times, or through telescope/eyepiece combinations that yield magnifications between 30× and 50×. Remember, this dark patch isn't small. It covers an area almost as large as two Full Moons.

This object's catalog name is one you may not have heard of. In 1977, Claes Bernes of the Stockholm Observatory compiled a new catalogue of bright nebulae in dense dust clouds. He found 160 such objects in 80 different dark clouds when he searched on Palomar Sky Survey plates. That survey allowed him to find nebulae to a declination of –46°. For more southerly objects, Bernes turned to the just-completed European Southern Observatory "blue" survey. Most of these objects are reflection nebulae, and all are star-forming regions Bernes suggested astronomers target for future study by radio and infrared observations.

| OBJECT #580 | NGC 6749 |
|---|---|
| Constellation | Aquila |
| Right ascension | 19h05m |
| Declination | 1°54′ |
| Magnitude | 12.4 |
| Size | 6.3′ |
| Type | Globular cluster |

If you're a beginning observer, I hope you didn't flip through this book at random, looking to observe your first object, and land on this page. If so, please accept my apology, and pick something else. Save this challenge for a time when you're more seasoned.

You'll find this large-telescope target 5.2° west-southwest of magnitude 3.4 Delta ($\delta$) Aquilae. Don't be discouraged if you don't spot it immediately because this is one of the least-concentrated globular clusters known. It lies at a distance of 24,000 light-years, so material in the Milky Way's plane dims its light by more than 4 magnitudes. Its brightest stars glow at magnitude 16.

Through a 12-inch telescope, you'll detect NGC 6749 as a soft haze. Step up to a 20-inch scope, and NGC 6749 still appears faint. You will, however, be able to distinguish between the slightly brighter central region and the oh-so-faint halo. Observers rate this globular as one of the two most difficult to spot in Dreyer's *New General Catalogue*. The other is NGC 6380 (Object #489).

| OBJECT #581 | NGC 6751 |
|---|---|
| Constellation | Aquila |
| Right ascension | 19h06m |
| Declination | –6°00′ |

| (continued) | |
|---|---|
| Magnitude | 11.9 |
| Size | 20″ |
| Type | Planetary nebula |

Our next object sits 1.1° south of magnitude 3.4 Lambda (λ) Aquilae. Through a 14-inch telescope, you'll see magnitude 12.9 GSC 5140:3169 shining less than 1′ to the east and a magnitude 13.2 star half that distance to the west. At 250×, the tiny planetary looks slightly uneven in its overall illumination.

If you can't detect the magnitude 14.5 central star at this magnification, increase it past 350×, and you'll have no problem. For a different perspective, back the power down past 150×, and examine this object against the rich Milky Way background. At what magnification can you no longer tell NGC 6751 is non-stellar?

| OBJECT #582 | NGC 6752 |
|---|---|
| Constellation | Pavo |
| Right ascension | 19h11m |
| Declination | −59°59′ |
| Magnitude | 5.5 |
| Size | 20.4′ |
| Type | Globular cluster |
| Other names | The Pavo Globular, the Starfish, the Windmill, Caldwell 93 |

You'll spot one of the sky's brightest globular clusters — tied for 4th-brightest with NGC 6397 in Ara (Object #493) — easily with your naked eyes from a dark site. Look for it 3.2° northeast of magnitude 4.2 Lambda (λ) Pavonis. This cluster appears big and bright for the reason you'd think — it's close. NGC 6752 lies only 13,000 light-years from Earth.

Through any optics, this is a spectacular sight. A 6-inch telescope reveals hundreds of stars orbiting a broad, concentrated core. The brightest star sits in the foreground. That's magnitude 7.4 SAO 254482, and it sits just 4′ south-southwest of the cluster's center. Several — no, many — starry chains radiate in various directions from the center, giving NGC 6752 two of its popular names.

One of those names, the Starfish, appears two other times in this book. Amateur astronomers refer to another globular, NGC 6544 (Object #521), and open cluster M38 (Object #948) both as the Starfish Cluster.

| OBJECT #583 | NGC 6755 |
|---|---|
| Constellation | Aquila |
| Right ascension | 19h08m |
| Declination | 4°14′ |
| Magnitude | 7.5 |
| Size | 14′ |
| Type | Open cluster |

Our next object lies 2.9° east of magnitude 4.5 Theta (θ) Serpentis. Through a 6-inch telescope at 125×, you'll have no trouble picking the cluster out from the Milky Way background, although it is loose. Note the two distinct groups of 11th- and 12th-magnitude stars split by a dark lane that stretch from east to west. The southern half of the cluster slightly outshines the northern section. You'll also find the cluster's two brightest stars, which shine at magnitude 10.4, south of center. Through 12-inch and larger telescopes, look around for many yellow and orange stars that populate this region and offer a nice color contrast.

| OBJECT #584 | NGC 6756 |
|---|---|
| Constellation | Aquila |
| Right ascension | 19h09m |
| Declination | 4°41′ |
| Magnitude | 10.6 |
| Size | 4′ |
| Type | Open cluster |

You'll find this target 0.5° north-northeast of our previous item, NGC 6755 (Object #583). Through a 6-inch telescope at 125×, look for a grouping of about 10 stars that concentrate over an ephemeral background glow. A 14-inch scope resolves the background glow, but not into multitudes of stars, only about two dozen of them.

| OBJECT #585 | NGC 6760 |
|---|---|
| Constellation | Aquila |
| Right ascension | 19h11m |
| Declination | 1°02′ |
| Magnitude | 9.1 |
| Size | 6.6′ |
| Type | Globular cluster |

Look for this object 1.8° west of the magnitude 5.1 star 23 Aquilae. A 10-inch telescope at 200× reveals a broadly concentrated disk with even illumination but little stellar resolution. A 14-inch scope does a bit better, resolving a handful of faint points and also bringing out the halo's mottled structure.

| OBJECT #586 | NGC 6765 |
|---|---|
| Constellation | Lyra |
| Right ascension | 19h11m |
| Declination | 30°33′ |
| Magnitude | 12.9 |
| Size | 38″ |
| Type | Planetary nebula |

This faint, small planetary nebula lies 0.8° south of the magnitude 5.9 star 19 Lyrae. Through large telescopes at magnifications above 250×, it appears elongated in a northeast-southwest direction. An OIII filter helps a lot, but don't expect to see much structure. Without a filter, try to spot the 14th-magnitude star that sits off NGC 6765's northeast side. Two 12th-magnitude stars sit 3′ and 4′ south, respectively. Visually, the nebula's relatively high surface brightness swamps the 16th-magnitude central star.

| OBJECT #587 | NGC 6772 |
|---|---|
| Constellation | Aquila |
| Right ascension | 19h15m |
| Declination | −2°42′ |
| Magnitude | 12.7 |
| Size | 62″ |
| Type | Planetary nebula |

Scan 3° northeast of magnitude 3.4 Lambda ($\lambda$) Aquilae to locate the large planetary nebula NGC 6772. It's a tough catch through a 6-inch telescope, but a 12-inch instrument at 250× shows it as circular with a hazy edge.

Larger scopes and higher magnifications show a slight north-south elongation. Several faint stars are superimposed on this object's face, but they add to the view rather than detracting from it. A nebula filter is a must for this object.

**Object #588** M56 Anthony Ayiomamitis

| OBJECT #588 | M56 (NGC 6779) |
|---|---|
| Constellation | Lyra |
| Right ascension | 19h17m |
| Declination | 30°11′ |
| Magnitude | 8.3 |
| Size | 7.1′ |
| Type | Globular cluster |

Lying 45 percent of the way from Albireo ($\beta$) Cygni to Gamma ($\gamma$) Lyrae, the 56th entry in Charles Messier's catalog is easy to find. From a dark site, you'll spot it through binoculars.

The density of stars in M56 increases dramatically as you move toward its core. And, because the individual cluster stars aren't all that bright, you'll resolve them best through 8-inch or larger telescopes and at magnifications exceeding 150×. When you're done examining the inner workings of M56, back off the power, and enjoy the star field in which this cluster sits.

| OBJECT #589 | IC 1297 |
|---|---|
| Constellation | Corona Australis |
| Right ascension | 19h17m |
| Declination | −39°37′ |

| (continued) | |
|---|---|
| Magnitude | 10.7 |
| Size | 7″ |
| Type | Planetary nebula |

Our next object lies 1.4° east of magnitude 4.1 Beta (β) Coronae Australis. Through an 8-inch telescope at 200×, this planetary still is tiny. At 300×, it appears as a square with rounded edges. A nebula filter helps a lot, but be sure to view it without the filter so you can see the blue, bluish-green, or greenish-blue color. Exactly which hue you see depends on your eyes' color receptors. Through a 20-inch telescope at magnifications above 400×, look for a slight north-south elongation.

| OBJECT #590 | NGC 6781 |
|---|---|
| Constellation | Aquila |
| Right ascension | 19h18m |
| Declination | 6°33′ |
| Magnitude | 11.4 |
| Size | 109″ |
| Type | Planetary nebula |

Although Aquila ranks 22nd in size among the 88 constellations, it contains no Messier objects or emission nebulae and few bright star clusters. Still, you'll want to point your telescope 3.8° north-northwest of magnitude 3.4 Delta (δ) Aquila to observe NGC 6781.

NGC 6781 is an almost perfect bubble of gas cast off by a single, formerly Sun-like, star that has died. The bubble continues to expand; it measures some 2 light-years across. Energetic photons from nearby bright stars decompose the gas bubble. A similar process occurs within the Eagle Nebula (M16).

Through a 6-inch telescope at 100×, NGC 6781 stands out well against a rich, star-filled background. The disk appears soft, irregular, and oval-shaped with a slightly darker center. If the seeing at your observing site is good, look for small, dark blotches over NGC 6781's face.

If you're able to observe this planetary through a 16-inch scope, you'll see lots of structure in NGC 6781's thick ring. This feature's southern rim appears brightest. The northern edge is broken and gradually fades into the background. To see the ring best, use a nebula or an OIII filter. Through this size or larger telescope, you might detect the central star, a bluish white dwarf, which glows weakly at magnitude 16.2.

| OBJECT #591 | NGC 6791 |
|---|---|
| Constellation | Lyra |
| Right ascension | 19h21m |
| Declination | 37°51′ |
| Magnitude | 9.5 |
| Size | 15′ |
| Type | Open cluster |

NGC 6791 is a pretty cluster that lies less than 1° east-southeast of magnitude 4.4 Theta (θ) Lyrae. Its size — nearly half the diameter of the Full Moon — means that, even at magnitude 9.5, NGC 6791 appears faint through small telescopes. In fact, you may be fooled into thinking it's a globular cluster.

Through 16-inch and larger instruments, NGC 6791 begins to strut its stuff. Dozens of faint cluster stars begin to resolve into a fine, evenly distributed pile of diamond-dust.

| OBJECT #592 | Collinder 399 |
|---|---|
| Constellation | Vulpecula |
| Right ascension | 19h25m |
| Declination | 20°11′ |
| Magnitude | 3.6 |
| Size | 60′ |
| Type | Asterism |
| Other names | The Coathanger, Brocchi's Cluster, Al Sufi's Cluster |

To find out next object, use binoculars or a finder scope and extend a line southward from magnitude 3.0 Albireo (Beta [β] Cygni) through magnitude 4.4 Alpha (α) Vulpeculae. That distance is roughly 3°. Head 4.5° farther south, and you'll encounter Collinder 399.

This group was the 399th entry in a catalog of open clusters compiled by Swedish astronomer Per Arne Collinder. His catalog contains 471 objects. It's most common name, the Coathanger, comes from its shape.

Its other common names derive from astronomers. Persian astronomer Al Sufi (A.D. 903–986) was discovered this object and included it in his *Book of Fixed Stars*, which he published in A.D. 964.

American Dalmiro Francis Brocchi (1871–1955) was an amateur astronomer and chartmaker for the American Association of Variable Star Observers (AAVSO). In the 1920 s, he created a map of this region so that astronomers could calibrate their photometers.

Because it's so big, the Coathanger looks best at magnifications of 20× or less. Ten stars glow brighter than 7th magnitude, so the group appears distinct to the naked eye on dark nights. The brightest are 4 Vulpeculae, at magnitude 5.1, 5 Vulpeculae, at magnitude 5.6, and 7 Vulpeculae, which shines at magnitude 6.3.

| OBJECT #593 | NGC 6800 |
|---|---|
| Constellation | Vulpecula |
| Right ascension | 19h27m |
| Declination | 25°08′ |
| Magnitude | 9.0 |
| Size | 5′ |
| Type | Open cluster |

Our next object lies only 35′ northwest of magnitude 4.4 Alpha (α) Vulpeculae. This small cluster looks best through 8-inch or larger telescopes. Expect to see 50 stars scattered haphazardly across the field of view. Two stars, magnitude 7.1 SAO 87256, and magnitude 8.0 SAO 87200 flank the cluster on its east and west sides, respectively.

| OBJECT #594 | NGC 6802 |
|---|---|
| Constellation | Vulpecula |
| Right ascension | 19h31m |
| Declination | 20°16′ |
| Magnitude | 8.8 |
| Size | 3.2′ |
| Type | Open cluster |

You'll find NGC 6802 at the east end of the Coathanger (Object #592). Through a small telescope, or at low power, it appears as a bright but unresolved haze. Large scopes don't reveal much more. Through a 12-inch scope at 200×, you'll barely count two dozen stars.

| OBJECT #595 | Beta ($\beta$) Cygni |
|---|---|
| Constellation | Cygnus |
| Right ascension | 19h31m |
| Declination | 27°58′ |
| Magnitudes | 3.1/5.1 |
| Separation | 34″ |
| Type | Double star |
| Other name | Albireo |

When you're observing faint objects in the fall, take a break. Point a small telescope at Albireo (Beta [$\beta$] Cygni). Most northern observers consider this the sky's finest double star.

Although the first two letters of Albireo are "al," according to Richard Hinckley Allen, this star's name is not Arabian. He says the now universal title, "apparently was first applied to the star from a misunderstanding as to the words *ab ireo* in the description of the constellation in the 1515 *Almagest.*" The Arabians called Albireo *Al Minhar al Dajajah*, the Hen's Beak.

Astronomers label Albireo's two components $\beta^1$ Cygni and $\beta^2$ Cygni. $\beta^1$ shines at magnitude 3.4, while $\beta^2$ is fainter, coming in at magnitude 5.1. But it's not their magnitudes that make these stars a terrific sight: It's their colors.

Before I describe the colors of these two stars, let me reiterate that no two human eyes see precisely the same hues. That said, most observers "see" $\beta^1$ as golden and $\beta^2$ as sapphire blue. Whether you see gold and blue, blue and white, yellow and green, or any number of other combinations, it all adds up to a strikingly different pair. Don't miss it.

| OBJECT #596 | NGC 6804 |
|---|---|
| Constellation | Aquila |
| Right ascension | 19h32m |
| Declination | 9°13′ |
| Magnitude | 12.0 |
| Size | 31″ |
| Type | Planetary nebula |

Our next target lies almost 2° north-northwest of magnitude 4.5 Mu ($\mu$) Aquilae. Although its magnitude suggests that it's faint, a 6-inch telescope easily reveals its disk. At 200×, the planetary appears diffuse. A nebula filter improves the view, especially through 12-inch or larger telescopes. You'll see the incomplete ring structure around the 14th-magnitude central star. It's brightest north and south of the star. Two other stars appear at the planetary's western and northeastern edges.

| OBJECT #597 | M55 (NGC 6809) |
|---|---|
| Constellation | Sagittarius |
| Right ascension | 19h40m |
| Declination | −30°58′ |
| Magnitude | 6.3 |
| Size | 19.0′ |
| Type | Globular cluster |

Our next treat comes from Charles Messier's catalog. M55 is a superb globular you may just spot with your naked eye from an ultra-dark site. Observers describe this cluster as "highly resolved." That means its core doesn't appear packed with stars, which often gives a "burned out" effect.

A 12-inch telescope at 300× will break it into hundreds of 11th and 12th-magnitude stars. Through an eyepiece with a small field of view, M55 may appear more like a rich, open cluster.

You'll find M55 8° east of magnitude 2.6 Ascella (Zeta [ζ] Sagittarii).

| OBJECT #598 | Barnard 143 |
|---|---|
| Constellation | Aquila |
| Right ascension | 19h41m |
| Declination | 11°01′ |
| Size | 40′ |
| Type | Dark nebula |
| Other name | Barnard's E |

Our next treat is one of my all-time favorite binocular objects. Barnard's E, a combo of two dark nebulae from Edward Emerson Barnard's famous catalog, lies against the rich Milky Way in Aquila. Start at yellow, magnitude 2.7 Tarazed (Gamma [γ] Aquilae). If you center Tarazed, you shouldn't have to move your binoculars at all. Barnard's E lies 1.4° west-northwest of the star.

Barnard 143 (often designated B143) is easiest of the pair to spot. It's a narrow bar about 15′ long, oriented east-west. Two, slightly less distinct dark bars connect to it and form a U-shape. Just to the south lies Barnard 142 (B142), another dark nebula not quite as long and only one-third as wide, making it more difficult to see. Behind all these dark clouds, you'll see thousands of unresolved points of light.

| OBJECT #599 | NGC 6810 |
|---|---|
| Constellation | Pavo |
| Right ascension | 19h44m |
| Declination | −58°40′ |
| Magnitude | 11.4 |
| Size | 3.8′ by 1.2′ |
| Type | Spiral galaxy |

To find our next treat, first locate magnitude 1.9 Peacock (Alpha [α] Pavonis). Then move not quite 6° west-southwest. NGC 6810 is an edge-on spiral that doesn't show much detail. Through a 12-inch telescope, all you'll see is a bright streak twice as long as it is wide, elongated north-south. Its bright central region appears evenly illuminated. At magnifications above 250×, you may spot the thin halo around the core.

| OBJECT #600 | NGC 6811 |
|---|---|
| Constellation | Cygnus |
| Right ascension | 19h37m |
| Declination | 46°23′ |
| Magnitude | 6.8 |
| Size | 15′ |
| Type | Open cluster |

I've tried to see this object with my naked eye, but the star field it's in is too rich. It's easy through binoculars, however, it's best at magnifications above 100×. A 4-inch telescope will reveal 50 stars, and through an 8-inch scope, you'll see twice that many distributed unevenly across NGC 6811's face.

To find this cluster, move 1.8° northwest from magnitude 2.9 Delta (δ) Cygni.

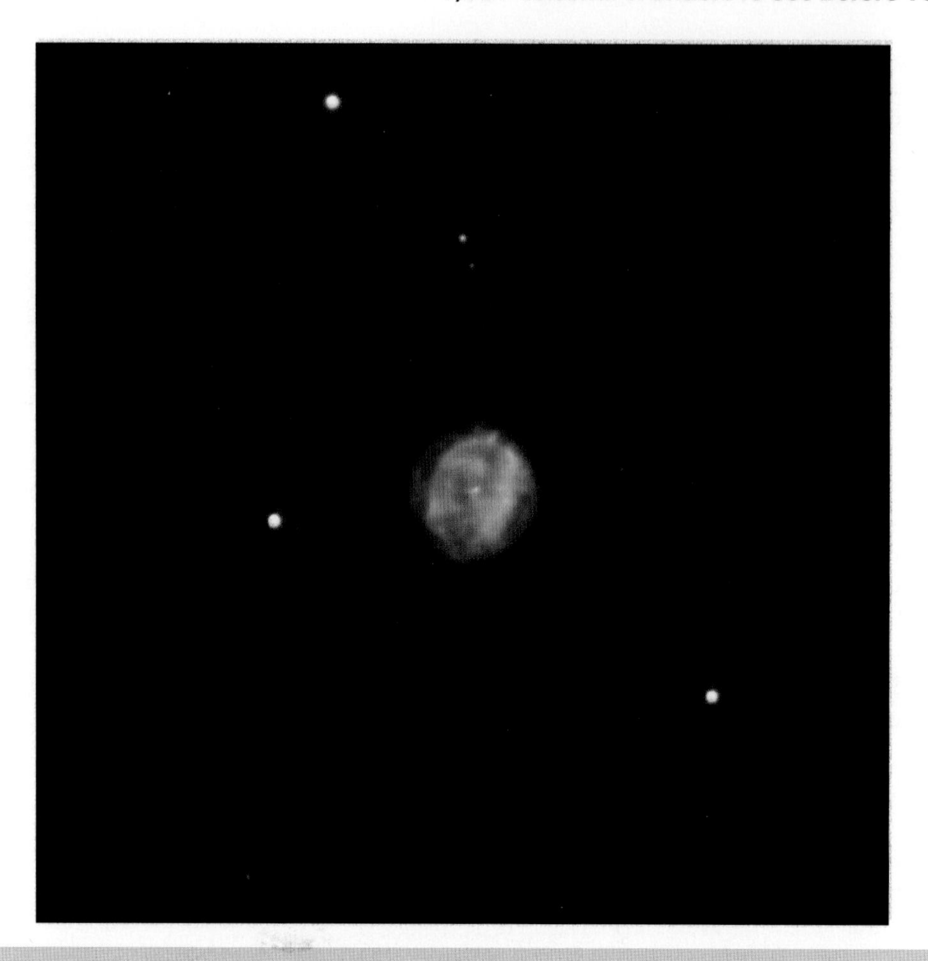

**Object #601** The Little Gem (NGC 6818) Mitch and Michael Dye/Adam Block/NOAO/AURA/NSF

| OBJECT #601 | NGC 6818 |
|---|---|
| Constellation | Sagittarius |
| Right ascension | 19h44m |
| Declination | −14°09′ |
| Magnitude | 9.3 |
| Size | 48″ |
| Type | Planetary nebula |
| Other names | The Little Gem, the Green Mars Nebula |

This treat sits in a no man's land bereft of bright stars near the Archer's northern border with Aquila. Look for this small planetary nebula 9° due west of 3rd-magnitude Beta ($\beta$) Capricorni. Although stellar luminaries in this part of the sky are scarce, this is a great area for deep-sky treats. The Little Gem lies only 0.7° north-northwest of Barnard's Galaxy (NGC 6822). Talk about two objects that are on the opposite ends of the surface brightness spectrum!

As luck would have it, the Little Gem is the bright one. It shines at magnitude 9.3 and measures roughly 0.8′ across from north to south and a bit less from east to west. The combination of brightness and size means NGC 6818's surface brightness is high, and you can really crank up the magnification.

The greenish-blue color most observers see appears best at around 100×. Above that power, look for this object's ever-so-slightly darker inner half.

Observers began to refer to this planetary as the Green Mars Nebula when they noticed its size through a small telescope (22″, less than half its listed diameter) equaled that of Mars when the Red Planet approaches closest to the Sun.

| OBJECT #602 | NGC 6819 |
|---|---|
| Constellation | Cygnus |
| Right ascension | 19h41m |
| Declination | 40°11′ |
| Magnitude | 7.3 |
| Size | 5′ |
| Type | Open cluster |
| Other names | The Fox Head Cluster, the Octopus Cluster |

Our next target lies 5° south of magnitude 3.0 Delta (δ) Cygni. It's a small, but bright open cluster that takes magnification well. Through a 4-inch telescope at 150×, you'll see more than two dozen stars. They're not evenly distributed because the northern half of the cluster appears brighter than the southern half. Double the aperture to 8′, and you'll count 50 stars. Double it again, and more than 100 stars are yours for the picking.

Many observers see a "V" in the cluster's brightest stars (usually at low magnification). Because a fox has a triangular head, amateur astronomers began calling NGC 6819 the Fox Head Cluster. *Astronomy* magazine Contributing Editor Stephen James O'Meara goes a step farther, however. Rather than a V, O'Meara sees a weak Greek letter Chi (χ). At low powers, he maintains the X shape has angular extensions that give the cluster a spiral shape, which he likens to that of an octopus.

| OBJECT #603 | NGC 6820 |
|---|---|
| Constellation | Vulpecula |
| Right ascension | 19h43m |
| Declination | 23°17′ |
| Size | 40′ by 30′ |
| Type | Emission nebula |

Our next target lies 3.5° east-southeast of magnitude 4.4 Alpha (α) Vulpeculae. Screw a nebula filter into your eyepiece to dim the stars of NGC 6820's associated cluster, NGC 6823. Any trace of nebulosity you see through a 6-inch telescope is a great catch. Usually, a 12-inch scope is the minimum required.

**Object #604** Barnard's Galaxy (NGC 6822) Julie and Jessica Garcia/Adam Block/NOAO/AURA/NSF

| OBJECT #604 | NGC 6822 |
|---|---|
| Constellation | Sagittarius |
| Right ascension | 19h45m |
| Declination | −14°48′ |
| Magnitude | 8.8 |
| Size | 19.1′ by 14.9′ |
| Type | Irregular galaxy |
| Other names | Barnard's Galaxy, Caldwell 57 |

If you use an 8-inch or larger telescope, insert the eyepiece that gives you the widest field of view, and look for Barnard's Galaxy, also known as NGC 6822. This object lies in Sagittarius 1.5° north-northeast of 5th-magnitude 55 Sagittarii. In 1881, American astronomer Edward Emerson Barnard discovered this object through a 6-inch refractor.

Barnard's Galaxy shines at magnitude 9.3 — pretty bright for a galaxy. Unfortunately, its light spreads out over an area 16′ by 14′, so its overall surface brightness is low.

Look for a dim haze roughly twice as long as it is wide. Note the slightly brighter streak that spans NGC 6822's long axis.

Larger scopes show several star-forming regions along the galaxy's northern end. To see just these, use a nebula filter. Without a filter, and through a 12-inch or larger scope, look for individual supergiant stars. They're faint — the brightest glow at only 14th magnitude — but they reveal themselves by lending a granular appearance to the galaxy.

| OBJECT #605 | NGC 6823 |
|---|---|
| Constellation | Vulpecula |
| Right ascension | 19h43m |
| Declination | 23°18′ |
| Magnitude | 7.1 |
| Size | 12′ |
| Type | Open cluster |

This object is the open cluster associated with emission nebula NGC 6820. Through a 4-inch telescope, you'll see 20 stars but no trace of the nebula. In fact, without a nebula filter in place, the cluster's brightness overwhelms that of the nebulosity at every aperture. At the center of the cluster lies a tight group of stars you'll have trouble resolving. You'll find this cluster 3.5° east-southeast of magnitude 4.4 Alpha (α) Vulpeculae.

| OBJECT #606 | NGC 6826 |
|---|---|
| Constellation | Cygnus |
| Right ascension | 19h45m |
| Declination | 50°31′ |
| Magnitude | 8.8 |
| Size | 25″ |
| Type | Planetary nebula |
| Other names | The Blinking Planetary, Caldwell 15 |

Our next object is a fun one to show to other amateur astronomers at star parties. Through 8-inch and smaller telescopes, NGC 6826 appears to blink when an observer views it with direct, and then averted, vision.

The first astronomers to notice this effect were James Mullaney and Wallace McCall at Allegheny Observatory in Pittsburgh. They described the effect in the August 1963 issue of *Sky & Telescope*, page 91. They used the observatory's 13-inch refractor and a magnification of "about 200×."

From a dark site, I've used a 6-inch telescope at about 100× to make this object blink. With direct vision, you'll spot the 11th-magnitude central star easily, but the nebula fades. Look a bit to the side (averted vision), and the nebula pops back into view. Furthermore, its apparent brightness under averted vision swamps the stars light. So, by looking back and forth with averted and then direct vision, you can make this object "blink."

You'll find the Blinking Planetary 0.5° east of the magnitude 6.0 star 16 Cygni.

| OBJECT #607 | IC 4889 |
|---|---|
| Constellation | Telescopium |
| Right ascension | 19h45m |
| Declination | –54°20′ |
| Magnitude | 11.3 |
| Size | 2.6′ by 1.8′ |
| Type | Spiral galaxy |

This elliptical galaxy lies 2° north-northwest of magnitude 5.3 Nu (ν) Telescopii. Through an 8-inch telescope you'll see a featureless oval. A 16-inch scope will divide the bright central region and the faint halo that surrounds it. Through that size scope, look for magnitude 14.3 IC 4888, which lies 8′ south.

| OBJECT #608 | NGC 6830 |
|---|---|
| Constellation | Vulpecula |
| Right ascension | 19h51m |
| Declination | 23°04′ |
| Magnitude | 7.9 |
| Size | 12′ |
| Type | Open cluster |

Our next target lies 0.5° north of the magnitude 4.9 star 12 Vulpeculae. Through a 6-inch scope, you'll see two dozen stars, the brightest of which form a distinct X shape. A 12-inch scope boosts the star count by a dozen.

| OBJECT #609 | NGC 6834 |
|---|---|
| Constellation | Cygnus |
| Right ascension | 19h52m |
| Declination | 29°25′ |
| Magnitude | 7.8 |
| Size | 6′ |
| Type | Open cluster |

You'll find this small cluster 2.9° east-southeast of magnitude 4.7 Phi (*Φ*) Cygni. At first glance, it seems a bit unbalanced because the southern half appears brighter than the northern part. At the center of the cluster you'll find its brightest star, magnitude 9.7 HIP 97785.

| OBJECT #610 | Harvard 20 |
|---|---|
| Constellation | Sagitta |
| Right ascension | 19h53m |
| Declination | 18°20′ |
| Magnitude | 7.7 |
| Size | 9′ |
| Type | Open cluster |

You'll need to look carefully to discern the stars of this cluster. They're the 10 brightest of all the ones you can see in your eyepiece's field of view. What's more, they appear to sit in front of a dizzying array of fainter background stars. The brightest star in the cluster, SAO 105381, shines at magnitude 8.9 and lies in the northeastern quadrant. The easiest way to find Harvard 20 is by dropping 0.5° south-southwest of M71 (our next celestial treat).

This object's catalog identifier comes from a list of star clusters compiled by American astronomer Harlow Shapley (1885–1972) in 1930.

**Object #611** M71 Anthony Ayiomamitis

| OBJECT #611 | M71 (NGC 6838) |
|---|---|
| Constellation | Sagitta |
| Right ascension | 19h54m |
| Declination | 18°47′ |
| Magnitude | 8.0 |
| Size | 7.2′ |
| Type | Globular cluster |

Our next target, the loose globular cluster M71, resides midway between magnitude 3.8 Delta ($\delta$) and magnitude 3.5 Gamma ($\gamma$) Sagittae, about 1.5° from each. Through a 4-inch telescope at low power, you'll see the bright core surrounded by a fuzz of barely resolved stars. Crank the magnification up to 200×, and M71's brightest stars will pop into view. Through a 12-inch scope, the star count passes 50.

| OBJECT #612 | 57 Aquila |
|---|---|
| Constellation | Aquila |
| Right ascension | 19h55m |
| Declination | –8°14′ |
| Magnitudes | 5.8/6.5 |
| Separation | 36″ |
| Type | Double star |

You'll find 57 Aql a bit more than 7° northwest of Alpha$^1$ ($\alpha^1$) Capricorni. The primary shines twice as brightly as the secondary, and both stars are white.

**Object #613** The Dumbbell Nebula (M27) Anthony Ayiomamitis

| OBJECT #613 | M27 (NGC 6853) |
|---|---|
| Constellation | Vulpecula |
| Right ascension | 20h00m |
| Declination | 22°43′ |
| Magnitude | 7.3 |
| Size | 348″ |
| Type | Planetary nebula |
| Other names | The Dumbbell Nebula, the Apple Core Nebula, the Diablo Nebula, the Double-Headed Shot |

Our next object is perfect for those of you who own a small telescope. The Dumbbell Nebula, also known as M27, rides high this month.

Good luck making the shape of a fox out of the stars of the constellation Vulpecula, the star figure that contains the Dumbbell Nebula. The faint star pattern's brightest star is magnitude 4.4 Alpha (α) Vulpeculae.

The Dumbbell Nebula owes its common names to a double-lobe shape common among planetary nebulae. Even through binoculars, this object is easy to spot. To see details in it, set up your telescope.

Small telescopes show the two bright lobes and several stars scattered across M27's face. This object responds well to high magnifications because it has a high surface brightness. Through a large telescope, use an OIII filter, and really crank up the magnification.

| OBJECT #614 | NGC 6857 |
|---|---|
| Constellation | Cygnus |
| Right ascension | 20h02m |
| Magnitude | 11.4 |
| Declination | 33°31′ |
| Size | 38.0″ |
| Type | Emission nebula |

This small, round emission nebula looks like a planetary nebula, and, in fact, that's how astronomers classified it until 1969. Because of its small size, NGC 6857 has a relatively high surface brightness. If you observe it through a 14-inch or larger telescope, look for an evenly illuminated central region with a faint edge that rapidly fades.

**Object #615** M75 Ken Siarkiewicz/Adam Block/NOAO/AURA/NSF

| OBJECT #615 | M75 (NGC 6864) |
|---|---|
| Constellation | Sagittarius |
| Right ascension | 20h06m |
| Declination | −21°55′ |
| Magnitude | 8.5 |
| Size | 6′ |
| Type | Globular cluster |

Our next object is globular cluster M75, which lies 60,000 light-years away. Because of its vast distance, you'll have trouble resolving even its brightest stars through a 12-inch telescope. The core appears bright and stellar at low magnification. Although M75 is in Sagittarius, it's outside the Milky Way's band, so the field is bereft of foreground stars. You'll find it 5° north-northeast of magnitude 4.7 Omega (ω) Sagittarii, next to that constellation's border with Capricornus.

| OBJECT #616 | NGC 6866 |
|---|---|
| Constellation | Cygnus |
| Right ascension | 20h04m |
| Declination | 44°09′ |
| Magnitude | 7.6 |
| Size | 7′ |
| Type | Open cluster |
| Other names | The Frigate Bird Cluster |

You'll find our next target 3.4° east-southeast of magnitude 2.9 Delta (δ) Cygni. Even through a 4-inch telescope, this cluster appears rich. A 10-inch scope at 150× shows a "river" of stars oriented east-west running through the cluster's center. Through neither aperture did I see or imagine a frigate bird, but your imagination might be better than mine.

So, where did the common name I list originate? If you guessed it comes from someone who definitely brings a great imagination and a flair for writing about it to the eyepiece, you're right. *Astronomy* magazine Contributing Editor Stephen James O'Meara sees the unmistakable outline of a frigate bird in the stars of NGC 6866.

| OBJECT #617 | NGC 6868 |
|---|---|
| Constellation | Telescopium |
| Right ascension | 20h10m |
| Declination | −48°23′ |
| Magnitude | 10.6 |
| Size | 4.0′ by 3.3′ |
| Type | Spiral galaxy |

With our next target, you get three for the price of one. Elliptical NGC 6868 is the brightest member of the Telescopium group of galaxies, a small cluster containing 10 members. Through a 12-inch telescope, you'll see an oval elongated east-west. Its bright center fades into a dim halo. Just 6′ to the north-northeast lies the magnitude 12.4 spiral galaxy NGC 6870. The third galaxy is NGC 6861, an elliptical that glows at magnitude 11.1 and lies 25′ west of NGC 6868.

You'll find this trio 4.8° west-southwest of magnitude 3.1 Alpha (α) Indi.

| OBJECT #618 | NGC 6871 |
|---|---|
| Constellation | Cygnus |
| Right ascension | 20h06m |
| Declination | 35°47′ |

| (continued) | |
|---|---|
| Magnitude | 5.2 |
| Size | 30′ |
| Type | Open cluster |

Our next object lies 2° east-northeast of magnitude 3.9 Eta ($\eta$) Cygni. This cluster contains perhaps 15 stars, but lies in front of an extremely rich background, so smaller telescopes work best. The magnitude 5.4 on the north end of the cluster isn't part of the group. Instead, look for several nice double stars within the confines of NGC 6871.

| OBJECT #619 | NGC 6885 |
|---|---|
| Constellation | Vulpecula |
| Right ascension | 20h12m |
| Declination | 26°28′ |
| Magnitude | 8.1 |
| Size | 20′ |
| Type | Open cluster |
| Other names | The 20 Vulpeculae Cluster, Caldwell 37 |

Our next target surrounds the magnitude 5.9 star 20 Vulpeculae, but, as you may have inferred from the cluster's brightness, doesn't include that luminary. At 100× through a 6-inch telescope, you'll count three dozen stars (not including 20 Vul). A 12-inch scope will net you 75 stars. NGC 6885 is loose and has a roughly triangular shape, best seen at low power. You'll find it 7.6° north-northeast of magnitude 3.5 Gamma ($\gamma$) Sagittae, the star that marks the tip of the arrow.

| OBJECT #620 | NGC 6876 |
|---|---|
| Constellation | Pavo |
| Right ascension | 20h18m |
| Declination | –70°52′ |
| Magnitude | 10.8 |
| Size | 3.7′ by 3.4′ |
| Type | Elliptical galaxy |

Our next target lies 2.5° northeast of the magnitude 4.0 Epsilon ($\varepsilon$) Pavonis. This massive, but featureless, elliptical lies at the center of a rich galaxy cluster. A 10-inch telescope brings several fainter companions into view.

Magnitude 12.2 elliptical galaxy NGC 6877 lies 1.5′ to the east-northeast. The spiral NGC 6880, which glows at magnitude 12.3, sits 6′ to the east of NGC 6876. Finally, move 9′ northwest from our starting galaxy to find magnitude 11.7 NGC 6872, a barred spiral galaxy with thin arms. You'll need a 20-inch scope to pull details out of this object.

| OBJECT #621 | Melotte 227 |
|---|---|
| Constellation | Octans |
| Right ascension | 20h12m |
| Declination | –79°18′ |
| Magnitude | 5.3 |
| Size | 50′ |
| Type | Open cluster |

For those of you who live in, or travel to, the Southern Hemisphere, here's an object that's not too far from the South Celestial Pole. And, although its discoverer, Belgian-born British astronomer Philibert Jacques Melotte, classified it as an open cluster in 1915, we now know this object is only a random alignment of stars.

Use your lowest power eyepiece, and look for 15 stars brighter than magnitude 10. To find Melotte 227, look 4.8° southwest of magnitude 3.7 Nu (ν) Octantis.

| OBJECT #622 | NGC 6882 |
|---|---|
| Constellation | Vulpecula |
| Right ascension | 20h12m |
| Declination | 26°33′ |
| Magnitude | 8.1 |
| Size | 20′ |
| Type | Open cluster |

Our next target combines two open cluster designations, NGC 6882 and NGC 6885. Are they the same? Probably.

Through a 4-inch telescope, you'll first see the magnitude 5.9 star 20 Vulpeculae surrounded by 20 cluster members. An 8-inch scope at 200× doesn't add more cluster stars, but it brings out a nice set of fainter background stars.

To find this object, move almost 10° east-northeast from magnitude 4.4 Alpha (α) Vulpeculae.

| OBJECT #623 | NGC 6886 |
|---|---|
| Constellation | Sagitta |
| Right ascension | 20h13m |
| Declination | 19°59′ |
| Magnitude | 11.4 |
| Size | 4″ |
| Type | Planetary nebula |

You'll find our next object 1.8° east of magnitude 5.1 Eta (η) Sagittae. To view this tiny planetary as anything more than a "star," you'll need a magnification above 300×. Don't expect much, however. All you'll notice is a circular disk with a bright center.

**Object #624** The Crescent Nebula (NGC 6888) Adam Block/NOAO/AURA/NSF

| OBJECT #624 | NGC 6888 |
|---|---|
| Constellation | Cygnus |
| Right ascension | 20h13m |
| Declination | 38°21′ |
| Size | 18′ by 13′ |
| Type | Emission nebula |
| Other names | The Crescent Nebula, Caldwell 27 |

Our next object is a bubble of gas carved out of the interstellar medium by an incredibly energetic star known as a Wolf-Rayet star, after the two astronomers who first identified the type. You'll see the W-R star easily. It shines at 7th magnitude and lies at NGC 6888's center.

Although you can detect the Crescent Nebula through small telescopes, 8-inch and larger instruments begin to show some of the structure. The slightly curved northwest edge is the brightest, but a short line of bright nebulosity also lies at the southwest edge. Larger telescopes also will show a thick nebulous patch that runs from the westernmost edge to the central star.

An Oxygen-III filter really helps to bring out the contrasting sections of this object. Such a filter also increases its overall visibility by dimming the vast number of background stars.

To find the Crescent Nebula, look 1.2° west-northwest of the magnitude 4.8 star 34 Cygni.

| OBJECT #625 | 31 Cygni |
|---|---|
| Constellation | Cygnus |
| Right ascension | 20h14m |
| Declination | 46°44′ |
| Magnitudes | 3.8/6.7/4.8 |
| Separations | 107″/337″ |
| Type | Double star |

To find 31 Cyg, look 5° west-northwest of 1st-magnitude Deneb (Alpha [α] Cygni). Sometimes called a triple-star system, but the third star — 30 Cygni — isn't associated with the pair. The brighter star shines yellow, while both the secondary and 30 Cyg are blue. This is a wide system, so use low power to view it.

| OBJECT #626 | NGC 6891 |
|---|---|
| Constellation | Delphinus |
| Right ascension | 20h15m |
| Declination | 12°42′ |
| Magnitude | 10.5 |
| Size | 14″ |
| Type | Planetary nebula |

Our next target lies 4.6° west-northwest of magnitude 4.0 Epsilon (ε) Delphini. Through even a 4-inch telescope, you'll see a bright, circular disk with a sharp edge that's distinctly blue. Step up to a 14-inch scope, and use a magnification of 500× to see the inner region as lens-shaped.

**Object #627** NGC 6894 Adam Block/NOAO/AURA/NSF

| OBJECT #627 | NGC 6894 |
|---|---|
| Constellation | Cygnus |
| Right ascension | 20h16m |
| Declination | 30°34′ |
| Magnitude | 12.3 |
| Size | 42″ |
| Type | Planetary nebula |

Our next object is faint, but well worth your time to search out. If your telescope doesn't have a go-to drive, draw a line between 39 Cygni (magnitude 4.4) and 21 Cygni (magnitude 5.2). NGC 6894 lies just slightly to the west of the midway point between the two stars.

Through a medium-size telescope, you'll see a faint, round disk. Step up the size of your instrument, and details start to appear. Through a 12-inch scope, look for a slightly darker central region that spans about half of NGC 6894's diameter. Now you're beginning to see this planetary nebula's ring structure. A 20-inch instrument will show a faint star that sits northwest of the object's center.

| OBJECT #628 | Alpha (α) Capricorni |
|---|---|
| Constellation | Capricornus |
| Right ascension | 20h18m |
| Declination | −12°33′ |
| Magnitudes | 3.6/4.2 |
| Separation | 378″ |
| Type | Double star |
| Other name | Algedi |

Our next target is an easy-to-find wide binary. You won't even need optical aid to split it, but binoculars help bring out the colors. The slightly brighter primary shines yellow, and the secondary is orange.

The Arabians gave this star its common name — actually, two of them. They called it Prima Giedi and Secunda Giedi. *Giedi* comes from the Arabian name of Capricornus, *Al Jady*, which means the goat (or ibex).

| OBJECT #629 | Beta (β) Capricorni |
|---|---|
| Constellation | Capricornus |
| Right ascension | 20h21m |
| Declination | −14°47′ |
| Magnitudes | 3.4/6.2 |
| Separation | 206″ |
| Type | Double star |
| Other name | Dabih |

Our next target sits 2.3° south-southeast of our last one, Alpha (α) Capricorni. Beta Cap also is a wide double star. Both components are yellow, but the primary outshines the secondary by some 13 times.

According to Richard Hinckley Allen, *Dabih* comes from the Arabian *Al Sa'd al Dhabih*, the Lucky One of the Slaughterers. This name refers to the sacrifice celebrated by the Arabians when Capricornus first became visible in the eastern morning sky just before sunrise.

| OBJECT #630 | NGC 6905 |
|---|---|
| Constellation | Delphinus |
| Right ascension | 20h22m |
| Declination | 20°07′ |
| Magnitude | 11.1 |
| Size | 39″ |
| Type | Planetary nebula |
| Other name | The Blue Flash |

You'll find this superb target 5.8° northwest of magnitude 3.8 Sualocin (Alpha [α] Delphini). Through an 8-inch telescope at 200×, you'll see a slightly oval shape with even illumination. A magnitude 10.3 star sits just off NGC 6905's northern tip, and a magnitude 11.4 star lies 1′ to the east.

Through a 16-inch scope at 300× or more, you'll barely see the central star. Look for two small regions that trail off from the north and south edges. The one to the south is easier to see.

The Blue Flash Planetary got its name when amateur astronomers using telescopes having less than 8′ of aperture saw a faint blue object pop in and out of view.

| OBJECT #631 | NGC 6907 |
|---|---|
| Constellation | Capricornus |
| Right ascension | 20h25m |
| Declination | −24°49′ |

| (continued) | |
|---|---|
| Magnitude | 11.1 |
| Size | 3.2′ by 2.3′ |
| Type | Barred spiral galaxy |

Call me strange. My nickname for this object is the "Giant Behemoth Galaxy." I think NGC 6907 looks like the prehistoric monster that terrorized England in the 1959 movie of the same name. The galaxy's central region is the behemoth's body, and the one thick spiral arm is its long neck and head, curving backward to strike terror in the masses.

You'll find NGC 6907 4.8° due west of magnitude 4.1 Psi ($\psi$) Capricorni. The galaxy's bright bar orients east-west, and the one visible arm curves northward from its east end.

| OBJECT #632 | NGC 6910 |
|---|---|
| Constellation | Cygnus |
| Right ascension | 20h23m |
| Declination | 40°47′ |
| Magnitude | 7.4 |
| Size | 10′ |
| Type | Open cluster |

You'll easily find this bright cluster 0.5° north-northeast of magnitude 2.2 Sadr (Gamma [$\gamma$] Cygni). A 4-inch telescope will reveal two dozen stars, and larger scopes will allow you to see as many as 50 stars. The brightest star here is magnitude 7.0 SAO 49563, which sits at the cluster's eastern edge. A 12-inch or larger scope equipped with a nebula filter will dim the stars in this region and reveal a large complex of nebulosity.

| OBJECT #633 | M29 (NGC 6913) |
|---|---|
| Constellation | Cygnus |
| Right ascension | 20h24m |
| Declination | 38°32′ |
| Magnitude | 6.6 |
| Size | 6′ |
| Type | Open cluster |
| Other name | The Cooling Tower |

Although M29 is a Messier object, it's one of the most difficult to identify. The reason is that M29 is a loose open cluster of about two dozen stars lying in front of a rich Milky Way star field.

To find M29, look 1.8° south of magnitude 2.2 Sadr (Gamma [$\gamma$] Cygni). A small telescope works best on this cluster because it won't reveal the multitude of surrounding stars. To prove this to myself, I once made a cardboard insert for the front of a 12-inch telescope. The insert had a 3-inch diameter hole drilled in it. I viewed M29 (and many other objects) with and without the insert, and the cluster was, indeed, easier to pick out when the insert was in place.

British amateur astronomer Jeff Bondono gave M29 its common name. He thought two curving lines of its stars looked like the concave sides of a cooling tower in operation at a nuclear power plant.

| OBJECT #634 | IC 5013 |
|---|---|
| Constellation | Microscopium |
| Right ascension | 20h29m |
| Declination | –36°02′ |
| Magnitude | 11.7 |
| Size | 1.8′ by 0.6′ |
| Type | Barred spiral galaxy |

IC 5013 combines with IC 5011 for our next treat. IC 5013 is the brighter and bigger of the two, and is a fat, lens-shaped spiral. IC 5011, which glows feebly at magnitude 14.0, is an elliptical galaxy measuring only 0.7′ long. It's round, and it appears attached to IC 5013's southwest edge.

You'll find this pair 4.9° west-southwest of magnitude 4.9 Alpha (α) Microscopii.

| OBJECT #635 | Omicron (ο) Capricorni |
|---|---|
| Constellation | Capricornus |
| Right ascension | 20h30m |
| Declination | −18°35′ |
| Magnitudes | 5.9/6.7 |
| Separation | 22″ |
| Type | Double star |

Look for Omicron Cap 4.4° south-southeast of magnitude 3.1 Dabih (Beta [β] Capricorni). It's the southernmost and easternmost of a small stellar triangle. The other two stars are Pi (π) and Rho (ρ) Capricorni. Most observers see both components of this system as bluish-white.

| OBJECT #636 | NGC 6920 |
|---|---|
| Constellation | Octans |
| Right ascension | 20h44m |
| Declination | –80°00′ |
| Magnitude | 12.4 |
| Size | 1.8′ by 1.5′ |
| Type | Spiral galaxy |

You'll find our next target only 10° from the South Celestial Pole, 3.8° southwest of magnitude 3.7 Nu (ν) Octantis. Although spiral, NGC 6920 doesn't hint at spiral structure through amateur telescopes. You'll see only a circular haze with a bright central region.

| OBJECT #637 | NGC 6925 |
|---|---|
| Constellation | Microscopium |
| Right ascension | 20h34m |
| Declination | –1°59′ |
| Magnitude | 11.3 |
| Size | 4.7′ by 1.3′ |
| Type | Spiral galaxy |

This lens-shaped spiral orients north-south and appears nearly three times as long as it is wide. The central region is broad and bright. Through a 12-inch telescope at 300×, you'll spot some of the slightly darker areas that hint at tightly wound spiral arms.

You'll find this object 3.7° west-northwest of magnitude 4.9 Alpha (α) Microscopii.

M.E. Bakich, *1,001 Celestial Wonders to See Before You Die*, Patrick Moore's Practical Astronomy Series,
DOI 10.1007/978-1-4419-1777-5_8, 

**Object #638** NGC 6934 Dale Niksch/Adam Block/NOAO/AURA/NSF

| OBJECT #638 | NGC 6934 |
|---|---|
| Constellation | Delphinus |
| Right ascension | 20h34m |
| Declination | 7°24′ |
| Magnitude | 8.7 |
| Size | 5.9′ |
| Type | Globular cluster |
| Other name | Caldwell 47 |

Our next target lies 3.9° south of magnitude 4.0 Epsilon (ε) Delphini. Because it lies some 55,000 light-years away, you'll have trouble resolving individual stars through a small telescope. Through a 10-inch scope, you'll start to pick out stars in the halo region, but not near the core. NGC 6934 is bright, so crank the magnification above 300×. The magnitude 9.2 star GSC 522:2249 sits 2′ west of the cluster's center.

| OBJECT #639 | NGC 6939 |
|---|---|
| Constellation | Cepheus |
| Right ascension | 20h31m |
| Declination | 60°38′ |
| Magnitude | 7.8 |
| Size | 7′ |
| Type | Open cluster |

Now here's a nice target. Even through a 4-inch telescope, you'll be able to count 50 stars in this rich cluster. Use low power, and contrast the cluster with the nice Milky Way background. A 10-inch telescope will reveal 100 stars spread fairly evenly across the field. Your eye will be drawn to the large number of patterns the stars contribute to.

You can find NGC 6939 2° southwest of magnitude 3.4 Eta (η) Cephei.

**Object #640** Mothra (NGC 6940) Anthony Ayiomamitis

| OBJECT #640 | NGC 6940 |
|---|---|
| Constellation | Vulpecula |
| Right ascension | 20h35m |
| Declination | 28°18′ |
| Magnitude | 6.3 |
| Size | 31′ |
| Type | Open cluster |
| Other name | Mothra |

You'll find our next object 2.3° south-southeast of the magnitude 4.0 star 41 Cygni. Through an 8-inch telescope, you'll see 100 stars brighter than magnitude 14. This cluster reminds me of M37 in Auriga. You can compare them in the fall from locations north of the equator.

*Astronomy* magazine Contributing Editor Stephen James O'Meara thought this cluster looked like a moth. Not any moth, mind you. He pictures Mothra, the Japanese kaiju (fantastic creature) that made its cinematic debut in 1961. If you're familiar with that creature (as I am), O'Meara goes on to say that the double star near the cluster's center represents the Cosmos, the pair of tiny telepathic priestesses who can summon Mothra when the need arises. I'm not certain which pair of stars O'Meara is referring to, but through even a 4-inch telescope, you'll have your choice of many in the 11th- and 12th-magnitude range.

**Object #641** NGC 6946 Anthony Ayiomamitis

| OBJECT #641 | NGC 6946 |
|---|---|
| Constellation | Cygnus |
| Right ascension | 20h35m |
| Declination | 60°09′ |
| Magnitude | 9.0 |
| Size | 11.5′ by ′9.8 |
| Type | Spiral galaxy |
| Other name | Caldwell 12 |

How can a galaxy be bright and faint simultaneously? The answer hinges on the object's surface brightness. In the case of NGC 6946, its overall brightness may be high, leading to a showy 9th magnitude. Unfortunately, however, that light spreads over an area 11′ across.

Another hindrance to NGC 6946's brightness is its location near the Milky Way's plane in Cepheus. So, the overall impression of NGC 6946, unless you're using a large telescope, is of a faint object.

Sir William Herschel discovered NGC 6946 September 9, 1798. For some unknown reason, this galaxy is a hotbed of supernova activity. In the last century, eight supernovae have appeared, in 1917, 1939, 1948, 1968, 1969, 1980, 2002, and 2004. The most recent was the brightest, reaching magnitude 12.3 September 30.

Although small telescopes will allow you to spot NGC 6946, you won't see much detail with apertures less than 12 inches. The galaxy's core appears bright and spans roughly 10% of its diameter. Two spiral arms are visible through a 12-inch scope, and you'll see four through a 16-inch or larger instrument.

To find NGC 6946, look 2.1° southwest of magnitude 3.4 Eta ($\eta$) Cephei.

**Object #642** NGC 6951 Cam and Connie Baher/Adam Block/NOAO/AURA/NSF

| OBJECT #642 | NGC 6951 |
|---|---|
| Constellation | Cepheus |
| Right ascension | 20h37m |
| Declination | 66°06′ |
| Magnitude | 10.7 |
| Size | 3.7′ by 3.3′ |
| Type | Spiral galaxy |

Our next celestial treat lies 5.7° northwest of magnitude 2.5 Alderamin (Alpha [α] Cephei). You'll need at least a 10-inch telescope to pick out any detail in this object. At magnifications above 250×, try to spot the thin spiral arm toward the southeast. Look for the magnitude 12.2 star GSC 4258:1945 superposed on the galaxy's western edge.

**Object #643** Gyulbudaghian's Nebula Adam Block/Mount Lemmon SkyCenter/University of Arizona

| OBJECT #643 | Gyulbudaghian's Nebula |
|---|---|
| Constellation | Cepheus |
| Right ascension | 20h46m |
| Declination | 67°58′ |
| Size | 30″ |
| Type | Reflection nebula |
| Notes | Illuminated by the star PV Cephei |

Say what? Our next target is Gyulbudaghian's Nebula. If you want to impress advanced amateur astronomers, get the pronunciation of this object right (gyool bu day' gee an's). The nebula bears the name of Armen Gyulbudaghian, the Russian astronomer who discovered it in 1977. The nebula has the catalog number GM 1–29.

This variable reflection nebula lies 6° north of magnitude 3.4 Eta ($\eta$) Cephei. It is a wedge-shaped object illuminated by the variable star PV Cephei. On photographs, Gyulbudaghian's Nebula appears blue because its gas scatters much of the star's light, and blue light scatters the most.

The nebula changes its brightness and shape on a non-regular basis. Use at least a 12-inch telescope at 150× to look for it. Make repeated observations (monthly?), and keep good notes so you'll have a record of its changes.

| OBJECT #644 | Gamma ($\gamma$) Delphini |
|---|---|
| Constellation | Delphinus |
| Right ascension | 20h47m |
| Declination | 16°07′ |
| Magnitudes | 4.3/5.2 |
| Separation | 10′ |
| Type | Double star |

You'll find this nice double star 15° northwest of Tarazed (Gamma [$\gamma$] Aquilae). The yellow and white stars have a subtle color contrast I find quite pleasing. Any size telescope at 100× will split this star.

| OBJECT #645 | Aquarius Dwarf |
|---|---|
| Constellation | Aquarius |
| Right ascension | 20h47m |
| Declination | −2°51′ |
| Magnitude | 13.9 |
| Size | 2.3′ by 1.2′ |
| Type | Dwarf elliptical galaxy |

The only reason to observe the challenging target called the Aquarius Dwarf is because it belongs to the Local Group. A 12-inch telescope will let you just define its outline. Through a 30-inch telescope under perfect conditions, I saw an oval, evenly illuminated smudge at 284×.

| OBJECT #646 | NGC 6958 |
|---|---|
| Constellation | Microscopium |
| Right ascension | 20h49m |
| Declination | −8°00′ |
| Magnitude | 11.3 |
| Size | 2.5′ by 2.1′ |
| Type | Elliptical galaxy |

You'll find our next target 4.2° south of magnitude 4.9 Alpha ($\alpha$) Microscopii. This fat oval orients east-west, but doesn't show much detail. The concentrated central region takes up three-quarters of the galaxy's area. Outside is a faint halo you may spot through a 12-inch telescope.

| OBJECT #647 | IC 5067 |
|---|---|
| Constellation | Cygnus |
| Right ascension | 20h51m |
| Declination | 44°21′ |
| Size | 25′ by 10′ |
| Type | Emission nebula |
| Other name | The Pelican Nebula |

The wonderful Pelican Nebula sits 1.5° west-southwest of its larger neighbor the North America Nebula (Object #652). The Pelican isn't as bright or easy to see, but that doesn't imply it's faint. To see this object's distinctive shape, use at least a 6-inch telescope with an eyepiece that gives 100×.

While unfiltered views of NGC 7000 may show detail, the same cannot be said about the Pelican Nebula. A nebula filter such as an Oxygen-III (OIII) will help quite a bit. You also won't see the Pelican with your unaided eyes. Use an 8-inch or larger telescope, and look for a dark break in the overall nebulous glow that signals the pelican's bill.

**Object #648** NGC 6960 (top) and NGC 6992/5 Adam Block/Mount Lemmon SkyCenter/University of Arizona

| OBJECT #648 | NGC 6960/92/95 |
|---|---|
| Constellation | Cygnus |
| Right ascension | 20h51m |
| Declination | 31°03′ |
| Size | 3° |

| (continued) | |
|---|---|
| Type | Supernova remnant |
| Other names | Bridal Veil Nebula, the Cygnus Loop, the Filamentary Nebula, the Network Nebula, the Witch's Broom, NGC 6960/74/79/92/95, Caldwell 33 (NGC 6992/5), and Caldwell 34 (NGC 6990) |

More than 15,000 years ago, the gaseous filaments in the Veil belonged to a massive star on the brink of blowing itself to bits. When it exploded as a supernova, the star would have shone brighter than the crescent Moon. Unfortunately, no record of this event exists.

The Veil Nebula contains two main segments. NGC 6960 wends its way past the star 52 Cygni, which shines at magnitude 4.2. The star is a foreground object unconnected to the supernova remnant. NGC 6960 tapers to a sharp point at the north end of a degree-long glowing strip. As it broadens to the south and passes 52 Cyg, a dark lane splits the nebulosity.

The Veil Nebula's brighter segment, NGC 6992/5, lies 2.7° northeast of 52 Cyg. At medium magnification (around 100×), the nebula breaks into numerous strands.

More nebulosity lies between the northern ends of the Veil's two main sections. Look for NGC 6979, the northernmost triangular bright patch. American astronomer Williamina Fleming (1857–1911) discovered this nebulous region in 1904 while working at Harvard Observatory. She named it Pickering's Triangle in honor of the observatory's director, Edward Charles Pickering (1846–1919). The nebula also goes by Pickering's Wedge and Fleming's Wisp.

For best results when observing the Veil Nebula, use a 10-inch or larger telescope and the eyepiece that gives the lowest magnification. Insert a nebula filter, disconnect the scope's drive (so you can move it freely), and pan this area. Take your time. There's a lot to see.

| OBJECT #649 | Abell 3716 |
|---|---|
| Constellation | Indus |
| Right ascension | 20h52m |
| Declination | –52°42′ |
| Size | 52′ |
| Type | Galaxy cluster |

Our next object is the distant galaxy cluster Abell 3716. This group contains more than 60 members in an area located 1° south of magnitude 5.1 Iota (ι) Indi. The brightest of these glows dimly at magnitude 13.6 and measures just 1′ across. Through a 16-inch telescope you'll spot a dozen members brighter than magnitude 15. I've counted two dozen galaxies in Abell 3716 through a 30-inch reflector.

| OBJECT #650 | M72 (NGC 6981) |
|---|---|
| Constellation | Aquarius |
| Right ascension | 20h54m |
| Declination | –2°32′ |
| Magnitude | 9.3 |
| Size | 5.9′ |
| Type | Globular cluster |

Our next object, M72, isn't all that easy to find if you're star-hopping. First, find magnitude 3.8 Albali (Epsilon [ε] Aquarii). Then move 3.3° southeast.

This is a classic Messier object because, through a small telescope (especially one with the low quality of Charles Messier's), it has a hazy appearance reminiscent of a comet.

Through a modern 8-inch telescope, the view is somewhat better. Most of M72's stars lie close to the core, which takes up some three-quarters of the cluster's diameter. The remaining outliers are difficult to resolve at magnifications under 200×.

The brightest star near the cluster is GSC 5765:1129. It shines at magnitude 9.6 and lies 5′ to the east.

| OBJECT #651 | M73 (NGC 6994) |
|---|---|
| Constellation | Aquarius |
| Right ascension | 20h59m |
| Declination | −2°38′ |
| Magnitude | 8.9 |
| Size | 2.8′ |
| Type | Cluster of four stars |

The fact that M73 is a Messier object shows how primitive Charles Messier's telescope was. This weak cluster of four stars sits not quite 3° west-southwest of magnitude 4.5 Nu (ν) Aquarii and only 1.2° east of one of Messier's faint globulars, M72. At least M72 is a globular.

M73 appears as an equilateral triangle of stars with the fourth star lying just to the triangle's north-northwest. My most recent observation of this cluster brought to mind the phrase "Nothing to see here. Move along."

**Object #652** The North America Nebula (NGC 7000) Adam Block/NOAO/AURA/NSF

| OBJECT #652 | NGC 7000 |
|---|---|
| Constellation | Cygnus |
| Right ascension | 20h59m |
| Declination | 44°20′ |
| Size | 120′ by 100′ |
| Type | Emission nebula |
| Other names | The North America Nebula, Caldwell 20 |

Few deep-sky objects with "common" names truly look like their namesakes. Not so with the North America Nebula. A quick glance at this object reveals the California coast, Mexico, the Gulf of Mexico, and even a faint Florida. It took more than 100 years, however, for people to see the similarity.

To find the North America Nebula, look 3° east of Deneb (Alpha [α] Cygni). The North America Nebula measures 2° across, so if you're using a telescope, start with the eyepiece that gives you the widest field of view.

Sir William Herschel discovered the North America Nebula October 24, 1786. German astronomer Maximilian Franz Joseph Cornelius Wolf (1863–1932) was the first to photograph this object, on December 12, 1890. He dubbed it the America Nebula, from which its current moniker derives.

From a dark site, sharp-eyed observers can see NGC 7000 with their naked eyes. Look roughly 3° east of Deneb (Alpha [α] Cygni). The whole continent will not be apparent at first glance. Find the brightest part — Mexico — then patiently, and with averted vision, try to see the rest. Don't give up on trying to find it with your eyes until you've searched through a nebula filter.

Low-power telescope views that provide at least a 2° field of view are best. And don't forget the nebula filter.

| OBJECT #653 | Equuleus |
|---|---|
| Type | Constellation |
| Right ascension (approx.) | 21h02m |
| Declination (approx.) | 16°11′ |
| Size (approx.) | |

The constellation Pegasus represents the mythological winged horse. But did you know that the sky holds a second horse? On Pegasus' southwest border sits Equuleus — and, by the way, the official pronunciation is ek woo oo' le us — the Foal.

By any reasonable standard, this is a weak constellation. It contains none of the 200 brightest stars, has no named star, meteor shower, or Messier object, and it ranks 87th in size out of the 88 constellations that cover the sky. It was, however, one of the original 48 constellations of the Greeks. Ptolemy, who lived from 73 to 151 A.D., mentioned Equuleus in his great work, *Almagest*. And your ability to identify it will set you apart from your observing buddies.

To see the Foal, or at least the faint stars that form its figure, first find the magnitude 2.4 star Epsilon (ε) Pegasi. We also will use this star later in this book to find the bright globular cluster M15. Just look 7° due west of Epsilon. Don't expect to see much. Equuleus' brightest star — Alpha (α) — glows at a disappointing magnitude 3.9. The fainter stars just north and just east of Alpha form the rest of the constellation.

| OBJECT #654 | Epsilon (ε) Equulei |
|---|---|
| Constellation | Equuleus |
| Right ascension | 20h59m |
| Declination | 4°18′ |
| Magnitudes | 6.0/7.1 |
| Separation | 10.7″ |
| Type | Double star |

You'll find Epsilon Equulei (also known as 1 Equ) 4.3° west-southwest of magnitude 3.9 Kitalpha (Alpha [α] Equulei). This easily split binary yields a primary that shines yellow-white and a secondary with an off-white hue.

| OBJECT #655 | NGC 7006 |
|---|---|
| Constellation | Delphinus |
| Right ascension | 21h02m |
| Declination | 16°11′ |
| Magnitude | 10.5 |
| Size | 2.8′ |
| Type | Globular cluster |
| Other name | Caldwell 42 |

For a nice large telescope target, I invite you to turn your attention to globular cluster NGC 7006. This magnitude 10.6 object resides in Delphinus the Dolphin.

An 8-inch scope can't hope to resolve the stars in this tightly packed mass. The problem here is distance — NGC 7006 lies approximately 140,000 light-years away.

To find NGC 7006, find Gamma (γ) Delphini, the easternmost star of Delphinus' crooked box (and the Dolphin's nose), and move 3.5° east.

When you observe at magnifications below about 250×, you'll see a comet-like haze 2′ across. If the seeing permits, crank the power above 300×, and NGC 7006 will appear slightly granular.

| OBJECT #656 | NGC 7008 |
|---|---|
| Constellation | Cygnus |
| Right ascension | 21h01m |
| Declination | 54°33′ |
| Magnitude | 10.7 |
| Size | 83′ |
| Type | Planetary nebula |
| Common names | The Fetus Nebula, the Coat Button Nebula |

The Fetus Nebula sits in a lonely region of northern Cygnus 5° north-northeast of the magnitude 5.4 star 51 Cygni. Through a 4-inch telescope, you'll see a small, uniformly illuminated disk. At high magnification through a 10-inch scope, NGC 7008 appears as an open, nebulous ring.

See it through a 16-inch or larger telescope, and you'll get two planetaries for the price of one. Crank up the magnification and try to spot a faint knot that lies 22″ northwest of the Fetus Nebula's central star. This object, which astronomers designate K 4–44 , is a distinct planetary nebula.

American amateur astronomer Eric Honeycutt gets the credit for bestowing the "Fetus Nebula" moniker onto NGC 7008. That's the shape he saw when he viewed this object through his 22-inch reflector. The first written mention appeared in *Amateur Astronomy* magazine #30, summer 2001.

*Astronomy* magazine Contributing Editor Stephen James O'Meara called it the Coat Button Nebula. He based that name on NGC 7008's photographic appearance, which reminded him of a large button on a winter coat. He added that the irregular shape of the annulus' center also looks like two buttonholes.

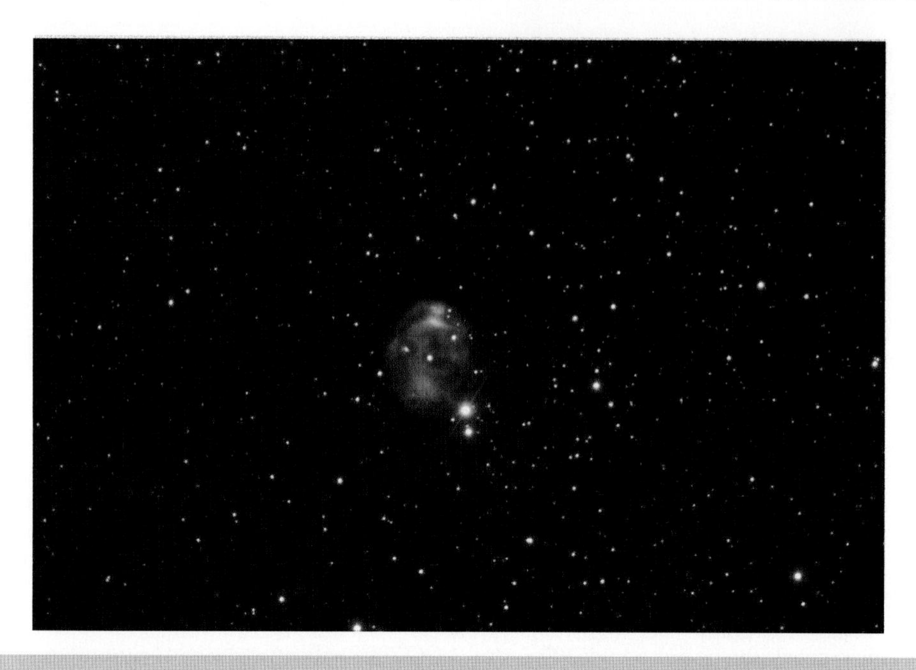

**Object #657** The Fetus Nebula (NGC 7008) Donn and Aaron Starkey/Adam Block/NOAO/AURA/NSF

| OBJECT #657 | NGC 7009 |
|---|---|
| Constellation | Aquarius |
| Right ascension | 21h04m |
| Declination | –1°22′ |
| Magnitude | 8.3 |
| Size | 25″ |
| Type | Planetary nebula |
| Other names | The Saturn Nebula, Caldwell 55 |

The Saturn Nebula, also known as NGC 7009, is a sky wonder you simply must seek out. This planetary nebula's common name arises from thin extensions on either end of its disk. These extensions, called ansae (singular = ansa), resemble Saturn's rings and represent material ejected from the nebula in two directions. Astronomers call objects like NGC 7009 bipolar planetaries.

Because of its similarity to the ringed planet, William Parsons, 3rd Earl of Rosse (1800–1867) first called NGC 7009 the Saturn Nebula.

The Saturn Nebula lies slightly more than 1° west of the magnitude 4.5 star Nu ($\nu$) Aquarii. Through an 8-inch or larger telescope, view NGC 7009 with magnifications above 200×. Its oval disk measures 25′ in its long dimension. The extensions each protrude another 15″. At the end of the extensions are fainter bulbs, but you'll need at least a 12-inch scope to pick them out.

Finally, what color is the Saturn Nebula to you? Whether you see it as mostly blue or mostly green depends on your color perception. And, just so you know, there's no right answer.

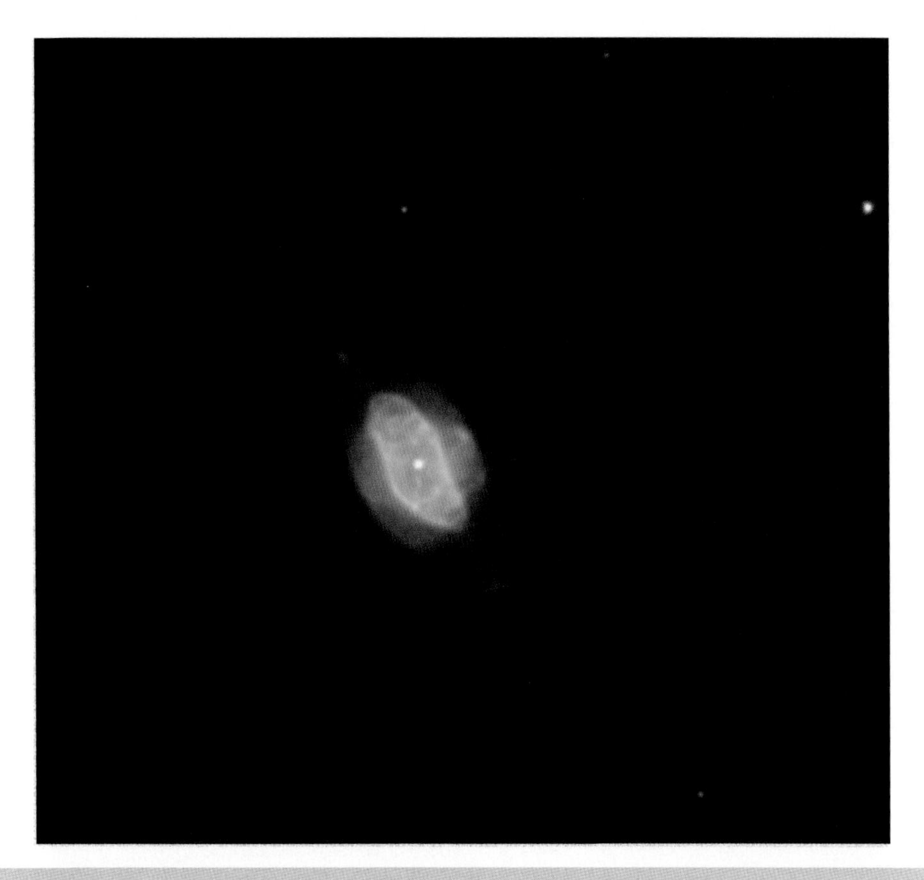

**Object #658** The Saturn Nebula (NGC 7009) Brad Ehrhorn/Adam Block/NOAO/AURA/NSF

| OBJECT #658 | NGC 7020 |
|---|---|
| Constellation | Pavo |
| Right ascension | 21h11m |
| Declination | −4°03′ |
| Magnitude | 11.8 |
| Size | 3.8′ by 1.7′ |
| Type | Spiral galaxy |

You'll find our next target 2.1° northwest of magnitude 4.2 Gamma ($\gamma$) Pavonis. NGC 7020 appears lenticular, twice as long as it is wide. Its illumination is even across its face.

| OBJECT #659 | NGC 7023 |
|---|---|
| Constellation | Cepheus |
| Right ascension | 21h01m |
| Declination | 68°10′ |
| Size | 10′ by 8′ |
| Type | Emission nebula |
| Common names | The Iris Nebula, Caldwell 4 |

Imagine a cloud of countless particles, each measuring less than a millionth of a millimeter wide. Together, however, these particles function as a special type of cosmic mirror we know as the Iris Nebula.

Sir William Herschel discovered the Iris Nebula in 1794. He described it as, "A star of 7th magnitude. Very much affected with nebulosity, which more than fills the field. It seems to extend to at least a degree all around; stars, such as 9th or 10th magnitude, of which there are many, are perfectly free from this appearance."

You'll find this object 3.4° southwest of magnitude 3.2 Alfirk (Beta [$\beta$] Cephei). Through a 10-inch telescope, NGC 7023 appears as a uniformly bright, circular haze about 1′ across surrounding the 7th-magnitude star HD 200775. Look closely to see a small, detached portion of the nebula just to the south.

Through 16-inch and larger scopes, the nebula's faint outer halo appears. You'll see about half of the 18′ diameter astroimagers capture.

**Object #660** The Iris Nebula (NGC 7023) Adam Block/NOAO/AURA/NSF

| OBJECT #660 | PK 80–6.1 |
|---|---|
| Constellation | Cygnus |
| Right ascension | 21h02m |
| Declination | 36°42′ |
| Magnitude | 13.5 |
| Size | 16.0″ |
| Type | Planetary nebula |
| Other name | The Egg Nebula |

To be fair, the Egg Nebula is a much more interesting object scientifically than visually. It is a bipolar protoplanetary nebula with highly polarized outflow.

And, how, you ask, did this object get its common name? In May 1974, Mike Merrill of the University of California at San Diego observed this object with the 60-inch infrared telescope atop Mount Lemmon near Tucson, Arizona. While examining photographic prints of the National Geographic Society Palomar Sky Survey, Merrill found that PK 80–6.1's position coincided with that of a small oval nebulosity. He called that interstellar cloud the Egg Nebula.

You can spot the Egg Nebula through a 10-inch telescope if your sky is dark and steady. It lies 0.7° due east of magnitude 6.0 SAO70794. Crank the power to 200×, and look for a slightly elongated "star". Through at least a 16-inch scope, insert and rotate a polarizing filter, and you'll see the nebula dim quite a lot.

**Object #661** The Egg Nebula (PK 80–6.1) Eric Africa/Adam Block/NOAO/AURA/NSF

| OBJECT #661 | NGC 7026 |
|---|---|
| Constellation | Cygnus |
| Right ascension | 21h06m |
| Declination | 47°51′ |
| Magnitude | 10.9 |
| Size | 21″ |
| Type | Planetary nebula |
| Other names | The Cheeseburger Nebula; the Tiny Dumbbell Nebula |

NGC 7026 is a tiny planetary nebula, but its surface brightness is high, so you can spot it even through a 4-inch telescope. The problem with that size instrument, however, is identifying the planetary from surrounding stars. Alternately moving an Oxygen-III filter in front of your eyepiece will make the stars fainter but leave the planetary undimmed.

Through a 12-inch or larger telescope, you'll see two nebulous lobes. Under a magnification less than 200×, the lobes will appear to touch. The eastern lobe glows slightly brighter.

American amateur astronomer Jay McNeil (who later found McNeil's Nebula in M78) dubbed this object the Cheeseburger Nebula in 1999.

| OBJECT #662 | NGC 7027 |
|---|---|
| Constellation | Cygnus |
| Right ascension | 21h07m |
| Declination | 42°14′ |
| Size | 25′ |
| Magnitude | 8.8 |
| Type | Planetary nebula |
| Other name | The Magic Carpet Nebula |

You'll find our next object 2.1° east-northeast of magnitude 3.9 Nu ($\nu$) Cygni. Through a telescope between 4 and 8 inches in aperture and at magnifications below 75×, NGC 7027 appears stellar, but as you increase the power, you'll begin to see the planetary's oval shape.

Through larger scopes and magnifications above 150×, you'll begin to lose the oval shape and gain a rectangle. An Oxygen-III filter really helps here. Without the filter, you'll see the nebula's faded green color through 14-inch instruments.

American amateur astronomer Kent Wallace gave this object its common name in 2000. He saw an article that described NGC 7027 as a hot coal on a carpet, and the rest is history. He says he actually prefers the "Green Rectangle" as a common name because of the appearance of this planetary through his 20-inch reflector at high powers.

| OBJECT #663 | 61 Cygni |
|---|---|
| Constellation | Cygnus |
| Right ascension | 21h07m |
| Declination | 38°45″ |
| Magnitudes | 5.2/6.0 |
| Separation | 28″ |
| Type | Double star |

The star 61 Cygni is famous for a reason you won't be able to observe. It has the largest proper motion of any star visible without optical aid. Also, in 1838, 61 Cyg became the first star to have its distance determined directly. In that year, German astronomer Friedrich Wilhelm Bessel measured its stellar parallax.

You'll find 61 Cyg in a busy region of the northern Milky Way 1.7° west-northwest of magnitude 3.8 Tau ($\tau$) Cygni. Any telescope will split the two stars. Both shine with an orange light, although the brighter component appears ever-so-slightly more yellow.

| OBJECT #664 | NGC 7041 |
|---|---|
| Constellation | Indus |
| Right ascension | 21h17m |
| Declination | –8°22′ |
| Magnitude | 11.2 |
| Size | 3.3′ by 1.4′ |
| Type | Spiral galaxy |

Our next target appears lens-shaped, more than twice as long as it does wide. NGC 7041 is evenly illuminated but high magnification does reveal a bright, stellar nucleus. You'll find it 6.6° east-southeast of magnitude 3.1 Alpha ($\alpha$) Indi. It makes a nice wide-field pair with our next object.

| OBJECT #665 | NGC 7049 |
|---|---|
| Constellation | Indus |
| Right ascension | 21h19m |
| Declination | −8°34′ |
| Magnitude | 10.3 |
| Size | 4.3′ by 3.2′ |
| Type | Spiral galaxy |

You'll find this object 0.5° east-southeast of NGC 7041, with which it forms a nice galactic pair. NGC 7049 appears as a circular haze with a bright central region. Through a 14-inch or larger telescope, look for magnitude 14.7 NGC 7041A, which lies midway between NGC 7041 and NGC 7049.

| OBJECT #666 | NGC 7062 |
|---|---|
| Constellation | Cygnus |
| Right ascension | 21h23m |
| Declination | 46°23′ |
| Magnitude | 8.3 |
| Size | 5′ |
| Type | Open cluster |

You'll find our next target 2° west-northwest of magnitude 4.0 Rho ($\rho$) Cygni. Through a 4-inch telescope, you'll spot two dozen stars scattered in a rough crescent shape. Larger scopes don't really reveal that many more stars.

| OBJECT #667 | NGC 7063 |
|---|---|
| Constellation | Cygnus |
| Right ascension | 21h24m |
| Declination | 36°29′ |
| Magnitude | 7.0 |
| Size | 9′ |
| Type | Open cluster |

You'll find our next object 2.5° southeast of magnitude 3.7 Tau ($\tau$) Cygni. This is a scattered cluster through any telescope with the northern side appearing richer than the southern side. A 6-inch scope reveals about two dozen stars. The brightest, magnitude 8.9 HIP 105673, sits at the cluster's southern edge.

| OBJECT #668 | Beta ($\beta$) Cephei |
|---|---|
| Constellation | Cepheus |
| Right ascension | 21h29m |
| Declination | 70°34′ |
| Magnitudes | 3.2/7.9 |
| Separation | 13.3″ |
| Type | Double star |
| Other name | Alfirk |

Here's a binary that's easy to find. It's the easternmost of the five stars that make up Cepheus' "house" asterism. This is an easy double star to split. Both stars shine with bluish light, but the magnitude difference (the primary outshines the secondary by 63 times) makes the brighter star appear blue-white.

This star's common name comes from Arabia. In *Star-Names and Their Meanings*, Richard Hinckley Allen tells us "Alfirk" once described Alpha (α) Cephei (now known as Alderamin). The name derives from *Al Kawakib al Firk*, the Stars of the Flock.

**Object #669** M15 Adam Block/NOAO/AURA/NSF

| OBJECT #669 | M15 (NGC 7078) |
|---|---|
| Constellation | Pegasus |
| Right ascension | 21h30m |
| Declination | 12°10′ |
| Magnitude | 6.3 |
| Size | 12.3′ |
| Type | Globular cluster |

For Northern Hemisphere observers, M15 ranks as autumn's showpiece globular cluster. From a dark site, sharp-eyed observers can spot magnitude 6.3 M15 with their naked eyes. Don't be confused by the magnitude 6.1 star only 17′ to the east. Confirm your sighting through your telescope.

And speaking of telescopes, a 4-inch scope will resolve dozens of stars around M15's strikingly bright core. Look for the chains of stars that wind out from its central region. Through a small telescope, these star patterns cause some observers to describe M15 as slightly oval.

Finding M15 is pretty easy. Use Theta (θ) and Epsilon (ε) Pegasi as pointers. Just draw a line from Theta through Epsilon, and continue another 4°.

With a 10-inch or larger telescope, try for the challenge object within M15 — Pease 1, the first planetary nebula found within a globular cluster. In 1928, American astronomer Francis Gladheim Pease discovered this object when he noticed an unusually bright "star" on a photographic plate taken with the 100-inch Hooker Telescope on Mount Wilson.

Look for this planetary nebula through an eyepiece that yields 200× or so. Use a nebula filter. The filter will suppress the brightness of the myriad stars surrounding Pease 1. Be aware of the sky conditions. You'll need good seeing to spot the tiny planetary, which lies about 1′ northeast of M15's core.

**Object #670** NGC 7082 Anthony Ayiomamitis

| OBJECT #670 | NGC 7082 |
|---|---|
| Constellation | Cygnus |
| Right ascension | 21h29m |
| Declination | 47°05′ |
| Magnitude | 7.2 |
| Size | 24′ |
| Type | Open cluster |

When you first look at NGC 7082, you might think, "Oh, what a rich cluster!" Not really. Here's a good example of a poor star cluster superimposed on a rich Milky Way background star field.

Through a 4-inch telescope, you'll spot 20 or so cluster members ranging in brightness from magnitudes 8–10. Larger apertures don't really add that many more stars.

You'll find this object 1.7° north-northwest of magnitude 4.0 Rho ($\rho$) Cygni.

| OBJECT #671 | NGC 7086 |
|---|---|
| Constellation | Cygnus |
| Right ascension | 21h31m |
| Declination | 51°35′ |
| Magnitude | 8.4 |
| Size | 12′ |
| Type | Open cluster |

Our next object lies 1.9° west-northwest of magnitude 4.7 Pi[1] ($\pi^1$) Cygni. You'll count more than 50 stars of similar brightnesses through an 8-inch telescope from a dark site. The brightest star near the cluster's center is magnitude 10.3 HIP 106175.

**Object #672** M2 Pat and Chris Lee/Adam Block/NOAO/AURA/NSF

| OBJECT #672 | M2 (NGC 7089) |
|---|---|
| Constellation | Aquarius |
| Right ascension | 21h34m |
| Declination | –0°49′ |
| Magnitude | 6.6 |
| Size | 12.9′ |
| Type | Globular cluster |

A small telescope will reveal the beautiful globular cluster M2 in the northern part of Aquarius. This deep-sky object is a real showpiece, even though it doesn't have a common name. M2 shines at magnitude 6.6, making it the 18th-brightest globular cluster.

To find M2, scan roughly 4.5° due north of Beta Aquarii. If you have sharp eyes, you might spot the cluster without optical aid from a dark site. Through your telescope, you'll notice that M2 appears slightly elliptical. It's one of the sky's richest and most compact globulars.

| OBJECT #673 | NGC 7090 |
|---|---|
| Constellation | Indus |
| Right ascension | 21h37m |
| Declination | –4°33′ |
| Magnitude | 10.7 |
| Size | 8.1′ by 1.4′ |
| Type | Spiral galaxy |

Our next object is another celestial needle. This one appears more than five times as long as it is wide, and it orients northwest to southeast. You won't see much detail until you observe through a 12-inch or larger telescope. The galaxy's central region appears broad and evenly illuminated, but crank up the magnification, and you'll see the northern edge appears irregularly lit. By contrast, the southern side is a bright, straight edge.

Look for NGC 7090 4.4° west-northwest of magnitude 4.7 Epsilon (ε) Indi.

**Object #674** M39 Anthony Ayiomamitis

| OBJECT #674 | M39 (NGC 7092) |
|---|---|
| Constellation | Cygnus |
| Right ascension | 21h32m |
| Declination | 48°26′ |
| Magnitude | 4.6 |
| Size | 31′ |
| Type | Open cluster |

Although Charles Messier included this open cluster as the 39th entry into his now-famous catalog, it's not at all impressive. You'll find it 2.8° north of magnitude 4 Rho (ρ) Cygni. In *Cycle of Celestial Objects*, Smyth described it as, "A loose cluster, or rather splashy galaxy field of stars, in a very rich vicinity between the Swan's tail and the Lizard."

Maybe it's the large size of M39 — the object covers an area equal to that of the Full Moon — coupled with the rich Milky Way background star field that makes it seem lost. Still, look carefully, and you'll count two dozen stars brighter than 12th magnitude. Use the lowest-power eyepiece you have.

| OBJECT #675 | NGC 7095 |
|---|---|
| Constellation | Octans |
| Right ascension | 21h52m |

| (continued) | |
|---|---|
| Declination | −1°32′ |
| Magnitude | 12.2 |
| Size | 2.8′ by 2.7′ |
| Type | Barred spiral galaxy |

Our next target lies far, far south, only 2° west of magnitude 4.1 Beta (β) Octantis. Through an 8-inch telescope, this galaxy appears faint, but you will detect the bar, which tilts northeast to southwest. Through a 16-inch scope, the thick spiral arms appear, defined by the dark zones across NGC 7095's face. The arms tightly wind around the central region. Wait for a moment of good seeing, and you might detect the magnitude 13.9 star at the galaxy's northern edge.

| OBJECT #676 | Struve 2816 |
|---|---|
| Constellation | Cepheus |
| Right ascension | 21h39m |
| Declination | 57°29′ |
| Magnitudes | 5.6/7.7/7.8 |
| Separations | 11.7′/20′ |
| Type | Double star |

Our next target gives you two treats for the price of one. Although you'll observe the triple-star system Struve 2816, you won't be able to miss open cluster Trumpler 37, sometimes called the Misty Clover Nebula. That's because the star lies pretty much dead-center within the cluster.

Look for these objects 5.7° south-southeast of magnitude 2.5 Alderamin (Alpha [α] Cephei), or 1.4° south-southwest of magnitude 4.2 Herschel's Garnet Star (Mu [μ] Cephei). Trumpler 37 contains some 50 stars and measures nearly 1° across. Struve 2816's closer pair displays a yellow primary and a blue secondary. The tertiary star, not quite twice as far away as the main pair's separation, also is blue.

**Object #677** IC 1396 Anthony Ayiomamitis

| OBJECT #677 | IC 1396 |
|---|---|
| Constellation | Cepheus |
| Right ascension | 21h39m |
| Declination | 57°30′ |
| Size | 170′ by 140′ |
| Type | Emission nebula |

Scanning Cepheus leads us to one of the sky's largest emission nebulae — IC 1396. But this region also contains dark nebulae and a bright star cluster. Get ready to spend lots of time observing this wonder.

In images of IC 1396, the most noticeable part is the Elephant Trunk Nebula. This region of sinuous bright and dark nebulosity conceals a new star-forming region. Because such a region resembles a comet, it's earned the name cometary globule. In addition to the gas the globules contain, they have dusty heads and elongated tails.

The brightest star near IC 1396 is Mu ($\mu$) Cephei, also known as Herschel's Garnet Star. This intensely red luminary varies in magnitude from 3.4 to 5.1 over a 730-day period. Mu is a cool, red supergiant that dredges up carbon compounds from its interior. The carbon temporarily settles on the surface, dimming and reddening the star's light. Eventually, the dark material absorbs enough energy to escape into space. Because of its proximity to Mu, some early nineteenth-century astronomers referred to IC 1396 as the Garnet Star Nebula. That name is now obsolete.

If you choose to observe IC 1396 without a nebula filter, use at least a 6-inch telescope from a dark site. Although you'll detect it through smaller instruments, you won't see much detail. If your field of view measures at least 2°, you'll see a circular mist crossed by many dark lanes. The most prominent lane points from the nebula's northwestern edge to the center.

At the nebula's heart, you'll find the naked-eye open cluster Trumpler 37, a 1°-wide collection of more than 50 stars between magnitudes 10 and 12. Increase the magnification past 150× to view the triple star at the cluster's center — a magnitude 6 sun flanked by two 8th-magnitude companions.

| OBJECT #678 | NGC 7098 |
|---|---|
| Constellation | Octans |
| Right ascension | 21h44m |
| Declination | –5°07′ |
| Magnitude | 11.4 |
| Size | 4.0′ by 2.3′ |
| Type | Spiral galaxy |

Our next object lies in a nondescript star field 2.3° north of magnitude 3.7 Nu ($\nu$) Octantis. Through an 8-inch telescope, the galaxy orients northeast to southwest. It has a bright, broad central region, and a halo that's easy to spot once you increase the magnification past 200×.

| OBJECT #679 | M30 (NGC 7099) |
|---|---|
| Constellation | Capricornus |
| Right ascension | 21h40m |
| Declination | –3°11′ |
| Magnitude | 7.3 |
| Size | 11′ |
| Type | Globular cluster |

Our next treat is another from Charles Messier's catalog, and it lies 3.2° east-southeast of magnitude 3.7 Zeta ($\zeta$) Capricorni. You can also use the magnitude 5.2 star 41 Capricorni as a guide. M30 lies only 23′ west of that star.

I've tried several times to see M30 without optical aid, but I haven't been successful. Perhaps it's because on those occasions the cluster sat low in the sky.

Through a 4-inch telescope, you'll see a bright, broad core you won't resolve surrounded by myriad stars you can resolve. Use a 12-inch scope, however, and crank the magnification to 300× or beyond, and the core will explode with detail. Note the magnitude 8.6 star SAO 190531 less than 6′ to the west-southwest of M30's center.

| OBJECT #680 | NGC 7103 |
|---|---|
| Constellation | Capricornus |
| Right ascension | 21h40m |
| Declination | −2°28′ |
| Magnitude | 12.6 |
| Size | 1.4′ by 1.2′ |
| Type | Elliptical galaxy |

Our next target is a group of galaxies that lies 3° east of magnitude 3.7 Zeta (ζ) Capricorni. The brightest, and that's not saying much, is NGC 7103. You'll need at least a 14-inch or larger telescope and a magnification of 250×. With that, you'll see NGC 7103 as slightly oval with a brighter central region. Magnitude 13.8 NGC 7104 lies 4′ to the northeast, and magnitude 14.8 IC 5122 lies 4′ to the north-northwest.

| OBJECT #681 | NGC 7128 |
|---|---|
| Constellation | Cygnus |
| Right ascension | 21h44m |
| Declination | 53°43′ |
| Magnitude | 9.7 |
| Size | 4′ |
| Type | Open cluster |

Our next target is one of the smallest open clusters in the list. At low power, it gives the appearance of a starry, fat crescent Moon. A 6-inch telescope at 120× reveals 15 stars randomly spread across the field of view.

You can find NGC 7128 2.5° north of magnitude 4.7 Pi$^1$ ($\pi^1$) Cygni.

**Object #682** The Small Cluster Nebula (NGC 7129) Anthony Ayiomamitis

| OBJECT #682 | NGC 7129 |
|---|---|
| Constellation | Cepheus |
| Right ascension | 21h43m |
| Declination | 66°06′ |
| Magnitude | 11.5 |
| Size | 7′ |
| Type | Open cluster |
| Other name | The Small Cluster Nebula |

Our next object is a star cluster, but it also includes a star-forming region that contains emission and reflection nebulosity, which explains this object's common name. The cluster carries the designation NGC 7129. The three nebulous regions within this area are IC 5132, IC 5133, and IC 5134.

A 6-inch telescope will reveal the cluster, a rather poor collection of a dozen or so stars. To see the nebulae, step up to at least a 10-inch scope. Most of the nebulosity is of the reflection type, so no filter helps.

You'll find NGC 7129 2.6° northwest of magnitude 4.4 Xi ($\xi$) Cephei.

**Object #683** Mu Cephei Anthony Ayiomamitis

| OBJECT #683 | Mu Cephei |
|---|---|
| Constellation | Cepheus |
| Right ascension | 21h44m |
| Declination | 58°47′ |
| Magnitude | 3.4 (at maximum) |
| Period | 730 days (variable) |
| Type | Variable star |
| Other name | Herschel's Garnet Star |

Look for Herschel's Garnet Star almost 5° southeast of Cepheus' brightest star, magnitude 2.5 Alderamin (Alpha [α] Cephei). Mu's color won't be immediately apparent to your unaided eye because it isn't bright enough to trigger your color receptors. The best approach to seeing the garnet color is to ever-so-slightly defocus a telescope. The tiny disk you'll create will then display its copper, ruby, rose, or garnet color. Which is it? That depends on your eye's color receptors.

| OBJECT #684 | Piscis Austrinus |
|---|---|
| Right ascension (approx.) | 22h14m |
| Declination (approx.) | –1° |
| Size (approx.) | 245.37 square degrees |
| Type | Constellation |

As its name indicates (if you know Latin, that is), the constellation Piscis Austrinus represents the Southern Fish. This refers to its position relative to another watery constellation, Pisces, which lies a bit to the north.

Piscis Austrinus would be a standard, middle-of-the-road constellation except for one thing — Fomalhaut, the constellation's Alpha star and the sky's 18th-brightest star. This luminary shines at magnitude 1.16, but it dominates its part of the sky because there are no nearby bright stars. The nearest reasonably bright star — Al Nair (Alpha [α] Gruis) — sits some 20° to the south-southwest and shines at magnitude 1.7.

Unfortunately, Piscis Austrinus contains no more of the 200 brightest stars, and none of its other stars have names. Also, the constellation can boast of no meteor shower or Messier object. Still, it's a constellation you should be able to identify.

You can find Piscis Austrinus most easily by using the Great Square of Pegasus. The Square's two westernmost stars — Markab (Alpha Pegasi) and Scheat (Beta Peg) — point southward to Fomalhaut. That being said, it's a long haul. Fomalhaut lies 45° — one-eighth of the way across the sky — south of Markab. After you find Fomalhaut, use a star chart to locate the rest of the figure, which lies to the west of that bright star.

| OBJECT #685 | NGC 7135 |
| --- | --- |
| Constellation | Piscis Austrinus |
| Right ascension | 21h50m |
| Declination | –4°53′ |
| Magnitude | 11.3 |
| Size | 3.0′ by 2.1′ |
| Type | Spiral galaxy |

Our next target lies 2.6° north-northwest of magnitude 3.0 Gamma (γ) Gruis. This small galaxy doesn't show any trace of spiral structure through amateur telescopes. Through 14-inch and larger telescopes, you'll note that the tiny central region has a thin halo surrounding it. A triangle of stars lies on NGC 7135's northwest side. The brightest (and also the farthest from the galaxy) is magnitude 9.5 SAO 213316.

| OBJECT #686 | NGC 7139 |
| --- | --- |
| Constellation | Cepheus |
| Right ascension | 21h46m |
| Declination | 63°49′ |
| Magnitude | 13.3 |
| Size | 78″ |
| Type | Planetary nebula |

Our next target requires at least a 12-inch telescope, although a few top-notch observers have viewed it through scopes with apertures as small as 8 inches. At magnifications above 250×, NGC 7139's disk appears circular with a diffuse edge. I think an OIII filter is a must when you observe this planetary.

You'll find it 2° west-southwest of magnitude 4.3 Kurhah (Xi [ξ] Cephei).

| OBJECT #687 | NGC 7142 |
| --- | --- |
| Constellation | Cepheus |
| Right ascension | 21h46m |
| Declination | 65°48′ |
| Magnitude | 9.3 |
| Size | 4.3′ |
| Type | Open cluster |

To find our next object, aim your telescope 2.3° west-northwest of magnitude 4.4 Kurhah (Xi [ξ] Cephei). This nice cluster presents an even distribution of similarly bright stars. Most shine at magnitude 12, but a few 10th-magnitude gems pepper the cluster. Through a 6-inch telescope, three dozen stars will pop into view. A 10-inch scope reveals 50 members.

| OBJECT #688 | Palomar 12 |
|---|---|
| Constellation | Capricornus |
| Right ascension | 21h47m |
| Declination | –1°15′ |
| Magnitude | 11.7 |
| Size | 2.9′ |
| Type | Globular cluster |

Now for something a bit different. Well, it's a celestial object from a different catalog, at least. Look 2.8° southeast of magnitude 4.5 Epsilon (ε) Capricorni for Palomar 12, a globular that lies more than 60,000 light-years away. The Palomar catalog of globular clusters contains 15 faint objects discovered on the Palomar Observatory Sky Survey plates

You'll need at least an 8-inch telescope to see this object. Even through a 20-inch scope, you'll see only an evenly illuminated glow. That's because the brightest stars in Palomar 12 glow around magnitude 15.

Oh, and in some catalogs you might see this object listed as the Capricornus Dwarf. In 1957, Swiss astronomer Fritz Zwicky cataloged it as a nearby dwarf galaxy, a member of the Local Group. Other astronomers later confirmed its globular cluster status.

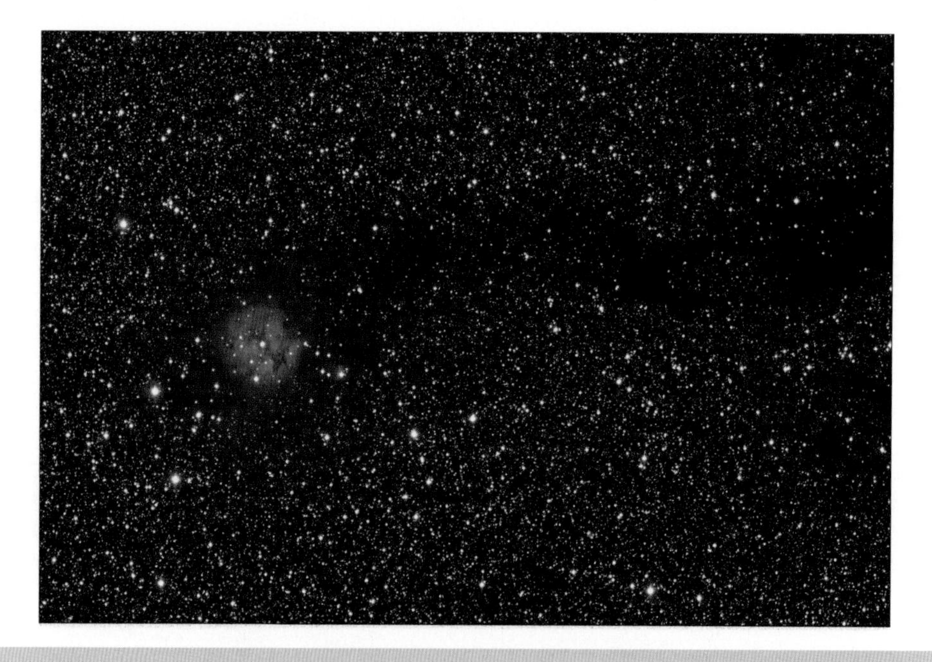

**Object #689** The Cocoon Nebula (IC 5146) Adam Block/NOAO/AURA/NSF

| OBJECT #689 | IC 5146 |
|---|---|
| Constellation | Cygnus |
| Right ascension | 21h53m |
| Declination | 47°16′ |
| Magnitude | 7.2 |
| Size | 12′ |
| Type | Emission nebula |
| Other names | The Cocoon Nebula, Caldwell 19 |

Some deep-sky objects stand on their own. Others pair up with notable companions. Occasionally, the surrounding star field helps elevate a rather ordinary object's status. Such is the case with the Cocoon Nebula.

The Cocoon Nebula overlaps the eastern edge of Barnard 168 (B168), one of the northern sky's finest dark nebulae. From a dark site, sharp-eyed observers can pick up this murky lane without optical aid, but such a sighting isn't easy. A better approach is to use binoculars or a telescope with an eyepiece that provides more than a 1° field of view.

British astronomer Thomas Henry Espinell Compton Espin (1858–1934) discovered the Cocoon Nebula August 13, 1899. German astronomer Maximilian Franz Joseph Cornelius Wolf (1863–1932) was the first to photograph it, in 1900. His description said, "It is placed centrally in a very fine lacuna, void of faint stars, which surrounds the luminous cloud like a trench. The most striking feature with regard to this object is that the star-void halo encircling the nebula forms the end of a long channel ... to a length of more than two degrees."

You can spot the Cocoon Nebula through a 4-inch telescope as a circular blur. Its light equals that of a 7th-magnitude star, but the nebula's surface brightness is low. Making matters worse, two magnitude 9.7 stars lie entangled within the glow. For best results, use at least an 8-inch telescope and an eyepiece equipped with a nebula filter. The filter will suppress the starlight while still passing most of the emission nebula's glow.

| OBJECT #690 | IC 5148 |
|---|---|
| Constellation | Grus |
| Right ascension | 22h00m |
| Declination | −9°23′ |
| Magnitude | 11.0 |
| Size | 120″ |
| Type | Planetary nebula |
| Other name | Spare Tire Nebula |

You'll find our next target 1.3° west of magnitude 4.5 Lambda ($\lambda$) Gruis. It's a gorgeous planetary that displays a thick ring, or annulus. The dark central zone seems small, but it spans one-quarter of IC 5148's diameter. The central star glows dimly at magnitude 15.5, but it's not hard to spot through a 14-inch telescope. Whatever scope you end up using, be sure to crank up the magnification. A magnitude 10.4 star, GSC 7986:150, lies only 2′ south-southwest of center.

According to American amateur astronomer Kent Wallace, the first written mention of this object's common name appeared in the July/August issue of *Southern Astronomy* magazine. Unfortunately, there was no mention of who coined the name.

| OBJECT #691 | NGC 7160 |
|---|---|
| Constellation | Cepheus |
| Right ascension | 21h54m |
| Declination | 62°36′ |
| Magnitude | 6.1 |
| Size | 5′ |
| Type | Open cluster |

You'll find our next target 1.8° north-northeast of magnitude 4.3 Nu ($\nu$) Cephei. This cluster, just barely visible to the naked eye from a dark site, shows 20 stars through a 4-inch telescope. The brightest, magnitude 7.0 SAO 19718, sits just to the northeast of NGC 7160's center.

| OBJECT #692 | NGC 7172 |
|---|---|
| Constellation | Piscis Austrinus |
| Right ascension | 22h02m |
| Declination | –1°52′ |
| Magnitude | 11.8 |
| Size | 2.8′ by 1.4′ |
| Type | Seyfert galaxy |

NGC 7172 forms part of the compact galaxy group Hickson 90. Even at magnitude 11.8, this spiral is the brightest member. The other three galaxies you can observe through a 12-inch or larger telescope from a dark site are NGC 7173 (13.2), NGC 7174 (13.1), and NGC 7176 (12.4). All three lie 7′ to the south of NGC 7172 in an area 2′ across.

NGC 7172 appears moderately bright and twice as long as it is wide. Telescopes with 20-inch and larger apertures may show a dark region halfway from the core to the edge on the galaxy's north side.

| OBJECT #693 | NGC 7184 |
|---|---|
| Constellation | Aquarius |
| Right ascension | 22h03m |
| Declination | –0°49′ |
| Magnitude | 11.2 |
| Size | 6.5′ by 1.4′ |
| Type | Spiral galaxy |

NGC 7184 sits in a desolate region of southwestern Aquarius near that constellation's border with Capricornus. To star-hop to its location, find 6th-magnitude 41 Aquarii, and move 2.7° west.

This spiral's long axis tilts roughly northeast to southwest (its position angle is 62°). It appears to point to a 12th-magnitude star just off its northeast end. Through most scopes NGC 7184 appears four times as long as it is wide.

If you observe NGC 7184 through a 12-inch or larger telescope, you'll spot a faint trio of galaxies in an ever-so-slightly curved line. NGC 7180 (magnitude 12.6), NGC 7185 (magnitude 12.2), and NGC 7188 (magnitude 13.2) lie 17′, 21′, and 32′ north, respectively.

| OBJECT #694 | Lacerta |
|---|---|
| Right ascension (approx.) | 22h25m |
| Declination (approx.) | 46° |
| Size | 200.69 square degrees |
| Type | Constellation |

One object I propose you seek out is the small, faint constellation Lacerta the Lizard. Talk about a loser! Lacerta contains none of the 200 brightest stars, no named star, no meteor shower, and no Messier object.

What's worse, no celestial map pictured Lacerta prior to 1690. In that year, Johannes Hevelius included it, and six other new constellations, in his star atlas *Sobiescianum, sive Uranographia, totum Coelum Stellatum*. Now that's a mouthful.

Look for Lacerta directly south of Cepheus and roughly 20° east of Deneb, Cygnus the Swan's brightest star. Don't get your hopes up. Lacerta's Alpha star glows only at magnitude 3.8.

| OBJECT #695 | IC 5152 |
|---|---|
| Constellation | Indus |
| Right ascension | 22h03m |
| Declination | –1°17′ |
| Magnitude | 10.6 |
| Size | 4.9′ by 3.0′ |
| Type | Irregular galaxy |

You'll find our next object, dwarf irregular galaxy IC 5152, in far eastern Indus, next to its border with Grus. Look 3.8° north of magnitude 4.4 Delta (δ) Indi. The first thing you'll notice when you observe this galaxy is the magnitude 8.2 foreground star that sits on its northwest edge. Because the star shines nine times brighter than the whole galaxy, it's a bit distracting.

Of course, I'm not saying don't observe it. Through an 8-inch telescope, you'll see it stretched in an east-west orientation twice as long as it is wide. With magnifications above 300×, the ends appear to taper to points. The central region is broad and bright, but the outer halo is tough to catch, probably because of the bright star.

| OBJECT #696 | Xi (ξ) Cephei |
|---|---|
| Constellation | Cepheus |
| Right ascension | 22h04m |
| Declination | 64°38′ |
| Magnitudes | 4.4/6.5 |
| Separation | 7.7′ |
| Type | Double star |
| Other name | Kurhah |

To find Xi Cep, look for a star at the midway point between Alderamin (Alpha [α] Cephei) and Iota (ι) Cephei. Although many observers see both components as white, some describe the secondary as a dull yellow. This may be due to the magnitude difference of the two stars. Xi Cephei B shines only one-seventh as bright as the primary.

According to Richard Hinckley Allen, Persian astronomer and philosopher al Kazwani called this star Al Kurhah, an Arabic word that German astronomer Christian Ludwig Ideler (1766–846) translated as a white spot, or blaze, in the face of a horse.

**Object #697** NGC 7209 Anthony Ayiomamitis

| OBJECT #697 | NGC 7209 |
|---|---|
| Constellation | Lacerta |
| Right ascension | 22h05m |
| Declination | 46°30′ |
| Magnitude | 7.7 |
| Size | 25′ |
| Type | Open cluster |

You'll find this object hidden in a rich starfield 5.8° southwest of magnitude 3.8 Alpha (α) Lacertae. A 4-inch telescope at 100× shows 50 stars nicely strewn around the field of view.

The half-dozen brightest stars within NGC 7209 fall in the 9th-magnitude range. They appear to sit in the foreground and contrast nicely with a layer of fainter background stars. Those stars, in turn, lie atop a field of innumerable faint points of light.

About half the stars — including four of the bright ones — shine with a yellowish-white light. The rest are blue, making for a striking overall color contrast throughout this cluster.

| OBJECT #698 | NGC 7213 |
|---|---|
| Constellation | Grus |
| Right ascension | 22h09m |
| Declination | –7°10′ |
| Magnitude | 10.0 |
| Size | 4.8′ by 4.2′ |
| Type | Spiral galaxy |

Although astronomers catalog this object as a spiral galaxy, through amateur telescopes it gives all indications of being elliptical in structure. No aperture I've seen it through (up to 30 inches) even suggests spiral arms. What you can see, however, is a round object with an evenly illuminated central region that takes up one-third of NGC 7213's diameter. The haze surrounding the core isn't faint, and all you need is a magnification above 250× to see it.

You'll find this galaxy easily. It lies only 16′ southeast of magnitude 1.7 Al Nair (Alpha [α] Gruis).

| OBJECT #699 | NGC 7217 |
|---|---|
| Constellation | Pegasus |
| Right ascension | 22h08m |
| Declination | 31°22′ |
| Magnitude | 10.1 |
| Size | 3.5′ by 3.1′ |
| Type | Spiral galaxy |

Our next object lies 1.9° south-southwest of the magnitude 4.3 Pi$^2$ ($\pi^2$) Pegasi. Through an 8-inch telescope, you'll see a bright central region that spans half the galaxy's diameter. The halo is easy to see at high magnifications.

| OBJECT #700 | NGC 7235 |
|---|---|
| Constellation | Cepheus |
| Right ascension | 22h12m |
| Declination | 57°16′ |
| Magnitude | 7.7 |
| Size | 4′ |
| Type | Open cluster |

To find our next object, look 0.4° west-northwest of magnitude 4.2 Epsilon (ε) Cephei. A 4-inch telescope at 150× only reveals about a dozen stars haphazardly strewn across the field of view. Increasing the aperture to 8 inches shows 20 stars. Despite the small number of stars, the range of brightnesses spans seven magnitudes, from 9th to 15th.

**Object #701** NGC 7243 Anthony Ayiomamitis

| OBJECT #701 | NGC 7243 |
|---|---|
| Constellation | Lacerta |
| Right ascension | 22h15m |
| Declination | 49°53′ |
| Magnitude | 6.4 |
| Size | 21′ |
| Type | Open cluster |
| Other name | Caldwell 16 |

Our next target is a barely naked-eye open cluster that lies 2.6° west of magnitude 3.8 Alpha (α) Lacertae. Through a 4-inch telescope, you'll pick out three dozen stars randomly strewn across the field of view. An 8-inch scope increases the number to 50. And, unlike some open clusters, adding aperture really cranks up the number of stars, although the question remains as to whether these are true cluster members or just background stars. Through a 16-inch at a dark site, I once estimated more than 200 stars in NGC 7243. The brightest star in the cluster, GSC 3614:2189, glows at magnitude 8.0 dead-center.

| OBJECT #702 | IC 5201 |
|---|---|
| Constellation | Grus |
| Right ascension | 22h21m |

| (continued) | |
|---|---|
| Declination | –6°04′ |
| Magnitude | 11.0 |
| Size | 6.6′ by 4.4′ |
| Type | Barred spiral galaxy |

Our next target lies 0.4° west-southwest of magnitude 5.6 Pi$^2$ ($\pi^2$) Gruis. Through an 8-inch telescope at 200×, this thin galaxy appears twice as long as it is wide, oriented north-south. The bar is easy to see as the broad, bright central region. A haze around the bar hints at spiral arms, but you won't see anything even remotely resembling spiral structure through any scope 20′ in diameter or less.

| OBJECT #703 | Zeta (ζ) Aquarii |
|---|---|
| Constellation | Aquarius |
| Right ascension | 22h29m |
| Declination | –°01′ |
| Magnitudes | 4.3/4.5 |
| Separation | 2.0″ |
| Type | Double star |

You'll find this binary star 2.3° northeast of magnitude 3.8 Gamma (γ) Aquarii. Crank the magnification up to 150× or more to spot both stars. The primary is only slightly (20%) brighter than the secondary, and both shine with a white light.

| OBJECT #704 | NGC 7245 |
|---|---|
| Constellation | Lacerta |
| Right ascension | 22h15m |
| Declination | 54°20′ |
| Magnitude | 9.2 |
| Size | 5′ |
| Type | Open cluster |

Our next target lies 2.4° north-northwest of magnitude 4.4 Beta (β) Lacertae. A 6-inch telescope will show you 20 closely spaced stars, the brightest of which shines at magnitude 10.7.

| OBJECT #705 | NGC 7261 |
|---|---|
| Constellation | Cepheus |
| Right ascension | 22h20m |
| Declination | 58°05′ |
| Magnitude | 8.4 |
| Size | 6′ |
| Type | Open cluster |

Look 1.2° east of magnitude 3.4 Zeta (ζ) Cephei for this small cluster. Through a 6-inch telescope, about 15 stars pop into view. The brightest is magnitude 9.6 SAO 34332 at the cluster's southeastern edge.

| OBJECT #706 | Delta (δ) Cephei |
|---|---|
| Constellation | Cepheus |
| Right ascension | 22h29m |
| Declination | 58°25′ |
| Magnitudes | 3.9/6.3 |
| Separation | 41′ |
| Type | Double star |

Our next target is a wide binary you'll split through any size telescope. To find it, look 2.4° east of magnitude 3.4 Zeta (ζ) Cephei. The primary appears yellow, and the secondary is blue.

Delta Cephei A also is famous because it's the prototype of a class of variable stars called Cepheid variables. Such stars expand and contract at exact intervals. Delta Cep takes 5.366341 days to cycle through its period. During that time, its magnitude varies from 3.48 to 4.37.

Because Cepheid variables pulsate at such regular intervals (what astronomers define as the period-luminosity relationship), researchers can use them as "standard candles" to star clusters or galaxies the stars are in. Using Cepheid variables, American astronomer Edwin Hubble determined that the Andromeda Galaxy was not part of the Milky Way, but rather an entirely separate star system.

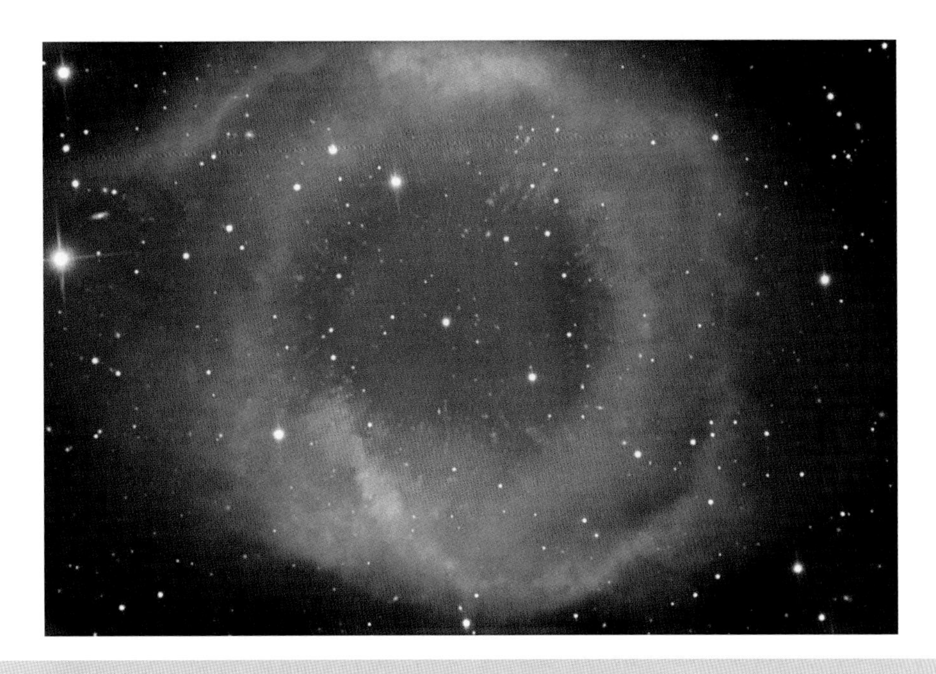

**Object #707** The Helix Nebula (NGC 7293) Adam Block/NOAO/AURA/NSF

| OBJECT #707 | NGC 7293 |
|---|---|
| Constellation | Aquarius |
| Right ascension | 22h30m |
| Declination | −0°48′ |
| Magnitude | 7.3 |
| Size | 13′ |
| Type | Planetary nebula |
| Other names | The Helix Nebula, the Sunflower Nebula, Caldwell 63 |

The Helix Nebula is one of the brightest hard-to-see objects in the sky. Although its total light output nearly reaches 7th magnitude, its surface brightness is disappointingly low. For best results, use binoculars that provide 7× to 15× magnification and have front lenses larger than 50 millimeters.

If your site is dark — and I mean really dark — try to spot the Helix with your naked eyes. A naked-eye sighting is a bit easier if you first find the Helix through binoculars. Fortunately, no stars of equivalent brightness lie in the Helix's immediate vicinity.

Through 12-inch and larger telescopes, and using a nebula filter, you can pick out brightness differences within the ring. Slightly brighter concentrations lie on the northern and southern edges. Also note that the central "dark" area appears brighter than the background sky outside the ring.

German astronomer Karl Ludwig Harding (1765–1834) published an account of his discovery of the Helix Nebula in the *Berliner Jahrbuch* in 1827. The Helix is one of the largest and nearest planetary nebulae.

You'll find this great object 1.2° west of magnitude 5.2 Upsilon (υ) Aquarii.

# September

| OBJECT #708 | 8 Lacertae |
|---|---|
| Constellation | Lacerta |
| Right ascension | 22h36m |
| Declination | 39°38′ |
| Magnitudes | 5.7/6.5 |
| Separation | 22.4″ |
| Type | Double star |

This binary lies in a rather star-poor region of the sky. One way to find it is to move 5.4° southwest from magnitude 3.6 Omicron ($o$) Andromedae. Any size telescope will split this pair. Both components are white.

| OBJECT #709 | NGC 7314 |
|---|---|
| Constellation | Piscis Austrinus |
| Right ascension | 22h36m |
| Declination | –26°03′ |
| Magnitude | 10.9 |
| Size | 4.2′ by 1.7′ |
| Type | Spiral galaxy |

This spiral is easy to spot through a 6-inch telescope because of its high surface brightness. It lies 1° due east of magnitude 6.4 Zeta ($\zeta$) Piscis Austrini. NGC 7314 has a position angle of 3°, which means the object pretty much appears on a north-south line.

While you're observing this galaxy, try to spot the smaller and fainter spiral galaxy NGC 7313 roughly 4′ to the southwest. Only one-sixth as long as NGC 7314, NGC 7313 glows at magnitude 14.2.

M.E. Bakich, *1,001 Celestial Wonders to See Before You Die*, Patrick Moore's Practical Astronomy Series,
DOI 10.1007/978-1-4419-1777-5_9, 

**Object #710** Stephan's Quintet Adam Block/NOAO/AURA/NSF

| OBJECT #710 | Stephan's Quintet |
|---|---|
| Constellation | Pegasus |
| Right ascension | 22h36m |
| Declination | 33°58′ |
| Magnitudes | 14.8, 14.6, 14.0, 14.4, 13.6 |
| Sizes | 1.1′ by 1.1′, 0.9′ by 0.9′, 1.9′ by 1.2′, 1.7′ by 1.3′, 2.2′ by 1.1′ |
| Type | Galaxy group |
| Notes | NGCs 7317, 7318A & B, 7319, 7320 |

One of the mantras of amateur astronomy is “aperture rules.” This means you'll see more detail as you observe through ever-larger telescopes. Few celestial objects demonstrate this better than Stephan's Quintet.

French astronomer Edouard Stephan discovered this group in 1877. The five galaxies now carry the designations NGC 7317, NGC 7318A, NGC 7318B, NGC 7319, and NGC 7320.

Four of these galaxies — the exception is NGC 7320 — form a compact galaxy group, the first ever discovered. NGC 7320 belongs to the Pegasus Spur, a group of about three dozen galaxies, the brightest of which is magnitude 9.5 NGC 7331.

Although you can “see” Stephan's Quintet through a 6-inch scope, 50× will show you only a faint, clumpy glow 3′ across. A 12-inch telescope, on the other hand, lets you identify the individual members.

At the Quintet's southwestern edge is NGC 7317, which lies next to a 13th-magnitude foreground star. The colliding pair NGC 7318A and NGC 7318B lies 2′ to the east. You'll need high magnification — above 200× — to “unmerge” them.

The brightest and largest member, NGC 7320, lies to the southeast and contains a 13th-magnitude foreground star in its halo. This galaxy shines at magnitude 12.5. NGC 7319, which sits at the group's northeastern edge, is the real test for visual observers and is the faintest galaxy in Stephan's Quintet. It glows softly at magnitude 15.3.

| OBJECT #711 | NGC 7329 |
|---|---|
| Constellation | Tucana |
| Right ascension | 22h40m |
| Declination | −66°29′ |
| Magnitude | 11.8 |
| Size | 3.2′ by 1.9′ |
| Type | Barred spiral galaxy |

Our next target lies 2° southeast of magnitude 4.5 Delta (δ) Tucanae. Although it lies 150 million light-years away, a 12-inch telescope will show some detail. Crank the power up to 250× or beyond, and you'll easily spot the bar. Then note the central bulge, slightly fatter than the bar. Finally observe the irregular haze surrounding the central region that gives away the galaxy's spiral structure.

**Object #712** The Deer Lick Group (NGC 7331) Adam Block/Mount Lemmon SkyCenter/University of Arizona

| OBJECT #712 | NGC 7331 |
|---|---|
| Constellation | Pegasus |
| Right ascension | 22h37m |
| Declination | 34°25′ |
| Magnitude | 9.5 |
| Size | 10.5′ by 3.7′ |
| Type | Spiral galaxy |
| Other names | The Deer Lick Group, Caldwell 30 |

Deep-sky objects have some fanciful names. Usually, however, they fit the view. For example, the Blue Snowball (Object #733) is blue and round, the Omega Nebula (Object #538) looks like that Greek letter, and Gomez's Hamburger (IRAS 18059–3211) looks like a sandwich.

But what's with a deep-sky object called the "Deer Lick Group"? Well, in the 1980s, American amateur astronomer Tom Lorenzin bestowed the common name on this galaxy group to honor the Deer Lick Gap, which lies in the mountains of North Carolina. Apparently, Tom had a memorable view of these galaxies from there.

The Deer Lick Group's brightest member is NGC 7331. From a dark sky, you can spot this magnitude 9.5 spiral galaxy through binoculars, but a telescope brings out a lot more detail. Through a 10-inch scope with a low-power eyepiece, you'll see three galaxies to NGC 7331's east that form an equilateral triangle. These galaxies are not NGC 7331's companions but lie much farther away.

At 200×, the galaxy shows a bright nucleus surrounded by a nebulous glow 3 times as long as it is wide. Larger scopes show the western edge ends abruptly at a dust lane. On nights of good seeing, look for a spiral arm shining beyond this lane.

| OBJECT #713 | NGC 7332 |
|---|---|
| Constellation | Pegasus |
| Right ascension | 22h37m |
| Declination | 23°48′ |
| Magnitude | 11.1 |
| Size | 3.7′ by 1.0′ |
| Type | Spiral galaxy |

I'm certain you'll enjoy our next object, or should I say pair of objects? NGC 7332 and NGC 7339, just 5′ to its east, form a gorgeous pair of lens-shaped galaxies just 2.1° west of magnitude 4.0 Lambda (λ) Pegasi.

Both objects appear more than three times as long as they are wide. Each has even illumination, but NGC 7332's core is a bit broader. NGC 7339 glows more faintly than its neighbor, at magnitude 12.2.

| OBJECT #714 | NGC 7361 |
|---|---|
| Constellation | Piscis Austrinus |
| Right ascension | 22h42m |
| Declination | –30°03′ |
| Magnitude | 12.2 |
| Size | 4.0′ by 0.9′ |
| Type | Spiral galaxy |

Just look 3.3° west of magnitude 1.2 Fomalhaut (Alpha [α] Piscis Austrini) to find our next target. Through a 10-inch telescope, you'll see a celestial sliver oriented north-south appearing five times as long as it is wide. At 300× through a 12-inch scope, look for a truncation at the northern tip.

**Object #715** NGC 7380 Kris Sandburg and Peter Jacobs/Adam Block/NOAO/AURA/NSF

| OBJECT #715 | NGC 7380 |
|---|---|
| Constellation | Cepheus |
| Right ascension | 22h47m |
| Declination | 58°06′ |
| Magnitude | 7.2 |
| Size | 20′ |
| Type | Open cluster |
| Other name | The Wizard Nebula |

Our next object lies 2.4° east of magnitude 4.1 Delta ($\delta$) Cephei. Through an 8-inch telescope, two dozen stars 10th-magnitude and fainter pop into view.

Now insert a nebula filter and observe emission nebula Sharpless 2–142. It appears as an unevenly bright haze with an irregular border 0.5° long stretching from north to south. The northern part of the nebula glows more brightly than the southern part.

Amateur astronomers began calling this object the Wizard Nebula in the early 2000s when astroimagers began to circulate long-exposures of it on the Internet.

| OBJECT #716 | NGC 7418 |
|---|---|
| Constellation | Grus |
| Right ascension | 22h57m |
| Declination | –37°02′ |
| Magnitude | 11.0 |
| Size | 4.2′ by 2.1′ |
| Type | Barred spiral galaxy |

You'll find this object 2.8° northwest of magnitude 5.6 Upsilon ($\upsilon$) Gruis. Through a 16-inch telescope at 350×, the galaxy appears face-on and irregularly bright with ever-so-faint traces of spiral structure. Look for the magnitude 11.7 spiral NGC 7421 just 20′ to the south.

| OBJECT #717 | Sharpless 2–155 |
| --- | --- |
| Constellation | Cepheus |
| Right ascension | 22h57m |
| Declination | 62°37′ |
| Magnitude | — |
| Size | 50′ by 30′ |
| Type | Emission nebula |
| Other names | The Cave Nebula, Caldwell 9 |

Our next object lies 3.7° south-southeast of magnitude 3.5 Iota ($\iota$) Cephei. To be honest, this is a better target for astroimagers than for visual observers, but a large telescope (16 inches of aperture or more) equipped with a nebula filter will bring it in for you. Look for a deep, dark indentation in the nebulosity that gives Sh 2–155 its common name.

This nebula's moniker arises from the wide, dark region easily visible on images. The dark nebulosity appears like the mouth of a cave, thus the name.

| OBJECT #718 | IC 1459 |
| --- | --- |
| Constellation | Grus |
| Right ascension | 22h57m |
| Declination | –36°28′ |
| Magnitude | 10.0 |
| Size | 4.9′ by 3.6′ |
| Type | Elliptical galaxy |

You'll find our next target a bit more than 3° northwest of magnitude 5.6 Upsilon ($\upsilon$) Gruis. Through a 12-inch telescope at 150, the galaxy inclines northeast to southwest and appears evenly illuminated. Double the magnification to 300×, and you'll see the thin faint halo that surrounds the extended central region. You'll also notice IC 1459 isn't quite round.

| OBJECT #719 | NGC 7457 |
| --- | --- |
| Constellation | Pegasus |
| Right ascension | 23h01m |
| Declination | 30°09′ |
| Magnitude | 11.2 |
| Size | 4.1′ by 2.5′ |
| Type | Spiral galaxy |

Our next object lies 2.1° north-northwest of magnitude 2.4 Scheat (Beta [$\beta$] Pegasi). Through an 8-inch telescope, it appears nearly rectangular, twice as long as it is wide oriented northwest to southeast.

| OBJECT #720 | NGC 7462 |
| --- | --- |
| Constellation | Grus |
| Right ascension | 23h03m |
| Declination | –40°50′ |
| Magnitude | 11.3 |
| Size | 5.1′ by 0.8′ |
| Type | Spiral galaxy |

Our next object lies 2.1° south-southwest of magnitude 5.6 Upsilon ($\upsilon$) Gruis. Another galactic "needle," NGC 7462 measures six times as long as it is wide. A magnitude 10.4 star lies at this galaxy's southwest tip.

| OBJECT #721 | NGC 7479 |
|---|---|
| Constellation | Pegasus |
| Right ascension | 23h05m |
| Declination | 12°19′ |
| Magnitude | 10.8 |
| Size | 4.0′ by 3.1′ |
| Type | Barred spiral galaxy |
| Other name | Caldwell 44 |

This showpiece galaxy (through a large scope) lies 2.9° south of magnitude 2.5 Markab (Alpha [α] Pegasi). A 10-inch telescope shows the odd spiral structure. At low power, you'll detect a bright core, the surrounding central bulge, and a bar elongated north-south. This galaxy's best feature is the single, tightly wound spiral arm curling to the west of the south end of the bar. The north end of the bar seems cut off. There's no trace of a spiral arm here.

| OBJECT #722 | IC 1470 |
|---|---|
| Constellation | Cepheus |
| Right ascension | 23h05m |
| Declination | 60°15′ |
| Size | 1.2′ by 0.8′ |
| Type | Emission nebula |

You'll find this nebula 5° east-northeast of magnitude 4.1 Delta (δ) Cephei. Through a 12-inch telescope at 200×, IC 1470 appears bright with a fainter extension to the southeast.

| OBJECT #723 | NGC 7492 |
|---|---|
| Constellation | Aquarius |
| Right ascension | 23h08m |
| Declination | −15°37′ |
| Magnitude | 11.4 |
| Size | 6.2′ |
| Type | Globular cluster |

Talk about low surface brightness! NGC 7492's diameter spans a quarter that of the Full Moon, but it shines more faintly than magnitude 11. In fact, even through a 12-inch telescope, you'll find it hard to believe you're observing a globular cluster.

For best results, head for a dark site, and crank the power past 200×. Even then, you only will spot 10 stars. NGC 7492 sits 3.3° east of magnitude 3.3 Skat (Delta [δ] Aquarii).

| OBJECT #724 | NGC 7510 |
|---|---|
| Constellation | Cepheus |
| Right ascension | 23h11m |
| Declination | 60°34′ |
| Magnitude | 7.9 |
| Size | 4′ |
| Type | Open cluster |

You'll find this nice cluster 5° west-northwest of magnitude 4.9 Tau (τ) Cassiopeiae. A 4-inch telescope at 150× will reveal three dozen stars. Back off the magnification to 75× or lower, and you'll see a bar-shaped stellar region stretching northeast to southwest through the center of NGC 7510.

**Object #725** NGC 7538 Fred Calvert/Adam Block/NOAO/AURA/NSF

| OBJECT #725 | NGC 7538 |
|---|---|
| Constellation | Aquarius |
| Right ascension | 23h14m |
| Declination | 61°31′ |
| Size | 9′ by 6′ |
| Type | Emission nebula |

Look for this nebula 5° northwest of magnitude 4.9 Tau (τ) Cassiopeiae. This object is part of a vast array of nebulosity known as the Cassiopeia complex. Through an 8-inch telescope, you'll see NGC 7538's outline, but it will be faint. A 12-inch scope will show a fat, nearly rectangular object half again as long as it is wide, oriented northeast to southwest.

| OBJECT #726 | NGC 7582 |
|---|---|
| Constellation | Grus |
| Right ascension | 23h18m |
| Declination | −42°22′ |
| Magnitude | 10.1 |
| Size | 6.9′ by 2.6′ |
| Type | Barred spiral galaxy |

Look for this galaxy, and several others, 2.4° east-northeast of magnitude 4.3 Theta (θ) Gruis. This galaxy stretches twice as long as it is wide and orients northwest to southeast. Through any telescope less than 30 inches in aperture, it appears evenly illuminated across its face. Only that giant scope could bring out the ever-so-faint haze that gave away the presence of tightly wound spiral arms.

Another spiral, magnitude 10.6 NGC 7552, sits 0.5° to the west-southwest. A pair of galaxies, magnitude 11.3 NGC 7590 and magnitude 11.5 NGC 7599, both spirals, lie 12′ to the east-northeast. Including NGC 7582, this foursome of galaxies is known as the Grus Quartet.

| OBJECT #727 | 94 Aquarii |
|---|---|
| Constellation | Aquarius |
| Right ascension | 23h19m |
| Declination | –13°28′ |
| Magnitudes | 5.3/7.3 |
| Separation | 12.7″ |
| Type | Double star |

Look for our next target 6.4° east-northeast of magnitude 3.3 Skat (Delta [δ] Aquarii). You'll find a pretty pair featuring a yellow primary shining 2 magnitudes brighter than its orange companion.

**Object #728** NGC 7606 Adam Block/Mount Lemmon SkyCenter/University of Arizona

| OBJECT #728 | NGC 7606 |
|---|---|
| Constellation | Aquarius |
| Right ascension | 23h19m |
| Declination | -8°29′ |
| Magnitude | 10.8 |
| Size | 4.4′ by 2.0′ |
| Type | Barred spiral galaxy |

Our next object lies slightly less than 1° southeast of magnitude 4.9 Chi (χ) Aquarii. NGC 7606 is an inclined spiral, but its arms lie tightly wound about the core, so you won't be observing them through normal-sized amateur telescopes. Look for a broadly concentrated core that occupies about half the overall length. The arms lie outside of that bright region, and then there's a rapid fade to the blackness of space.

| OBJECT #729 | NGC 7626 |
|---|---|
| Constellation | Pegasus |
| Right ascension | 23h21m |
| Declination | 8°13′ |

| (continued) | |
|---|---|
| Magnitude | 11.1 |
| Size | 2.4′ by 1.9′ |
| Type | Elliptical galaxy |

Head 2.6° northwest of magnitude 4.3 Theta ($\theta$) Piscium, just over the Pegasus border, to find this giant elliptical galaxy. It and its near twin, NGC 7619, which lies just 7′ to the west, are the brightest members of the Pegasus I galaxy cluster. Despite being nearly 200 million light-years away, the two appear quite bright. An 8-inch telescope shows each galaxy as a bright core immersed in a featureless halo. Both galaxies have a slightly oval appearance.

**Object #730** The Bubble Nebula (NGC 7635) Brad Ehrhorn/Adam Block/NOAO/AURA/NSF

| OBJECT #730 | NGC 7635 |
|---|---|
| Constellation | Cassiopeia |
| Right ascension | 23h21m |
| Declination | 61°12′ |
| Size | 15′ by 8′ |
| Type | Emission nebula |
| Other names | The Bubble Nebula, Caldwell 11 |

The bright open cluster M52 in Cassiopeia serves as a guide to the Bubble Nebula. Observing it will give you a glimpse of how a star interacts with what's around it.

The remarkably spherical bubble marks the boundary between an intense wind of particles from a massive, hot star, BD+602522, and the nebula's interior. The central star is 40 times more massive than the Sun and emits a stellar wind moving at 4 million mph (7 million km/h). The bubble's surface actually marks the leading edge of this wind's gust front, which slows as it plows into the denser surrounding material.

The bubble's surface is not uniform because, as the shell expands outward, it encounters gaseous regions of different densities that impede the expansion by differing amounts. More material lies to the northeast than to the southwest, so the wind progresses less in that direction, offsetting the central star from the bubble's center.

Sir William Herschel discovered the Bubble Nebula in 1787. Of it, he wrote, "A star of 9th magnitude with very faint nebulosity of small extent about it." The star measures a magnitude brighter than Herschel's estimate, but his assessment of the nebula was correct — it's faint.

An 8-inch telescope at a dark site barely shows NGC 7635 as a 3′ by 1′ arc surrounding the star. This wisp of light floats in a rich field of faint background stars. Through a 16-inch scope, you'll see the whole bubble. Try to detect a fainter haze north of the bright arc and separated from it by a dark lane. A nebula filter will help you see this object better.

| OBJECT #731 | NGC 7640 |
| --- | --- |
| Constellation | Andromeda |
| Right ascension | 23h22m |
| Declination | 40°51′ |
| Magnitude | 11.3 |
| Size | 10.0′ by 2.2′ |
| Type | Barred spiral galaxy |

Near Andromeda's border with Lacerta is where you'll find our next object. It lies 4° east-southeast of magnitude 3.6 Omicron (*ο*) Andromedae. An 8-inch telescope brings out the general lens shape, which is the stretched-out core and the first hint of spiral arms. Move up to a 12-inch scope, and you'll see faint extensions to the north and south. Large star-forming regions near the core give NGC 7640's surface an uneven texture.

**Object #732** M52 Anthony Ayiomamitis

| OBJECT #732 | M52 (NGC 7654) |
|---|---|
| Constellation | Cassiopeia |
| Right ascension | 23h24m |
| Declination | 61°35′ |
| Magnitude | 6.9 |
| Size | 12′ |
| Type | Open cluster |

Messier's 52nd entry is a fine open cluster observers with the sharpest eyes may just discern from the darkest locations. To find it, draw a line from magnitude 2.2 Alpha ($\alpha$) Cassiopeiae to magnitude 2.3 Beta ($\beta$) Cas. That distance is 5°. Now extend the line another 6°, and you'll land on M52.

Through an 8-inch telescope you'll see at least 75 stars ranging from 9th to 12th magnitude. The cluster appears well defined against the starry background, particularly on its western edge. A prominent clump of six stars lies on the eastern edge.

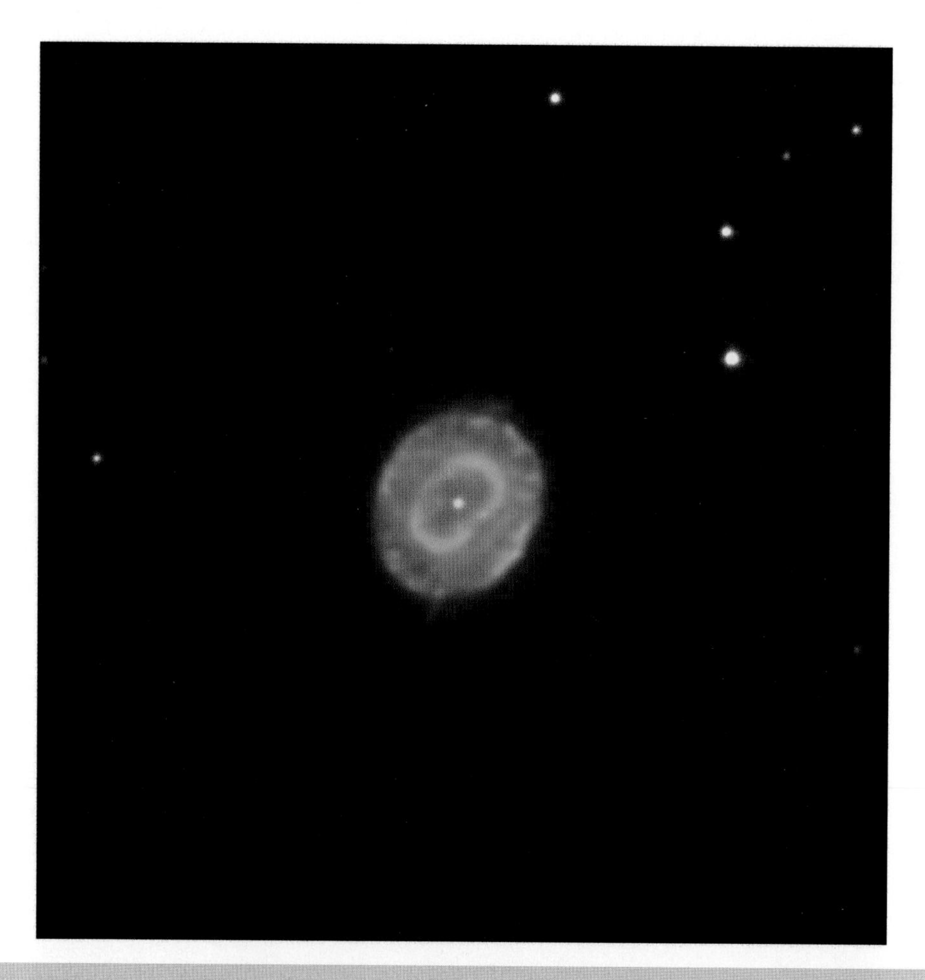

**Object #733** The Blue Snowball (NGC 7662) Adam Block/NOAO/AURA/NSF

| OBJECT #733 | NGC 7662 |
|---|---|
| Constellation | Andromeda |
| Right ascension | 23h26m |
| Declination | 42°33′ |
| Magnitude | 8.3 |
| Size | 12″ |
| Type | Planetary nebula |
| Other names | The Blue Snowball, Caldwell 22 |

If you have access to an 8-inch or larger telescope, look in the northern part of the constellation Andromeda for a planetary nebula called the Blue Snowball. You'll see immediately why astronomers gave it that name.

The Blue Snowball shines at magnitude 9. Luckily, its light isn't spread out over a large area. NGC 7662 — another designation for this object — measures only 2.2′ across. This small size concentrates the planetary's light, allowing it to trigger your eyes' color receptors. If you're looking for (or wanting to show somebody) color in a deep-sky object, look no further than the Blue Snowball.

That being said, different observers have described it as pale blue, faint blue, light blue, Robin's-egg blue, slightly blue, whitish-blue, and, occasionally, various shades of light green. What's more, nobody's wrong. Each of us has our own sense of color perception, and it may differ a little or a lot from the observer next in line.

Through an 8-inch scope, the Blue Snowball appears as a small, evenly illuminated disk. You won't see the 13th-magnitude central star in anything less than a 16-inch scope, so search for other details, like the nebula's rich inner structure.

Look for a bright ring of gas surrounding NGC 7662's hollow center. A fainter gas shell — tough to see — encompasses the ring. The ring's brightest areas lie to the northeast and southwest. At magnifications above 300×, the brightness of the shell drops quickly near its edge.

| OBJECT #734 | NGC 7678 |
|---|---|
| Constellation | Pegasus |
| Right ascension | 23h29m |
| Declination | 22°25′ |
| Magnitude | 11.3 |
| Size | 2.3′ by 1.7′ |
| Type | Spiral galaxy |

Here's another nice deep-sky treat that disproves the notion that nothing lies within Pegasus' Great Square. NGC 7678 lies 1.2° southeast of magnitude 4.4 Upsilon ($\upsilon$) Pegasi. This face-on spiral has tightly wrapped arms that only begin to reveal themselves through 14-inch telescopes at high power. Note the nice isosceles triangle of 12th-magnitude stars that frames the galaxy.

| OBJECT #735 | NGC 7686 |
|---|---|
| Constellation | Andromeda |
| Right ascension | 23h30m |
| Declination | 49°08′ |
| Magnitude | 5.6 |
| Size | 15′ |
| Type | Open cluster |

Our next object is one you'll be able to spot with your naked eyes from a dark site. Through a 4-inch scope, about 20 stars appear, ranging in magnitudes from 7.5 to 11. Look for this cluster 3° north-northwest of magnitude 3.8 Lambda ($\lambda$) Andromedae.

| OBJECT #736 | UGC 12613 |
|---|---|
| Constellation | Pegasus |
| Right ascension | 23h29m |
| Declination | 14°45′ |
| Magnitude | 12.6 |
| Size | 4.6′ by 2.8″ |
| Type | Dwarf irregular galaxy |
| Other name | The Pegasus Dwarf |

Our next target, the Pegasus Dwarf (UGC 12613), isn't bright, but it's one you should seek out. It lies just below the Great Square of Pegasus. The Pegasus Dwarf is one of the most distant members of our Local Group of galaxies. At a distance of 5.7 million light-years, it sits a bit more than twice as far from us as the Andromeda Galaxy (M31).

Through a 10-inch telescope, this dwarf irregular appears as a dim mist twice as long as it is wide. Don't expect to see many details. You'll find it 5.8° due east of Markab (Alpha [α] Pegasi).

**Object #737** Cederblad 211 Adam Block/Mount Lemmon SkyCenter/University of Arizona

| OBJECT #737 | Cederblad 211 |
|---|---|
| Constellation | Aquarius |
| Right ascension | 23h44m |
| Declination | −15°17′ |
| Size | 2′ by 1′ |
| Type | Emission nebula |

This ultra-faint patch of nebulosity surrounds the variable star R Aquarii. It's a challenge object even through a 24-inch telescope, so be warned.

You'll need an Oxygen-III filter and an eyepiece that will provide a magnification of 300× or more. Look for two tiny (10″-long) streaks that extend from R Aquarii to the west-southwest and the east-northeast. Observers have postulated that the nebula may be easier to see when R Aquarii is near its minimum brightness.

| OBJECT #738 | R Aquarii |
|---|---|
| Constellation | Aquarius |
| Right ascension | 23h44m |
| Declination | –15°17′ |
| Magnitude range | 6–12 |
| Period | 386.96 days |
| Type | Variable star |

This Mira-type variable star (for Mira, see Object #822) takes a bit more than a year to cycle through a 6-magnitude brightness range. When it's at maximum brightness, the star is usually barely visible without optical aid from a dark site. If you enjoy variable-star observing, R Aquarii is one to watch. Both its minimum and maximum brightnesses vary by more than a magnitude.

| OBJECT #739 | NGC 7741 |
|---|---|
| Constellation | Pegasus |
| Right ascension | 23h44m |
| Declination | 26°05′ |
| Magnitude | 11.3 |
| Size | 4.0′ by 2.7′ |
| Type | Barred spiral galaxy |

Our next object requires an 8-inch telescope to reveal its details. Through such an instrument at 100×, NGC 7741 appears as a mottled, round haze. Move up to a magnification of 250×, and you'll see the uniformly lit bar that extends from east to west across the glow. Apertures of 18 inches or more show hints of the faint surrounding spiral arms.

A nice double star consisting of magnitude 9.8 GSC 2254:1685 and magnitude 11.9 GSC 2254:1349 lies at the northern edge of the halo and points to the galaxy's core. Approximately 20″ separates the two stars.

You'll find NGC 7741 6.2° west-southwest from magnitude 2.0 Alpheratz (Alpha [α] Andromedae).

| OBJECT #740 | NGC 7762 |
|---|---|
| Constellation | Cepheus |
| Right ascension | 23h50m |
| Declination | 68°02′ |
| Magnitude | 10.0 |
| Size | 15′ |
| Type | Open cluster |

You'll find our next target a bit more than 6° east-northeast of magnitude 3.5 Iota (ι) Cephei. Look carefully. NGC 7762 has a tendency to fade into the background when viewed with medium to high magnifications. Back off the power initially, and then increase it in steps to see several dozen 11th- and 12th-magnitude stars across this cluster's face.

| OBJECT #741 | NGC 7788 |
|---|---|
| Constellation | Cassiopeia |
| Right ascension | 23h57m |

| (continued) | |
|---|---|
| Declination | 61°24′ |
| Magnitude | 9.4 |
| Size | 4.0′ |
| Type | Open cluster |

Our next celestial target, NGC 7788, is the most northwestern of a line of three open clusters. The brightest, magnitude 8.5 NGC 7790 (Object #743), lies 17′ to the southeast. Continue the line another 20′ to the southeast, and you'll encounter magnitude 9.7 Berkeley 58. To find this group, start at magnitude 2.3 Caph (Beta [$\beta$] Cassiopeiae), and move 2.5° northwest.

**Object #742** The Screaming Skull (NGC 7789) Anthony Ayiomamitis

| OBJECT #742 | NGC 7789 |
|---|---|
| Constellation | Cassiopeia |
| Right ascension | 23h57m |
| Declination | 56°44′ |
| Magnitude | 6.7 |
| Size | 15′ |
| Type | Open cluster |
| Other names | The Crab Cluster, Herschel's Spiral Cluster, the Screaming Skull |

You'll find this terrific open cluster midway between magnitude 5.0 Sigma ($\sigma$) Cassiopeiae and magnitude 6.0 Rho ($\rho$) Cas. Even through a 4-inch telescope, you'll see 50 stars evenly spread across this rich cluster's face. An 8-inch telescope shows more than a hundred members and the number just keeps increasing with aperture.

Because so many equally bright stars occupy your view, your eyes will have a tendency to create patterns. Some observers see dark lanes between lines of stars. I see a stellar pattern that resembles a pinwheel, or a face-on spiral galaxy with four distinct arms. Some observers note the apparent counter-clockwise spiral pattern of the stellar "arms" of this cluster. Because British astronomer Caroline Herschel discovered this object in 1783, this feature led to the nickname Herschel's Spiral Cluster.

NGC 7789 (long before it had an NGC number) was the last deep-sky object Admiral Smyth described in his 1844 epic *Cycle of Celestial Objects.* He said, "It is, indeed, a very glorious assemblage, both in extent and richness, having spangly rays of stars which give it a remote resemblance to a crab, the claws reaching the confines of the space in view, under an eyepiece magnifying 185 times. With this form in the mind's eye, the imagined head will be in the *np* [north preceding, i.e. northwest], the tail in the *sf* [south following, i.e. southeast], and where the eyes would be, is the minute close double star of the 11th and 12th magnitudes, above estimated. There are several other pairs in the figure, especially towards the tail. The crab itself is but a mere condensed patch in a vast region of inexpressible splendour, spreading over many fields."

*Astronomy* magazine Contributing Editor Stephen James O'Meara sees a skull with an open mouth. Most importantly, however, what do you see?

| OBJECT #743 | NGC 7790 |
|---|---|
| Constellation | Cassiopeia |
| Right ascension | 23h58m |
| Declination | 61°13′ |
| Magnitude | 8.5 |
| Size | 17′ |
| Type | Open cluster |

Our next target lies 2.5° northwest of magnitude 2.3 Caph (Beta [$\beta$] Cassiopeiae). Although you can see this cluster through binoculars, you'll need at least a 4-inch scope to resolve it well. You'll easily spot several dozen stars in an area half as wide as the Full Moon. The background stars here are dense, but the cluster stands out nicely. Two other open clusters flank NGC 7790: Magnitude 9.4 NGC 7788 lies 17′ to the northwest, and magnitude 9.7 Berkeley 58 sits 20′ to the southeast.

| OBJECT #744 | NGC 7793 |
|---|---|
| Constellation | Sculptor |
| Right ascension | 23h58m |
| Declination | –32°36′ |
| Magnitude | 9.0 |
| Size | 9.3′ by 6.3′ |
| Type | Spiral galaxy |
| Other name | Bond's Galaxy |

Northern Hemisphere observers would be a lot more familiar with this galaxy if it didn't lie in the southern constellation Sculptor. It's bright, relatively large, and it has good surface brightness across its face-on view.

To find it, look 4.9° south-southeast of magnitude 4.6 Delta ($\delta$) Sculptoris. NGC 7793 takes high magnification well, so crank the power past 250×. Only then do you have a chance to see the slightly darker markings in the many broad spiral arms.

American astronomer George Phillips Bond (1825–1865) discovered this galaxy in 1850 from Cambridge, Massachusetts. He found it while using a 4-inch f/8 refractor to search for comets.

| OBJECT #745 | Sigma ($\sigma$) Cassiopeiae |
|---|---|
| Constellation | Cassiopeia |
| Right ascension | 23h59m |

| (continued) | |
|---|---|
| Declination | 55°45′ |
| Magnitudes | 5.0/7.1 |
| Separation | 3″ |
| Type | Double star |

Look for Sigma Cas 3.7° south of magnitude 2.3 Caph (Beta [$\beta$] Cassiopeiae). The two components lie fairly close together, so use a magnification near 150× to get a nice split. Many observers describe both stars as white, but bluish-white seem to be the next most chosen colors.

**Object #746** NGC 7814 Adam Block/NOAO/AURA/NSF

| OBJECT #746 | NGC 7814 |
|---|---|
| Constellation | Pegasus |
| Right ascension | 0h03m |
| Declination | 16°09′ |
| Magnitude | 10.6 |
| Size | 6.0′ by 2.5′ |
| Type | Spiral galaxy |
| Other name | Caldwell 43 |

Our next target lies 2.6° west-northwest of magnitude 2.8 Algenib (Gamma [$\gamma$] Pegasi), and it's a good one: NGC 7814. Small telescopes reveal this object's football shape, but with more tapered ends. The central region spans a third of NGC 7814's length. This galaxy does have a prominent dust lane captured by many astroimagers. You'll need a huge telescope to have a chance to see it, however. Through a 20-inch scope, crank the magnification past 400×, and look for two thin lines that emanate from points outside the core, and cross the galaxy's long axis.

| OBJECT #747 | NGC 7822 |
|---|---|
| Constellation | Cepheus |
| Right ascension | 0h03m |
| Declination | 68°37′ |
| Size | 65′ by 20′ |
| Type | Emission nebula |

Our next target, emission nebula NGC 7822, is a challenge object through a 12-inch telescope. You'll find it not quite 7.5° east of magnitude 3.5 Iota (ι) Cephei. Its seemingly huge size only becomes apparent through imaging. What you'll see visually (and, I suggest a nebula filter) appears like a thin cloud half as wide and long as the above measurements. If you see NGC 7822, it's worth going after Cederblad 214, another faint nebula just 5′ to the northeast.

| OBJECT #748 | NGC 7840 |
|---|---|
| Constellation | Pisces |
| Right ascension | 0h07m |
| Declination | 8°23′ |
| Magnitude | 15.5 |
| Size | 0.4′ |
| Type | Elliptical galaxy |
| Notes | The last NGC object |

The last object in the NGC catalog isn't as easy to find as the first but it's worth the effort, I think, just to "bookend" the *New General Catalogue*. To locate it, first find the magnitude 13.4 spiral galaxy NGC 3. It lies 2.4° northeast of magnitude 4.0 Omega (ω) Piscium. NGC 7840 lies 5′ north-northwest of NGC 3, and it's really faint and small. It glows dimly at magnitude 15.5 and measures a scant 0.4′ across. You won't have much luck searching for this object through telescopes smaller than 16′ in diameter.

| OBJECT #749 | NGC 1 |
|---|---|
| Constellation | Pegasus |
| Right ascension | 0h07m |
| Declination | 27°43′ |
| Magnitude | 13.6 |
| Size | 1.7′ by 1.2′ |
| Type | Spiral galaxy |

Those of you with 8-inch or larger telescopes who observe under a dark sky can search for spiral galaxy NGC 1. I like pointing this object out to fellow amateur astronomers because few have seen it. Although this object lies in Pegasus, it sits really close to that constellation's border with Andromeda.

Look for this magnitude 12.8 object 1.4° south of 2nd-magnitude Alpheratz (Alpha [α] Andromedae). Don't expect to see much detail through anything smaller than a 20-inch telescope, but, hey! At least you can say you've observed the first object in the *New General Catalogue*.

As you observe NGC 1, avert your gaze less than 2′ south to spot another spiral galaxy, NGC 2. At magnitude 14.2, this object poses more of a challenge than NGC 1. NGC 2 measures 1′ by 0.6′.

| OBJECT #750 | NGC 40 |
|---|---|
| Constellation | Cepheus |
| Right ascension | 0h13.0m |
| Declination | 72°32′ |
| Magnitude | 12.4 |

| (continued) | |
|---|---|
| Size | 37″ |
| Type | Planetary nebula |
| Other names | The Bow-Tie Nebula, Caldwell 2 |

Our next celestial wonder sits unassumingly near the head of Cepheus the King. Despite its low overall magnitude, the Bow-Tie Nebula makes a fine target through all telescopes because its surface brightness is high. The Bow-Tie sits 5.5° south-southeast of the star marking the head of Cepheus the King, magnitude 3.2 Gamma ($\gamma$) Cephei.

As is the case with many of our celestial wonders, NGC 40 is a discovery of Sir William Herschel. He found it November 25, 1787.

An object's magnitude tells observers how bright the object appears. For objects that are not point-sources, like galaxies, integrated magnitude compares all the light from the object to the light of a single star. For NGC 40, then, magnitude 11 means its total light output equals that of a magnitude 11 star.

If you combine magnitude with surface brightness, you get a better description of how easily you'll see an astronomical object. Surface brightness is given in units of magnitude per square arcsecond. Let's compare NGC 40 with an equally bright galaxy that measures 6′ across. Because the galaxy covers an area 100 times greater than NGC 40, its surface brightness is only 1% of NGC 40's. So, the Bow-Tie Nebula will be much easier to see.

A dear friend, Alaskan amateur astronomer Jeff Medkeff (1968–2008) developed a simple rule of thumb to help him determine how difficult a nebula or galaxy will be to see. He multiplied the magnitude by the surface brightness. This gave him a number (no units, just a number). The higher that number, the tougher the object will be to observe.

A 4-inch telescope at a dark site will reveal an oval disk about one-third longer than it is wide. The magnitude 11.6 central star appears bright compared to the nebula. Through a 10-inch scope, NGC 40's disk shows several bright knots toward the southeast and northwest. Increase the magnification to 200× (if the seeing permits), and look for a dark cavity between the shell and the central star.

I rate this planetary as one of the most surprising in the sky. It never fails to delight when I show it to other observers.

| OBJECT #751 | NGC 45 |
|---|---|
| Constellation | Cetus |
| Right ascension | 0h14m |
| Declination | −23°10′ |
| Magnitude | 10.7 |
| Size | 8.5′ by 5.9′ |
| Type | Spiral galaxy |

Observing NGC 45 isn't that straightforward because of two stars in the field of view. Right in front of the galaxy sits magnitude 9.9 SAO 166133. Then, not even 5′ to the west-southwest, sits magnitude 6.5 GSC 6413:626.

Through an 8-inch telescope, you'll see the galaxy's condensed nucleus. Spotting the spiral arms takes at least 16 inches of aperture, and, even through that size scope, they're not all that bright.

You'll find NGC 45 about 8.5° southwest of magnitude 2.0 Diphda (Beta [$\beta$] Ceti).

| OBJECT #752 | NGC 55 |
|---|---|
| Constellation | Sculptor |
| Right ascension | 0h14.9m |
| Declination | −39°11′ |
| Magnitude | 8.1 |
| Size | 30′ by 6.3′ |
| Type | Barred spiral galaxy |
| Other names | Caldwell 72, the Southern Cigar Galaxy |

You'll find this magnificent object 3.7° northwest of magnitude 2.4 Ankaa (Alpha [$\alpha$] Phoenicis). Be prepared to spend some high-quality time observing it.

This galaxy's core lies distinctly offset to the west, giving the object a tapered, cigar-like appearance at low power and a two-part appearance at high magnifications. You'll see the faint dark lanes that divide the arms best if you use averted vision.

NGC 55 also is one of the sky's few galaxies that will benefit from a nebula filter. Because it's big and bright, an OIII filter will dim its stars, therefore increasing the contrast with star-forming regions of ionized hydrogen. Several of these are visible along the galaxy's arms through a 12-inch scope.

Finally, for those of you with large amateur telescopes, look for a magnitude 15.3 galaxy off NGC 55's eastern tip. That's PGC 599897. It measures approximately 2′ by 1′.

**Object #753** IC 10 Adam Block/NOAO/AURA/NSF

| OBJECT #753 | IC 10 |
|---|---|
| Constellation | Cassiopeia |
| Right ascension | 0h20m |
| Declination | 59°18′ |
| Magnitude | 11.3 |
| Size | 7.3′ by 6.4′ |
| Type | Irregular galaxy |

Pan 1.4° east of magnitude 2.3 Caph (Beta [$\beta$] Cassiopeiae, and you'll find a member of the Local Group of galaxies. Even though it's close (2 million light-years), don't get your hopes up. IC 10's not all that bright, and it's pretty big as galaxies go. Even professional astronomers didn't recognize this diffuse object as a galaxy until 1935.

Through a 12-inch telescope, IC 10 appears amorphous. Larger scopes reveal many star-forming regions that lie in the galaxy's plane.

# October

| OBJECT #754 | NGC 103 |
|---|---|
| Constellation | Cassiopeia |
| Right ascension | 0h25m |
| Declination | 61°19′ |
| Magnitude | 9.8 |
| Size | 5′ |
| Type | Open cluster |

The easiest way to find our next target is to first locate magnitude 4.2 Kappa ($\kappa$) Cassiopeiae. Then move 1.5° southwest to the star 12 Cas. From there, drop south 0.5°.

Through an 8-inch telescope, you'll spot about three dozen stars in a tight pattern. A magnification around 50× will set this object in a nice star field. Triple the power, and look for arcs and hooks made by the cluster's stars. Some observers have detected the Greek letter Psi ($\Psi$) formed by stars in the center of NGC 103. Don't get it confused with the Psi Cassiopeiae Cluster (NGC 457), however.

| OBJECT #755 | NGC 104 |
|---|---|
| Constellation | Tucana |
| Right ascension | 0h24m |
| Declination | −72°05′ |
| Magnitude | 3.8 |
| Size | 50′ |
| Type | Globular cluster |
| Other names | 47 Tucanae, Caldwell 106 |

M.E. Bakich, *1,001 Celestial Wonders to See Before You Die*, Patrick Moore's Practical Astronomy Series,
DOI 10.1007/978-1-4419-1777-5_10, 

One of the sky's greatest celestial pairings occurs in the far Southern Hemisphere sky, where 47 Tucanae and the Small Magellanic Cloud align. Among globular clusters, only Omega Centauri (NGC 5139) outshines NGC 104, and only Omega appears larger.

French astronomer Abbe Nicholas Louis de Lacaille discovered the deep-sky nature of NGC 104 and cataloged it in 1751. Before then, this globular cluster carried the stellar designation bestowed on it by German astronomer Johann Elert Bode who extended John Flamsteed's catalog to the Southern Hemisphere. He continued to number all bright stars within constellations and called this object 47 Tucanae.

To the naked eye, NGC 104 appears as a fuzzy "star." A 3-inch telescope will begin to resolve this cluster's stars, but they really put on a show when you use an 8-inch or larger scope. Through such instruments, the cluster is a ball of stars you can resolve nearly to the core. The 2′-wide core spikes in brightness because of thousands of unresolved stars. Be sure to note the many streams of stars that emanate from a region 6′ in diameter.

Most observers consider 47 Tucanae the sky's most impressive globular cluster. The reason involves its stars' evolutionary stage. You can resolve a globular cluster well when your telescope lets you see the cluster's horizontal branch stars. In a globular cluster, where all stars formed at the same time, the number of stars of a given magnitude takes a sudden leap at a point astronomers refer to as the horizontal branch. That's the stage in a star's life right after its red-giant stage. Globular 47 Tucanae's horizontal-branch magnitude, around 13, makes it easy to resolve.

| OBJECT #756 | NGC 129 |
|---|---|
| Constellation | Cassiopeia |
| Right ascension | 0h30m |
| Declination | 60°13′ |
| Magnitude | 6.5 |
| Size | 12′ |
| Type | Open cluster |

Here's an object sharp-eyed observers can detect with their naked eyes from a dark site. To find it, draw a line between magnitude 2.3 Caph (Beta [$\beta$] Cassiopeiae) and magnitude 2.2 Tsih (Gamma [$\gamma$] Cas). NGC 129 lies at the midpoint of this line.

Through an 8-inch telescope, you'll see a rich cluster in which you can resolve several dozen stars evenly spread over its diameter. Step up to a 12-inch scope, and 50 stars will pop into view.

| OBJECT #757 | NGC 133 |
|---|---|
| Constellation | Cassiopeia |
| Right ascension | 0h31m |
| Declination | 63°22′ |
| Magnitude | 9.4 |
| Size | 3′ |
| Type | Open cluster |

NGC 133 forms a nice triangle with NGC 146 and King 14. This trio of open clusters lies near magnitude 4.2 Kappa ($\kappa$) Cassiopeiae. This star sits just above the constellation's "W" pattern. NGC 133 looks like a hook of four equally bright stars. King 14 is a fairly loose cluster. NGC 146 shines at magnitude 9.1 and measures 5′ across.

| OBJECT #758 | NGC 134 |
|---|---|
| Constellation | Sculptor |
| Right ascension | 0h30.4m |
| Declination | −33°15′ |

| (continued) | |
|---|---|
| Magnitude | 10.4 |
| Size | 8.5′ by 1.9′ |
| Type | Barred spiral galaxy |

You'll find this attractive galaxy 0.5° east-southeast of magnitude 4.9 Eta ($\eta$) Sculptoris. Through an 8-inch telescope, you'll see a small bright nucleus surrounded by an elliptical haze. If you move to larger telescopes, the haze resolves into tightly wound spiral arms marked here and there by dark regions.

Finally, look 9′ west of NGC 134 for the magnitude 13.0 spiral galaxy NGC 131. It has nearly the same orientation as its larger companion.

**Object #759** NGC 147 Anthony Ayiomamitis

| OBJECT #759 | NGC 147 |
|---|---|
| Constellation | Cassiopeia |
| Right ascension | 0h33m |
| Declination | 48°30′ |
| Magnitude | 9.5 |
| Size | 15.0′ by 9.4′ |
| Type | Elliptical galaxy |
| Other name | Caldwell 17 |

Our next object is a satellite galaxy of the Andromeda Galaxy (M31), and so belongs to the Local Group of galaxies. To find it, move 1.9° west from magnitude 4.5 Omicron ($o$) Cassiopeiae.

Look carefully for this challenging object. NGC 147 is a dwarf elliptical galaxy without much apparent structure. When you do see it, you'll notice an oval halo a bit brighter than the background glow. Is the galaxy ever-so-slight brighter toward the center, or is that a foreground star? You decide.

Not quite 1° east of NGC 147, you'll find magnitude 9.2 NGC 185. It touts a higher surface brightness than its companion because it's slightly smaller, measuring 14′ by 12′. A 12-inch telescope reveals an oval halo with a bright core that spans two-thirds of the galaxy's diameter.

| OBJECT #760 | NGC 150 |
|---|---|
| Constellation | Sculptor |
| Right ascension | 0h34m |
| Declination | −27°48′ |
| Magnitude | 11.3 |
| Size | 3.4′ by 1.6′ |
| Type | Barred spiral galaxy |

Our next treat lies 5.5° west-northwest of magnitude 4.3 Alpha (α) Sculptoris. This galaxy has a bright concentrated core. Through a 10-inch telescope with an eyepiece yielding 150×, you'll see a ring of light surrounding the core, with a dark region between the core and the ring.

Through a 16-inch scope at 300× or more, you'll start to resolve NGC 150's ring into two large, curving spiral arms. Only through the largest amateur scopes will you begin to see the short bar that emanates from the core.

**Object #761** NGC 157 Erica and Dan Simpson/Adam Block/NOAO/AURA/NSF

| OBJECT #761 | NGC 157 |
|---|---|
| Constellation | Cetus |
| Right ascension | 0h35m |
| Declination | −8°24′ |
| Magnitude | 10.4 |
| Size | 4.0′ by 2.4′ |
| Type | Spiral galaxy |

I think you'll enjoy observing NGC 157. This galaxy lies 3.5° east of magnitude 3.5 Iota ($\iota$) Ceti. Through small telescopes, it appears almost rectangular with an even brightness distribution and a faint outer halo. If you aim a 14-inch or larger scope at it, however, NGC 157 will reveal clumps of star formation within its spiral arms. The brightest of these lie to the northwest and south of the galaxy's center.

**Object #762** NGC 188 Anthony Ayiomamitis

| OBJECT #762 | NGC 188 |
|---|---|
| Constellation | Cepheus |
| Right ascension | 0h44m |
| Declination | 85°20′ |
| Magnitude | 8.1 |
| Size | 13′ |
| Type | Open cluster |
| Other name | Caldwell 1 |

Our next treat is one of the sky's most northerly deep-sky targets and the first entry on Sir Patrick Caldwell Moore's list of 109 deep-sky objects. Because "M" was taken by Messier, Moore called his list the Caldwell catalog, and, thus, a "C" precedes each entry.

Open cluster NGC 188 (Caldwell 1) lies approximately 4° from Polaris (Alpha [$\alpha$] Ursae Minoris). Through an 8-inch telescope at 100×, you'll see about 50 magnitude 13 and fainter stars. A 12-inch scope at a dark site will allow you to count twice that number of star. Because the stars all appear about the same brightness, our eyes form them into patterns. Curved lines, hooks, and letters are the ones I most often see. As you focus on the stars, dark lanes seem to run through the cluster, adding to the visual appeal.

| OBJECT #763 | NGC 189 |
|---|---|
| Constellation | Cassiopeia |
| Right ascension | 0h40m |
| Declination | 61°05′ |
| Magnitude | 8.8 |
| Size | 5′ |
| Type | Open cluster |

NGC 189 lies a bit southwest of another nice open cluster NGC 225. Both are near Gamma ($\gamma$) Cassiopeiae, the middle star of the Queen's W pattern. Some observers have described NGC 189 as a figure U with a line thru it.

**Object #764** NGC 205 Adam Block/NOAO/AURA/NSF

| OBJECT #764 | NGC 205 |
|---|---|
| Constellation | Andromeda |
| Right ascension | 0h40m |
| Declination | 41°41′ |
| Magnitude | 8.1 |
| Size | 19.5′ by 12.5′ |
| Type | Elliptical galaxy |

Our next target is bright and easy to find. Just locate the Andromeda Galaxy, and look 0.6° northwest of its core. Elliptical galaxy NGC 205 shines about as brightly as M31's other easy-to-see companion, M32. NGC 205, however, is nearly three times larger. Even through large amateur telescopes, you won't see detail in this galaxy.

| OBJECT #765 | NGC 210 |
| --- | --- |
| Constellation | Cetus |
| Right ascension | 0h41m |
| Declination | −13°52′ |
| Magnitude | 10.9 |
| Size | 5′ by 3.3′ |
| Type | Spiral galaxy |

At first glance, you may initially think our next target is an elliptical galaxy, but that's only because NGC 210 has incredibly faint spiral arms. The extended central region appears oval with an even brightness across its surface. A magnitude 12.4 star lies slightly more than 1′ west-northwest of the nucleus. NGC 210 lies 4.2° north of magnitude 2.0 Diphda (Beta [$\beta$] Ceti).

**Object #766** M32 Adam Block/NOAO/AURA/NSF

| OBJECT #766 | M32 (NGC 221) |
| --- | --- |
| Constellation | Andromeda |
| Right ascension | 0h43m |
| Declination | 40°52′ |
| Magnitude | 8.1 |
| Size | 11.0′ by 7.3′ |
| Type | Elliptical galaxy |

Because M32 lies 0.4° due south of the heart of the Andromeda Galaxy (M31), I'll forgive you if you choose not to spend too much time observing it. Any size telescope will show this featureless elliptical.

**Object #767** The Andromeda Galaxy (M31) Adam Block/NOAO/AURA/NSF

| OBJECT #767 | M31 (NGC 224) |
|---|---|
| Constellation | Andromeda |
| Right ascension | 0h43m |
| Declination | 41°16′ |
| Magnitude | 3.4 |
| Size | 185′ by 75′ |
| Type | Spiral galaxy |
| Other name | The Andromeda Galaxy |

The sky's greatest galaxy gets its familiar name from the northern constellation where it resides, Andromeda the Princess. This star system is our nearest large spiral, and it sits at the far end of the Local Group of galaxies.

Observers have described M31 as something other than starlike as far back as 964. In that year, Persian astronomer Abdal-Rahman Al-Sufi called it a "little cloud" in his *Book of Fixed Stars*.

German astronomer Simon Marius (1573–1624) was the first to study it telescopically. He described it, “Like the flame of a candle seen through horn, and like a cloud consisting of three rays; whitish, irregular and faint; brighter toward the center.”

Messier cataloged it August 3, 1764: “The beautiful nebula of the belt of Andromeda, shaped like a spindle; it resembles two cones or pyramids of light, opposed at their bases.”

In 1888, British astronomer Isaac Roberts (1829–1904) became the first to photograph spiral structure in M31. American astronomer Vesto M. Slipher (1875–1969) first measured M31’s radial velocity (the speed of a celestial object toward or away from us) in 1912. Slipher found its velocity far surpassed that of any other object, and his measurement helped prove M31 lay far from the Milky Way. In 1923, Edwin Hubble measured Cepheid variable stars in M31 and confirmed its extragalactic nature.

Observers approach the Andromeda Galaxy in one of two ways. Some use low-power optics for an overall view, which includes M31’s nucleus, dust lanes, and two companion galaxies, M32 and NGC 205. If this is your approach, try 20×80 (or similar) binoculars from a dark site. Try to trace M31’s full length, which equals 6 Full Moons side by side.

Other amateur astronomers eschew wide-field views of the Andromeda Galaxy in favor of greatly magnified looks at small regions through large telescopes. If this plan appeals to you, use a 10-inch or larger scope and crank up the magnification to 300× and more. Scan M31’s spiral arms for bright clumps, which indicate star-forming regions.

| OBJECT #768 | NGC 225 |
|---|---|
| Constellation | Cassiopeia |
| Right ascension | 0h44m |
| Declination | 61°46′ |
| Magnitude | 7.0 |
| Size | 15′ |
| Type | Open cluster |
| Other name | The Sailboat Cluster |

NGC 225 sits not quite 2° northwest of magnitude 2.5 Gamma ($\gamma$) Cassiopeiae, the star at the middle of the W asterism. The cluster contains two groups of stars, one a bit fainter than the other.

American amateur astronomer Rod Pommier gave this object its common name in an article in the May 2000 issue of *Astronomy* magazine. Pommier called NGC 225 the ‘Sailboat Cluster’ because it has a distinctive four-star arc that outlines the leading edge of what could be a sail, inflated, he said, by an imaginary wind. A mast of three to four stars in a line supports the sail, and the line connects to a boat outlined by eight stars arranged in an elongated ellipse. Alas, as Earth turns, the sailboat slowly drifts backward across the sky. Pommier suggested switching your drive off and watching NGC 225 sail across the eyepiece on a star-filled sea.

**Object #769** NGC 246 Jeff Cremer/Adam Block/NOAO/AURA/NSF

| OBJECT #769 | NGC 246 |
|---|---|
| Constellation | Cetus |
| Right ascension | 0h47m |
| Declination | −11°53′ |
| Magnitude | 10.9 |
| Size | 225″ |
| Type | Planetary nebula |
| Other name | Caldwell 56 |

The easiest way to find our next object is to locate two stars. Magnitude 4.8 Phi$^1$ ($\Phi^1$) Ceti and magnitude 5.2 Phi$^2$ ($\Phi^2$) Ceti form an equilateral triangle with NGC 246. They sit 1.5° north-northwest and north-northeast of the planetary nebula, respectively.

A 6-inch telescope under a dark sky reveals a large disk with several stars across its face, including an obvious central star. Through a 12-inch scope, a hollow center and a bright, thin rim on the northeast appear. A narrowband nebula filter, such as an OIII, dramatically improves the visibility of the patchy inner structure. Look 0.4° north-northeast for the magnitude 11.8 spiral galaxy NGC 255.

| OBJECT #770 | NGC 247 |
|---|---|
| Constellation | Cetus |
| Right ascension | 0h47m |
| Declination | −20°46′ |
| Magnitude | 9.2 |
| Size | 19.0′ by 5.5′ |
| Type | Spiral galaxy |
| Other name | Caldwell 62 |

This is a big galaxy. It's so big, and its surface brightness is so low that, unless your sky is really dark, you'll have problems seeing it. You'll find it 2.9° south-southeast of Diphda (Beta [$\beta$] Ceti).

NGC 247 is one of the standout members of the Sculptor group of galaxies. The Sculptor group consists of a loose gathering of a couple dozen galaxies that lie between 50% farther away and twice as far as the most distant members of the Local Group. NGC 247 lies about 13 million light-years away.

Through a 10-inch telescope, NGC 247 has a concentrated, circular center with a tight, oval haze around it. The galaxy orients north-south. The magnitude 8.1 star GSC 5849:2326 sits at its tapered southern end. The northern edge of the galaxy is more rounded.

**Object #771** The Silver Coin Galaxy (NGC 253) Doug Matthews/Adam Block/NOAO/AURA/NSF

| OBJECT #771 | NGC 253 |
|---|---|
| Constellation | Sculptor |
| Right ascension | 0h48m |
| Declination | −25°17′ |
| Magnitude | 7.6 |
| Size | 30′ by 6.9′ |
| Type | Spiral galaxy |
| Other names | The Silver Coin Galaxy, the Sculptor Galaxy, Caldwell 65 |

The Silver Coin Galaxy makes every observer's top 10 list of galaxies. It's that good. This object's "fame quotient" is low, however, because from northern sites it lies low in the southern sky. That, and Messier didn't observe it.

British amateur astronomer Caroline Herschel (1750–1848), sister of Sir William Herschel, discovered NGC 253 September 23, 1783. Sir William added it to his catalog October 30.

The Silver Coin Galaxy — so named because of its appearance through small telescopes — is visible to sharp-eyed observers from a dark, southerly site. From the Northern Hemisphere, it's best to conduct this test of vision and sky quality in early October when NGC 253 lies due south, at its highest point.

NGC 253 belongs to the Sculptor Group of galaxies, the closest such system to our Local Group. This collection also includes NGC 55, NGC 247, and NGC 300.

Through an 8-inch or larger telescope, you'll detect the galaxy's mottled appearance. Although the core appears well-defined, it's only slightly brighter than NGC 253's outer regions. Dark dust lanes radiate from the center. Use a 12-inch scope to discern the spiral arms, one to the northeast and one to the southwest.

| OBJECT #772 | Eta ($\eta$) Cassiopeiae |
|---|---|
| Constellation | Cassiopeia |
| Right ascension | 0h49m |
| Declination | 57°49′ |
| Magnitudes | 3.4 and 7.5 |
| Separation | 12.9″ |
| Type | Double star |
| Other name | Achird |

This colorful binary lies 1.7° northeast of magnitude 2.2 Schedar (Alpha [$\alpha$] Cassiopeiae). Low power works best to bring out the yellow color of the primary and the reddish hue of the secondary. Magnifications above 100× accentuate the brightness difference between the two, making them appear more white and yellow.

The common name "Achird" is a twentieth-century addition. Richard Hinckley Allen mentions the star as a nice double but says nothing about the name in *Star Names and Their Meaning*, which appeared in 1899.

| OBJECT #773 | 65 Piscium |
|---|---|
| Constellation | Pisces |
| Right ascension | 0h50m |
| Declination | 27°43′ |
| Magnitudes | 6.3 and 6.3 |
| Separation | 4.6″ |
| Type | Double star |

Crank the power up past 100× to split this relatively close double. What you'll see is a pair of equally bright yellow stars. You'll find 65 Piscium 3.9° southeast of magnitude 3.3 Delta ($\delta$) Andromedae.

**Object #774** NGC 281 Anthony Ayiomamitis

| OBJECT #774 | NGC 281 |
|---|---|
| Constellation | Cassiopeia |
| Right ascension | 0h53m |
| Declination | 56°37′ |
| Size | 35′ by 30′ |
| Type | Emission nebula |
| Other name | The Pacman Nebula |

Our next target is a fine emission nebula that lies 1.7° east of magnitude 2.2 Schedar (Alpha [$\alpha$] Cassiopeiae) and only 1.3° south-southeast of magnitude 3.5 Eta ($\eta$) Cas. Astroimagers noticed the nebula's resemblance to the main character in the Namco video game Pac-Man, which Midway distributed throughout the U.S.

Through a 12-inch telescope equipped with a nebula filter, look for the dark lane that divides the brighter from the dimmer portion. The magnitude 7.4 star HD 5005 that forms the Pacman's eye shines through the brightest region of the nebula. This young, hot double star, along with open cluster IC 1590, provides most of the ultraviolet radiation that ionizes the nebula.

**Object #775** NGC 288 Pat and Chris Lee/Adam Block/NOAO/AURA/NSF

| OBJECT #775 | NGC 288 |
|---|---|
| Constellation | Sculptor |
| Right ascension | 0h53m |
| Declination | −26°35′ |
| Magnitude | 8.1 |
| Size | 13.8′ |
| Type | Globular cluster |

This nice globular lies 3° north-northwest of magnitude 4.3 Alpha ($\alpha$) Sculptoris. Through binoculars at a dark site, you'll easily spot NGC 288 and the Sculptor Galaxy (NGC 253) within the same field of view only 1.8° to the northwest.

NGC 288 has a broad central region characterized by a poor concentration of stars. But that just means you'll resolve many of them through an 8-inch or larger telescope. Magnifications above 200× will reveal more than 100 individual stars. As with other clusters whose stars are all about the same brightness, look for interesting patterns of stars.

| OBJECT #776 | Small Magellanic Cloud |
|---|---|
| Constellation | Tucana |
| Right ascension | 0h54m |
| Declination | −72°50′ |
| Magnitude | 2.7 |
| Size | 320′ by 185′ |
| Type | Irregular galaxy |
| Common names | SMC; Nubecula Minor |

The sky's second-largest galaxy has the word "small" in its name only because of its counterpart, the Large Magellanic Cloud appears larger.

Historians often credit Portuguese explorer Ferdinand Magellan (1480–1521) with the discovery of these objects, but that is incorrect. The SMC and the LMC had been visible to Southern Hemisphere dwellers throughout history. What Magellan did was to make both "clouds" known to Europe; thus, they bear his name.

To the naked eye, the SMC appears 50% longer than it is wide, oriented on a northeast-southwest line. The central region and a 1°-wide area on the galaxy's northeastern end look brightest.

After you've observed the SMC with your naked eyes and then through binoculars, point your telescope at this wondrous object — and the bigger the scope, the better. The SMC contains no less than two dozen NGC objects brighter than magnitude 12.6.

Harvard College astronomer Henrietta Swan Leavitt (1868–1921) first noticed something was going on with Cepheid variable stars within this galaxy. She discovered the direct relationship between the brightness of Cepheids and the rate at which they cyclically brightened and dimmed. Astronomers now use such stars as standard candles to determine galactic distances.

| OBJECT #777 | NGC 300 |
|---|---|
| Constellation | Sculptor |
| Right ascension | 0h55m |
| Declination | −37°41′ |
| Magnitude | 8.1 |
| Size | 20.0′ by 13.0′ |
| Type | Spiral galaxy |
| Other names | The Southern Pinwheel Galaxy, Caldwell 70 |

Perhaps the sky's most classic spiral galaxy inhabits a region in southeastern Sculptor. Along with the Silver Coin Galaxy (NGC 253), NGC 300 creates interest in a constellation whose brightest star glows at a feeble magnitude 4.3.

One glance at NGC 300 through an 8-inch telescope, and you'll know immediately why it's called the Southern Pinwheel Galaxy. The resemblance to the Pinwheel Galaxy (M33) is uncanny. NGC 300 has low surface brightness and a broadly concentrated central region that stretches nearly one-third this galaxy's length. The core, however, appears small, almost starlike.

Through 12-inch and larger scopes, examine NGC 300's two prominent spiral arms for dark lanes and bright spots signaling star-forming regions. Several foreground stars superimposed over the galaxy complete the picture.

| OBJECT #778 | NGC 346 |
|---|---|
| Constellation | Tucana |
| Right ascension | 0h59m |
| Declination | −72°11′ |
| Magnitude | 10.3 |
| Size | 5.2′ |
| Type | Open cluster |

This cluster lies within the borders of the Small Magellanic Cloud, just to the northeast of its center. This loose, wedge-shaped cluster ionizes a surrounding nebula, which you can spot even through small binoculars. To observers, the nebula resembles a barred spiral galaxy, but you'll need an 8-inch telescope to see it. The bar is about 4′ long and oriented roughly east-west. A thin, bright arm wraps around the south side from the eastern end of the bar, but the northern arm is more diffuse.

| OBJECT #779 | Sculptor Dwarf |
|---|---|
| Constellation | Sculptor |
| Right ascension | 1h00m |
| Declination | −33°42′ |
| Magnitude | 8.8 |
| Size | 1.1° by 0.8° |
| Type | Dwarf spheroidal galaxy |

You'll find our next target 2.3° south-southwest of magnitude 5.5 Sigma ($\sigma$) Sculptoris. Well, that's where to look for it, at least. Because the Sculptor Dwarf covers more area than 4 Full Moons its surface brightness is incredibly low. Eagle-eyed observers have spotted it through 6-inch telescopes, but your best bet is to head to an ultra-dark site, and insert your lowest-power eyepiece into at least a 12-inch scope. Then, ever-so-slowly scan the area. What you're looking for is a slight brightening of the background glow.

The Sculptor Dwarf was the first dwarf spheroidal galaxy discovered. In 1937, American astronomer Harlow Shapley found it on a photographic plate. The main difference between this type of galaxy and a dwarf elliptical is that a dwarf spheroidal galaxy has even lower surface brightness.

| OBJECT #780 | NGC 362 |
|---|---|
| Constellation | Tucana |
| Right ascension | 1h03m |
| Declination | −70°51′ |
| Magnitude | 6.5 |
| Size | 12.9′ |
| Type | Globular cluster |
| Other name | Caldwell 104 |

You'll have to head south to spot our next treat. The bright globular cluster NGC 362 lies on the northern edge of the Small Magellanic Cloud. It's not part of that galaxy, however. Rather, it lies seven times closer to us.

Sharp-eyed observers will see it without optical aid as a faint, extended "star." Through an 8-inch telescope, NGC 362 explodes with detail. Well, except for the core. You'll need a larger scope and high magnification to resolve any of the stars near the cluster's center.

| OBJECT #781 | IC 1613 |
|---|---|
| Constellation | Cetus |
| Right ascension | 1h05m |
| Declination | 2°07′ |
| Magnitude | 9.2 |
| Size | 18.8′ by 17.3′ |
| Type | Irregular galaxy |
| Other name | Caldwell 51 |

Our next target is the huge Local Group galaxy IC 1613. You'll find it 0.8° north-northeast of the magnitude 6.1 double star 26 Ceti, right next to that constellation's border with Pisces.

Although some observers have detected it through an 8-inch (or even smaller) telescope, I suggest at least a 12-inch instrument to guarantee success. Look for a faint, round, uniform haze. And, remember, its diameter is half as large as the Full Moon's. Large amateur telescopes — those above 20 inches in aperture — at high magnifications reveal an entirely different object. Instead of a softly glowing core, several dozen 17th-magnitude stars will make up the galaxy's central glow.

| OBJECT #782 | Psi[1] ($\Psi^1$) Piscium |
|---|---|
| Constellation | Pisces |
| Right ascension | 1h06m |
| Declination | 21°28′ |
| Magnitudes | 5.6/5.8 |
| Separation | 30″ |
| Type | Double star |

You can split this wide binary through any size telescope. Both stars appear pale blue or blue-white. Although similarly bright Psi[2] ($\Psi^2$) Piscium lies 0.9° to the southeast, you'll have no trouble distinguishing the binary Psi[1] ($\Psi^1$) from the equally bright Psi[2].

| OBJECT #783 | NGC 381 |
|---|---|
| Constellation | Cassiopeia |
| Right ascension | 1h08m |
| Declination | 61°35′ |
| Magnitude | 9.3 |
| Size | 7′ |
| Type | Open cluster |

You'll find our next target 1.7° east-northeast of magnitude 2.5 Tsih (Gamma [$\gamma$] Cassiopeiae). At low power through a 4-inch telescope, note how well the cluster blends into the rich background. Crank the magnification to 150×, however, and you'll see an even distribution of some three dozen stars.

**Object #784** Mirach's Ghost (NGC 404) Anthony Ayiomamitis

| OBJECT #784 | NGC 404 |
|---|---|
| Constellation | Andromeda |
| Right ascension | 1h09m |
| Declination | 35°43′ |
| Magnitude | 10.3 |
| Size | 6.1′ by 6.1′ |
| Type | Elliptical galaxy |
| Other names | Mirach's Ghost, the Lost Pearl Galaxy |

Our next deep-sky object is a perfect target for a clear Halloween night: Mirach's Ghost, also known as NGC 404. Amateur astronomers call this magnitude 10.3 elliptical galaxy Mirach's Ghost because it lies only 6.8′ from 2nd-magnitude Mirach (Beta [$\beta$] Andromedae). As you might imagine, a 10th-magnitude galaxy next to that bright a star is pretty difficult to see.

*Astronomy* magazine Contributing Editor Stephen James O'Meara christened it the Lost Pearl Galaxy because some star atlases do not plot this object. The reason is that the printed image of Mirach overlaps it. He says, "Imagine a loose pearl rolling across the deck of a pirate ship, until it wedges firmly against the ship's brass."

This S0 galaxy — a type that has the disk shape of a spiral galaxy but no spiral arms — lies roughly 30 million light-years from Earth. Use high magnification to increase the contrast between the galaxy and the bright star. NGC 404 looks round and bright with an intense center. And don't fret too much about the glare from Mirach. There's no detail to be seen in the galaxy.

| OBJECT #785 | NGC 428 |
|---|---|
| Constellation | Cetus |
| Right ascension | 1h13m |
| Declination | 0°59′ |
| Magnitude | 11.5 |
| Size | 4.6′ by 3.4′ |
| Type | Spiral galaxy |

You'll find our next object 1.6° south-southeast of magnitude 6.0 star 33 Ceti, right at the Pisces border. This spiral shows a broadly concentrated oval disk with a hazy glow around it. Through a 14-inch or larger telescope, challenge yourself to see NGC 428's magnitude 16.2 companion, UGC 772, which lies 13′ to the southeast.

| OBJECT #786 | Zeta ($\zeta$) Piscium |
|---|---|
| Constellation | Pisces |
| Right ascension | 1h14m |
| Declination | 7°35′ |
| Magnitudes | 5.6/6.5 |
| Separation | 23″ |
| Type | Double star |

You'll have to look closely at this binary to see a color difference. Most observers report pale yellow or yellow-white as the primary star's color and white for the secondary. The separation is large enough, however, that you'll have no trouble splitting it.

| OBJECT #787 | Kappa ($\kappa$) Tucanae |
|---|---|
| Constellation | Tucana |
| Right ascension | 1h16m |
| Declination | −68°53′ |

| (continued) | |
|---|---|
| Magnitudes | 5.1/7.3 |
| Separation | 5.4″ |
| Type | Double star |

This nice binary sits far below the horizon for most Northern Hemisphere observers. The primary appears yellow-white and the secondary yellow, although some observers see no color at all in the fainter star. You'll find this star 4.4° north-northeast of the Small Magellanic Cloud.

| OBJECT #788 | NGC 436 |
|---|---|
| Constellation | Cassiopeia |
| Right ascension | 1h16m |
| Declination | 58°49′ |
| Magnitude | 8.8 |
| Size | 5′ |
| Type | Open cluster |

You can find this cluster 1.9° southwest of magnitude 2.7 Ruchbah (Delta [$\delta$] Cassiopeiae). Actually, however, it's easier to find the bright Owl Cluster (NGC 457), and just move 0.7° northwest.

NGC 436 is a sparse cluster with two distinct stellar brightnesses. The brighter stars form a V shape whose point faces roughly eastward.

**Object #789** The Owl Cluster (NGC 457) Anthony Ayiomamitis

| OBJECT #789 | NGC 457 |
|---|---|
| Constellation | Cassiopeia |
| Right ascension | 1h19m |
| Declination | 58°20′ |
| Magnitude | 6.4 |

| (continued) | |
|---|---|
| Size | 13′ |
| Type | Open cluster |
| Other names | The Owl Cluster, the ET Cluster, the Psi Cassiopeiae Cluster, Caldwell 13 |

One of the sky's two celestial owls (the other is the Owl Nebula [M97]) takes silent flight in Cassiopeia's rich star fields. Magnitude 5 Phi ($\Phi$) Cassiopeiae lends its name to the cluster but doesn't travel with it through space.

Sir William Herschel discovered NGC 457 in 1787. Messier missed this cluster, although it outshines the two objects he included from this constellation — magnitude 6.9 M52 and magnitude 7.4 M103.

While observing this cluster in 1977, *Astronomy* magazine Editor David J. Eicher saw an owl figure made of the two brightest stars and the cluster's overall shape. He dubbed it the Owl Cluster, and it's carried that name ever since. Five years later, Universal Pictures released the movie *E.T.: The Extra-Terrestrial*. Some observers saw a resemblance between the alien character in the film and NGC 457, and they subsequently dubbed it the E.T. Cluster.

NGC 457 contains 25 stars brighter than 12th magnitude. Its most luminous star shines at magnitude 8.6. A 6-inch telescope at 50× shows nearly 75 cluster stars. Note the uniform background glow caused by distant, unresolved Milky Way stars.

| OBJECT #790 | NGC 488 |
|---|---|
| Constellation | Pisces |
| Right ascension | 1h22m |
| Declination | 5°15′ |
| Magnitude | 10.3 |
| Size | 5.5′ by 4.0′ |
| Type | Spiral galaxy |

This relatively bright galaxy lies 2.3° west-southwest of magnitude 4.8 Mu ($\mu$) Piscium. This bright spiral has a large core surrounded by bright halo that stretches north-south. Look for a line of four equally spaced 10th- and 11th-magnitude stars south of the galaxy.

**Object #791** NGC 520 Jeff Newton/Adam Block/NOAO/AURA/NSF

| OBJECT #791 | NGC 520 |
|---|---|
| Constellation | Pisces |
| Right ascension | 1h25m |
| Declination | 3°48′ |
| Magnitude | 11.4 |
| Size | 4.6′ by 1.9′ |
| Type | Spiral galaxy |

Here's an odd duck. Although it's cataloged as NGC 520, this object actually is a pair of interacting galaxies astronomers cataloged as a single object. Even a small scope will show its odd shape.

Through a 6-inch telescope at low power, you'll think you're looking at an edge-on spiral. Crank the magnification up to 150×, however, and the sharp northwest edge will pop into view. Through larger instruments, a dark lane that divides the two galaxies becomes visible.

| OBJECT #792 | NGC 524 |
|---|---|
| Constellation | Pisces |
| Right ascension | 1h25m |
| Declination | 9°32′ |
| Magnitude | 10.4 |
| Size | 2.8′ |
| Type | Spiral galaxy |

You'll find our next target nestled in a group of galaxies that lies 3.4° northeast of magnitude 5.2 Zeta (ζ) Piscium. Within 1° of NGC 524, you'll also spot magnitude 12.6 NGC 489, magnitude 12.7 NGC 502, and magnitude 13.1 NGC 532, all spirals. Those objects, although fainter than our initial target, appear more like traditional spirals than NGC 524. That's because it's a face-on spiral with little details visible. Through any size telescope, you'll see a bright central region surrounded by a much fainter haze.

| OBJECT #793 | NGC 559 |
|---|---|
| Constellation | Cassiopeia |
| Right ascension | 1h30m |
| Declination | 63°19′ |
| Magnitude | 9.5 |
| Size | 7′ |
| Type | Open cluster |
| Other name | Caldwell 8 |

To find our next target, look 2.8° west of magnitude 3.4 Epsilon (ε) Cassiopeiae. Although its magnitude doesn't promise much, this cluster looks great through any size telescope. From a dark observing location, even a 4-inch scope will reveal three dozen stars. Most of them shine around magnitude 12 and pack into a roughly triangular core. A 12-inch instrument will double that number, bringing into view many fainter members that appear as "background" to those you've seen through smaller scopes.

**Object #794** NGC 578 Adam Block/NOAO/AURA/NSF

| OBJECT #794 | NGC 578 |
|---|---|
| Constellation | Cetus |
| Right ascension | 1h30m |
| Declination | –22°40′ |
| Magnitude | 10.8 |
| Size | 4.8′ by 3′ |
| Type | Spiral galaxy |

You'll find our next target a bit more than 1° south-southeast of the magnitude 5.1 star 48 Ceti. Through a 12-inch telescope at 150× or higher, you'll see this galaxy's relatively wide core and hints (or stubs) of two of its spiral arms. Through a 30-inch scope at 450×, I detected four distinct spiral arms, which all contained bright star-forming knots.

**Object #795** M103 Anthony Ayiomamitis

| OBJECT #795 | M103 (NGC 581) |
|---|---|
| Constellation | Cassiopeia |
| Right ascension | 1h33m |
| Declination | 60°42′ |
| Magnitude | 7.4 |
| Size | 6′ |
| Type | Open cluster |

Our next object is small telescope target M103. This magnitude 7.4 open cluster is really easy to find. It lies 1° east-northeast of magnitude 2.7 Delta ($\delta$) Cassiopeiae.

M103 isn't a spectacular cluster, but you must seek it out because of its status as a Messier object. Its 40 bright stars stand out well from the rich fields of the Milky Way. Cluster members range from 8th through 13th magnitude and group tightly in a triangle 5′ on a side. Most observers report the best views when they use a magnification around 100×.

| OBJECT #796 | NGC 584 |
|---|---|
| Constellation | Cetus |
| Right ascension | 1h31m |
| Declination | −6°52′ |
| Magnitude | 10.5 |
| Size | 4.1′ by 2′ |
| Type | Elliptical galaxy |
| Other name | The Little Spindle Galaxy |

NGC 584 is a fat lens-shaped galaxy that doesn't show much detail. Through an 8-inch telescope, you'll see the broad, bright core take up three-quarters of the galaxy's length. A bright halo lies outside the core, but it quickly fades to the black of space.

Only 4′ east-southeast of NGC 584's core lies the magnitude 13.2 spiral galaxy NGC 586. Crank up the magnification past 200× to put some distance between the two objects before you observe the fainter galaxy.

*Astronomy* magazine Contributing Editor Stephen James O'Meara called this object the Little Spindle because of its resemblance to the Spindle Galaxy (Object #132).

| OBJECT #797 | NGC 596 |
|---|---|
| Constellation | Cetus |
| Right ascension | 1h33m |
| Declination | −7°02′ |
| Magnitude | 10.9 |
| Size | 3.2′ by 2′ |
| Type | Elliptical galaxy |

You'll find another relatively bright elliptical galaxy 2.5° east-northeast of magnitude 3.6 Theta ($\theta$) Ceti. Through scopes 8′ in diameter and less, NGC 596 looks circular. It's only when you view this object through larger instruments that its slightly oval halo appears.

| OBJECT #798 | Triangulum |
|---|---|
| Right ascension (approx.) | 1h34m |
| Declination (approx.) | 30°39′ |
| Size (approx.) | 131.85 square degrees |
| Type | Constellation |

If you're just starting out in the sky, an easy test is to locate the tiny constellation Triangulum the Triangle. And, when I say small, I'm not kidding. Triangulum ranks 78th in size out of the 88 constellations. It covers an area of 132 square degrees — only 0.3% of the sky.

Only 12 stars in this constellation shine more brightly than magnitude 5.5, but don't despair. Triangulum's small size actually makes it easier for you to locate.

To find this diminutive star pattern, look just above the horns of Aries the Ram for a small, thin, three-sided starry figure. Andromeda lies to Triangulum's upper right, and Perseus sits to its upper left.

**Object #799** The Pinwheel Galaxy (M33) Adam Block/NOAO/AURA/NSF

| OBJECT #799 | M33 (NGC 598) |
|---|---|
| Constellation | Triangulum |
| Right ascension | 1h34m |
| Declination | 30°39′ |
| Magnitude | 5.7 |
| Size | 67′ by 41.5′ |
| Type | Spiral galaxy |
| Other names | The Pinwheel Galaxy, the Triangulum Galaxy |

Triangulum the Triangle would be easy to miss except that it contains one of the sky's standout galaxies. The Pinwheel Galaxy offers targets within its borders that the constellation lacks. Be prepared to spend lots of time observing this wonder.

Giovanni Hodierna probably observed the Pinwheel Galaxy before 1654. Messier independently discovered it August 25, 1764: "The nebula is a whitish light of almost even brightness. However, along two-thirds of its diameter it is a little brighter. Contains no star. Seen with difficulty in a 1-foot telescope."

Although you can glimpse M33 with your naked eyes from a dark site, this galaxy's surface brightness is low. Binoculars and small telescopes will help you gauge its overall shape, but you'll want to do a lot better than that. The best approach is to observe M33 through a 10-inch or larger telescope.

At 50×, look for an S shape emanating from a slightly brighter center. Luminous knots around M33's main body are vast star-forming regions. Through a 12-inch or (preferably) larger scope, find emission nebula NGC 604, which sits at the tip of M33's northern spiral arm. Crank up the magnification and look for two tiny lobes in contact. A nebula filter will improve your view of this nebula, but worsen the view of the rest of the galaxy.

| OBJECT #800 | NGC 602 |
|---|---|
| Constellation | Hydrus |
| Right ascension | 1h30m |
| Declination | −73°33′ |
| Size | 34′ |
| Type | Emission nebula |

Our next object sits just outside the eastern border of the Small Magellanic Cloud. NGC 602 combines an emission nebula with a cluster of stars. A 12-inch telescope, an eyepiece that gives a magnification of 200×, and a nebula filter will reveal an oval haze divided into two parts, the eastern side a bit brighter. The magnitude 12.2 star GSC 9142:30 shines at the nebula's southwestern edge.

**Object #801** NGC 613 Fred Calvert/Adam Block/NOAO/AURA/NSF

| OBJECT #801 | NGC 613 |
|---|---|
| Constellation | Sculptor |
| Right ascension | 1h34m |
| Declination | −29°25′ |

| (continued) | |
|---|---|
| Magnitude | 10.1 |
| Size | 5.5′ by 4.1′ |
| Type | Spiral galaxy |

This nice spiral sits 0.6° northwest of magnitude 5.7 Tau (τ) Sculptoris. Through a small telescope, it appears as just an evenly illuminated oval glow. A 12-inch scope and 200× or more reveals a bright, extended central region from which short spiral arms emanate. Note the magnitude 9.6 star SAO 167149 only 2′ north-northeast of the galaxy's core.

| OBJECT #802 | M74 (NGC 628) |
|---|---|
| Constellation | Pisces |
| Right ascension | 1h37m |
| Declination | 15°47′ |
| Magnitude | 9.4 |
| Size | 11.0′ by 11.0′ |
| Type | Spiral galaxy |

Just move 1.3° east-northeast of magnitude 3.6 Eta (η) Piscium, and you'll find the gorgeous face-on spiral M74. At a distance of 24 million light-years, stellar associations and gas clouds in this galaxy stand out well, so it's one of the few that will benefit from a nebula filter. If you plan to observe this galaxy filtered, use at least a 12-inch telescope to assure good light throughput.

Through an 8-inch telescope, you can see the uneven spiral arms and the mottled halo that surrounds the bright core. The half dozen stars superimposed on the galaxy's glow make it a more impressive sight.

| OBJECT #803 | NGC 637 |
|---|---|
| Constellation | Cassiopeia |
| Right ascension | 1h43m |
| Declination | 64°02′ |
| Magnitude | 8.2 |
| Size | 3′ |
| Type | Open cluster |

To find this open cluster, move 1.3° west-northwest of magnitude 3.4 Epsilon (ε) Cassiopeiae. A 4-inch telescope at 100× will reveal approximately 25 stars, and larger scopes don't show all that many more.

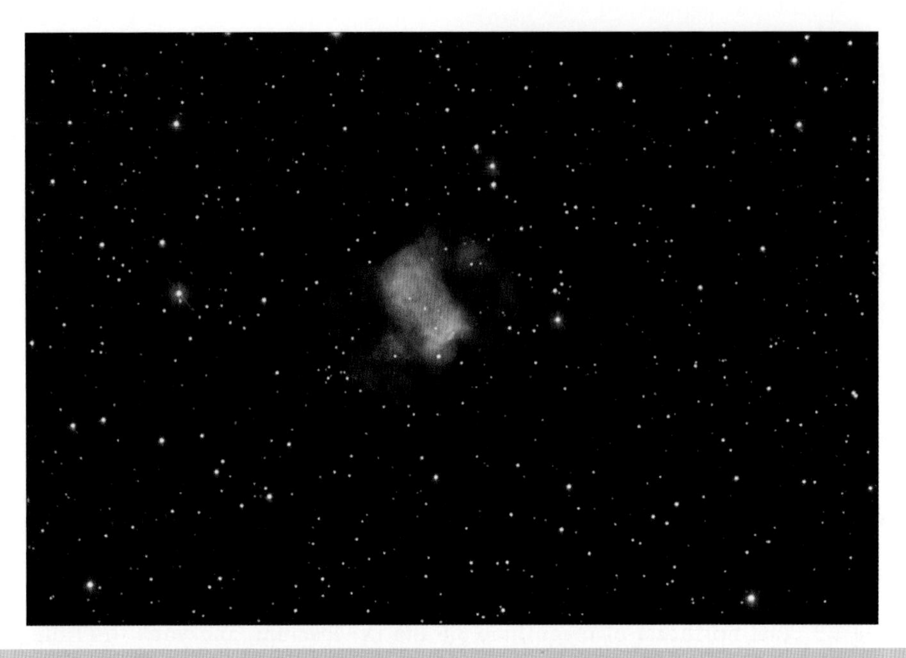

**Object #804** The Little Dumbbell Nebula (M76) Adam Block/NOAO/AURA/NSF

| OBJECT #804 | M76 (NGC 650) |
|---|---|
| Constellation | Perseus |
| Right ascension | 1h42m |
| Declination | 51°34′ |
| Magnitude | 10.1 |
| Size | 65″ |
| Type | Planetary nebula |
| Other names | The Little Dumbbell Nebula, the Barbell Nebula, the Cork Nebula |

The Little Dumbbell Nebula in Perseus is number 76 on Messier's famous list. This planetary nebula sits in far western Perseus, near its border with Cassiopeia and Andromeda. You'll find it 1° north of the 4th-magnitude star Phi ($\Phi$) Persei.

At magnitude 10.1 it's tied for the designation "Messier's faintest object" with galaxies M98 and M91. But don't let its magnitude fool you. M76 appears fairly bright because it isn't that large. Its elongated disk measures about 1′ across. And don't be afraid to crank up the magnification on this object — there's lots of detail to see.

An 8-inch telescope reveals the two lobes that gave the nebula its name, with the southwestern patch appearing a little brighter. A 16-inch scope brings out a faint strand extending west from the northeastern lobe. You might even catch a hint of the large, diffuse halo that surrounds the entire inner region.

| OBJECT #805 | NGC 654 |
|---|---|
| Constellation | Cassiopeia |
| Right ascension | 1h44m |
| Declination | 61°53′ |
| Magnitude | 6.5 |
| Size | 5′ |
| Type | Open cluster |

NGC 654 is a moderately rich open cluster that contains 40 stars shining between 7th and 12th magnitude. You'll find it by drawing a line between magnitude 2.7 Ruchbah (Delta ($\delta$) Cassiopeiae) and magnitude 3.4 Epsilon ($\varepsilon$) Cas. The cluster lies just to the east of the line's midpoint. Lots of open clusters populate this region of sky, so be careful with your identification.

| OBJECT #806 | NGC 659 |
|---|---|
| Constellation | Cassiopeia |
| Right ascension | 1h44m |
| Declination | 60°40′ |
| Magnitude | 7.9 |
| Size | 6′ |
| Type | Open cluster |
| Other name | The Yin-Yang Cluster |

Our next nice open cluster lies 2.3° east-northeast of magnitude 2.7 Ruchbah (Delta ($\delta$) Cassiopeiae) and only 10.5′ northeast of the magnitude 5.8 star 44 Cassiopeiae.

When you've finished observing NGC 659, look a bit less than 0.5° east, and have a look at the fainter open cluster IC 155.

*Astronomy* magazine Contributing Editor Stephen James O'Meara christened this collection of stars the Yin-Yang Cluster. He combines the stars in the core and a smaller, fainter group southwest of the core, with separate streams of stars. Separately, each of these, to O'Meara, looks like an apostrophe. Together, however, they appear as the Chinese symbol that combines Yin and Yang, the complementary opposite forces of life.

**Object #807** NGC 663 Peter and Suzie Erickson/Adam Block/NOAO/AURA/NSF

| OBJECT #807 | NGC 663 |
|---|---|
| Constellation | Cassiopeia |
| Right ascension | 1h46m |
| Declination | 61°15′ |
| Magnitude | 7.1 |
| Size | 16′ |
| Type | Open cluster |
| Other names | Caldwell 10 |

Here's a gorgeous cluster you'll enjoy through any size telescope, and perhaps even with the naked eye. Several sharp-eyed observers (I'm one of them) have seen this object without optical aid from a true-dark site. Some have even proposed that the cluster's "official" magnitude is too faint. Check it out, and see what you think.

NGC 663 breaks into some 40 stars through a 6-inch scope at 75×. It does so, however, in a twin waterfall pattern. Each waterfall (you might see these features as chains of faint stars) terminates at a pair of much brighter stars. At this magnification, the space between them appears dark.

A 10-inch will fill in that region and show you more than 75 stars within the cluster. You'll find NGC 663 just 45′ north-northeast of the magnitude 5.8 star 44 Cassiopeiae. Be sure to go past NGC 659, which lies only one-quarter of NGC 663's distance from the star.

**Object #808** NGC 672 Adam Block/NOAO/AURA/NSF

| OBJECT #808 | NGC 672 |
|---|---|
| Constellation | Triangulum |
| Right ascension | 1h48m |
| Declination | 27°26′ |
| Magnitude | 10.9 |
| Size | 6.6′ by 2.6′ |
| Type | Barred spiral galaxy |

Look for this seemingly rectangular object 2.4° south-southwest of magnitude 3.4 Mothallah (Alpha [α] Trianguli). This barred spiral galaxy is interacting with its neighbor, the magnitude 11.4 spiral IC 1727 only 8′ to the southwest. NGC 672 appears elongated roughly east-west. The eastern half looks slightly brighter, as does the central region.

| OBJECT #809 | NGC 676 |
|---|---|
| Constellation | Pisces |
| Right ascension | 1h49m |
| Declination | 5°54′ |
| Magnitude | 9.6 |
| Size | 4.6′ by 1.7′ |
| Type | Spiral galaxy |

NGC 676 is a lens-shaped spiral galaxy that doesn't reveal many details through any size telescope. Here's what every observer notices, however: the magnitude 9.4 star SAO 110143 placed squarely over NGC 676's center. The galaxy orients roughly north-south. Look for it not quite 2° east-northeast of magnitude 4.4 Nu (ν) Piscium.

| OBJECT #810 | Gamma (γ) Arietis |
|---|---|
| Constellation | Aries |
| Right ascension | 1h54m |
| Declination | 19°18′ |
| Magnitudes | 4.6/4.7 |
| Separation | 7.8″ |
| Type | Double star |
| Other name | Mesarthim |

I like to call this binary the "headlights" because the two components shine at nearly the same brightness, and both appear white. This is the double star I show first during autumn stargazes. Then I move on to the magnificent Albireo (Beta [β] Cygni).

Richard Hinckley Allen in *Star Names and Their Meanings* gives two possibilities for this star's common name. He said some of his contemporaries connected it with the Hebrew word Mesharetim, which means ministers. He thinks, however, that it's more likely an error made by celestial cartographer Johannes Bayer, who was unaware that the word Mesarthim referred to the lunar station to which this star and Sheratan (Beta [β] Arietis) belong.

| OBJECT #811 | NGC 720 |
|---|---|
| Constellation | Cetus |
| Right ascension | 1h53m |
| Declination | −13°44′ |
| Magnitude | 10.2 |
| Size | 4.7′ by 2.4′ |
| Type | Elliptical galaxy |

You'll find our next target 3° northeast of magnitude 3.5 Tau (τ) Ceti. It appears as a relatively bright oval of light with a broad central region. Through 10-inch and larger telescopes, you'll pick out the thin halo that quickly fades to black.

| OBJECT #812 | NGC 744 |
|---|---|
| Constellation | Perseus |
| Right ascension | 1h59m |
| Declination | 55°28′ |

| (continued) | |
|---|---|
| Magnitude | 7.9 |
| Size | 5′ |
| Type | Open cluster |

NGC 744 is a bright open cluster. The star count is relatively poor, however. A 4-inch telescope will reveal two dozen stars. So will a 12-inch scope. The brightest nearby star is magnitude 7.9 SAO 22809, which sits 7′ to the north-northeast. Look for this cluster a bit more than 3° west-southwest of the Double Cluster (NGC 869 and NGC 884).

| OBJECT #813 | NGC 752 |
|---|---|
| Constellation | Andromeda |
| Right ascension | 1h58m |
| Declination | 37°41′ |
| Magnitude | 5.7 |
| Size | 50′ |
| Type | Open cluster |
| Other name | Caldwell 28 |

Although this object lies in Andromeda, the best way to locate it is to look 3.7° northwest of Beta ($\beta$) Trianguli. NGC 752 is large and bright, and easily visible to the naked eye from a dark site.

NGC 752 is huge, so view it at magnifications below 50×. Through 10×70 binoculars, you'll count about three dozen magnitude 10 stars. An 8-inch telescope drives the star count above 100. A crooked line of four magnitude 7 and 8 stars crosses the central part of the cluster running east to west. A pair of 6th-magnitude stars lies at NGC 752's southern end.

Another cluster abuts the southern end of NGC 752, but it's another story altogether. Rather than a cluster of stars, Abell 262 is a cluster of galaxies containing several dozen 13th-magnitude and fainter objects.

| OBJECT #814 | Lambda ($\lambda$) Arietis |
|---|---|
| Constellation | Aries |
| Right ascension | 1h58m |
| Declination | 23°36′ |
| Magnitudes | 4.9/7.7 |
| Separation | 37″ |
| Type | Double star |

Here's a great target for a small telescope, and, because this binary has such a wide separation, you won't even need high magnification to view it. If you observe this pair with others, have some fun and ask everyone to view Lambda and tell you what colors they see. You'll get "yellow and blue," "white and blue," "orange and green," and lots of other combinations. Observations like these are proof positive that the color receptors in human eyes are unique to each of us.

**Object #815** NGC 772 Adam Block/NOAO/AURA/NSF

| OBJECT #815 | NGC 772 |
|---|---|
| Constellation | Aries |
| Right ascension | 1h59m |
| Declination | 19°01′ |
| Magnitude | 10.3 |
| Size | 7.3′ by 4.6′ |
| Type | Spiral galaxy |
| Other name | The Fiddlehead Galaxy |

A little more than a degree east of magnitude 4.5 Gamma (γ) Arietis you'll find spiral galaxy NGC 772. Through a 10-inch scope, it exhibits a bright center surrounded by a haze. A bit more than 3′ south-southwest of NGC 772's core lies magnitude 13.0 NGC 770. This pair of galaxies interacts gravitationally. Images of NGC 772 show the result — a distorted spiral structure.

This galaxy's common name comes from *Astronomy* magazine Contributing Editor Stephen James O'Meara, who noted that its brightest arm looked to him like a fiddlehead unfolding.

| OBJECT #816 | Alpha (α) Piscium |
|---|---|
| Constellation | Pisces |
| Right ascension | 2h02m |
| Declination | 2°46′ |
| Magnitudes | 4.2/5.1 |
| Separation | 1.7″ |
| Type | Double star |
| Other name | Al Rischa |

Most observers report a pale yellow and pale blue for the colors of the primary and secondary, respectively. If you see the "yellow" component as white and the "blue" component as green, don't worry. Human color receptors vary from one individual to the next. Do note that this is a close binary, so crank up the power past 150× to get a clean split.

This star's common name, Al Rischa, has an easy translation from Arabic — the cord. Its position marks the knot that ties the two ribbons of this constellation's fish together.

**Object #817** Gamma Andromedae Adam Block/NOAO/AURA/NSF

| OBJECT #817 | Gamma ($\gamma$) Andromedae |
|---|---|
| Constellation | Andromeda |
| Right ascension | 2h04m |
| Declination | 42°20′ |
| Magnitudes | 2.2/5.0 |
| Separation | 9.8″ |
| Type | Double star |
| Other name | Almach |

As I compiled this list, I wasn't sure if I wanted to put Gamma ($\gamma$) Andromedae in the small-scope category or present it as a large-scope challenge object. I'll explain below.

Gamma And is the 3rd-brightest star in the constellation, shining at 2nd magnitude. But its visual output actually combines the light from a colorful pair of stars: yellowish Gamma$^1$ ($\gamma^1$) Andromedae shines at magnitude 2.3, and bluish Gamma[2] ($\gamma^2$) glows at magnitude 3.6. About 10″ separate the two components, so any telescope will easily split this pair.

Why, then, did I consider describing it as a challenge object? The reason is that Gamma[2] is also a double, but one with a separation of only 0.4″. Gamma[2]'s components are just about the same brightness, at magnitudes 4.84 and 4.87. I split this star with a perfectly aligned and totally cooled-down 11-inch Schmidt-Cassegrain telescope in 2001. You won't split it with much less aperture.

*In Star Names and Their Meanings*, Richard Hinckley Allen explains that the common name Almach comes from "Al Anak al Ard," a small predatory animal found in Arabia. If you find yourself at a loss as to how to reconcile this meaning with the star marking the left foot of Andromeda the Princess, you're not alone. Allen can't either.

| OBJECT #818 | Iota ($\iota$) Trianguli |
|---|---|
| Constellation | Triangulum |
| Right ascension | 2h12m |
| Declination | 30°18′ |
| Magnitudes | 5.3/6.9 |
| Separation | 3.9″ |
| Type | Double star |

To find this binary, look 4.2° east of magnitude 3.4 Alpha ($\alpha$) Trianguli. At low power, you'll have trouble splitting Iota, so crank the magnification past 100×. Most observers see a yellow primary with a slightly fainter light-blue companion.

| OBJECT #819 | Stock 2 |
|---|---|
| Constellation | Cassiopeia |
| Right ascension | 2h15m |
| Declination | 59°16′ |
| Magnitude | 4.4 |
| Size | 60′ |
| Type | Open cluster |

A bit more than 2° north-northwest of the Double Cluster (NGC 869 and NGC 884) lies the nearby cluster Stock 2. This object appeared on a list of open clusters compiled by astronomer Jürgen Stock. This object spans a full degree, so small telescopes do a good job displaying it, if you keep the magnification below 50×. Expect to see a loose collection of about 50 stars between 8th and 10th magnitudes.

| OBJECT #820 | NGC 821 |
|---|---|
| Constellation | Aries |
| Right ascension | 2h08m |
| Declination | 11°00′ |
| Magnitude | 10.8 |
| Size | 2.4 by 1.7 |
| Type | Elliptical galaxy |

Although NGC 821 lies in Aries, it's easiest to find by starting at magnitude 4.4 Xi[1] ($\xi^1$) Ceti. From that star, move 2.4° north-northwest. The galaxy has an oval shape oriented roughly north-south. The evenly illuminated central region spans two-thirds of NGC 821's length. Outside is a thin halo visible through 12-inch and larger telescopes. You'll also spot the magnitude 9.2 star SAO 92805 on the galaxy's northwest edge.

**Object #821** NGC 869 and NGC 884 Fred Calvert/Adam Block/NOAO/AURA/NSF

| OBJECT #821 | NGC 869 and NGC 884 |
|---|---|
| Constellation | Perseus |
| Right ascension | 2h19m, 2h22m |
| Declination | 57°09′, 57°07′ |
| Magnitude | 5.3, 6.1 |
| Size | 29′, 29′ |
| Type | Open cluster |
| Also known as | The Double Cluster, Caldwell 14 |

The Double Cluster in Perseus is a real treat through low-power optics. NGC 869 is the richer of the two clusters. It contains nearly three dozen 9th- and 10th-magnitude stars in an area 10′ across.

NGC 884 appears less concentrated but holds more bright stars, with the densest group on the southwestern edge. Through a 4-inch telescope at 50×, you'll see an abundance of color in these stellar jewels. Look for red, yellow, and blue gems scattered among the abundant white stars.

Each cluster holds a treasure at its core. NGC 869 contains a dark, Y-shaped rift. NGC 884's stars surround RS Persei, a deep-red semiregular variable star.

By the way, the Double Cluster looks particularly nice through binoculars that magnify 15 or more times.

To find this wonderful pair, draw a line from magnitude 2.7 Ruchbah (Delta [$\delta$] Cassiopeiae) to magnitude 2.9 Gamma ($\gamma$) Persei. The Double Cluster lies at the midpoint of this line.

| OBJECT #822 | Omicron Ceti |
|---|---|
| Constellation | Cetus |
| Right ascension | 2h19m |
| Declination | −2°59′ |

(continued)

| | |
|---|---|
| Magnitude range | 2.0-10.1 |
| Period | 331.96 days |
| Type | Variable star |
| Other name | Mira |

The name Mira is Latin for "the Wonderful," and this luminary is the archetype for an important class of variable stars. Mira-type variables have long periods and large-amplitude variations in brightness. Mira itself takes 332 days to vary between 2nd and 10th magnitude, but it does so irregularly. In February 1997, Mira reached nearly 2nd magnitude, but at its next maximum at the beginning of 1998, the star disappointed observers by attaining only 4th magnitude. You can locate Mira slightly more than 7° southeast of magnitude 3.8 Al Rischa (Alpha [$\alpha$] Piscium).

# November

**Object #823** The Silver Sliver Galaxy (NGC 891) Adam Block/NOAO/AURA/NSF

M.E. Bakich, *1,001 Celestial Wonders to See Before You Die*, Patrick Moore's Practical Astronomy Series,
DOI 10.1007/978-1-4419-1777-5_11, © Springer Science+Business Media, LLC 2010

| OBJECT #823 | NGC 891 |
| --- | --- |
| Constellation | Andromeda |
| Right ascension | 2h22.6m |
| Declination | 42°21′ |
| Magnitude | 9.9 |
| Size | 13.0′ by 2.8′ |
| Type | Spiral galaxy |
| Other names | The Silver Sliver Galaxy, Caldwell 23 |

You know, second place is fine if people remember you. NGC 891's problem is that it's Andromeda's second-best galaxy. Not bad, except the constellation's top dog happens to be one of the sky's supreme wonders — the Andromeda Galaxy (M31).

Despite that, NGC 891 ranks as one of the sky's best edge-on galaxies. It inclines only 1.4° to our line of sight. Its more than four-to-one length to width ratio and 10th-magnitude brightness easily earned it the nickname the "Silver Sliver."

A 10-inch telescope reveals a symmetrical object about 10′ long with a noticeable but narrow central bulge. A dark dust lane bisects the galaxy and runs nearly its entire length. Dozens of foreground stars populate the field, adding a third dimension to the view.

At magnifications above 200×, note the sections of NGC 891's nucleus on each side of the dust lane. The western section glows slightly brighter. Likewise, the galaxy's disk to the southwest outshines its lesser half, which lies to the northeast.

**Object #824** NGC 896 Sean and Renee Stecker/Adam Block/NOAO/AURA/NSF

| OBJECT #824 | NGC 896 |
| --- | --- |
| Constellation | Cassiopeia |
| Right ascension | 2h25m |
| Declination | 61°54′ |
| Size | 20′ by 20′ |
| Type | Emission nebula |

To find this emission nebula, move 3.9° east-southeast from magnitude Epsilon (ε) Cassiopeiae. An 8-inch telescope and a nebula filter will reveal only the bright condensation on NGC 896's western edge. Just 8′ to the east-northeast lies the more impressive nebula IC 1795.

**Object #825** NGC 908 George and Laura Mishler/Adam Block/NOAO/AURA/NSF

| OBJECT #825 | NGC 908 |
|---|---|
| Constellation | Cetus |
| Right ascension | 2h23m |
| Declination | –21°14′ |
| Magnitude | 10.4 |
| Size | 5.9′ by 2.3′ |
| Type | Spiral galaxy |

To find spiral galaxy NGC 908, look 5.4° east of magnitude 4.0 Upsilon (υ) Ceti. Through an 8-inch telescope, it appears as an oval haze stretched roughly east-west. The core is quite a bit brighter, and ghostly hints of spiral structure appear at high magnifications.

Through a 16-inch scope, an arm that radiates to the north of the core and turns abruptly to the west is quite prominent. A fainter arm that can be seen with more difficulty extends due east and ends near a 15th-magnitude star. The core also appears irregular.

| OBJECT #826 | NGC 925 |
|---|---|
| Constellation | Triangulum |
| Right ascension | 2h27.3m |
| Declination | 33°35′ |
| Magnitude | 10.1 |
| Size | 12.0′ by 7.4′ |
| Type | Spiral galaxy |

NGC 925 is an attractive, nearly face-on spiral galaxy in the small constellation Triangulum. What makes NGC 925 special among spirals is the bar that projects from its core.

To find NGC 925, point your telescope 2° east of the magnitude 4 star Gamma (γ) Trianguli. Through a small scope, this galaxy's figure appears indistinct, but an 8-inch or larger instrument reveals the spiral arms that fold back abruptly from a long bar. At high magnification, say, above 250×, you'll spot NGC 925's stellar nucleus.

| OBJECT #827 | NGC 936 |
|---|---|
| Constellation | Cetus |
| Right ascension | 2h28m |
| Declination | –1°09′ |
| Magnitude | 10.2 |
| Size | 4.7′ by 4.1′ |
| Type | Barred spiral galaxy |

This relatively bright galaxy looks a bit oval through small telescopes, but use a 12-inch or larger instrument, and you'll instantly spot the central bar, which stretches east to west. NGC 936 lies 1.1° west from the magnitude 5.4 star 75 Ceti.

| OBJECT #828 | Alpha Ursae Minoris |
|---|---|
| Constellation | Ursa Minor |
| Right ascension | 2h32m |
| Declination | 89°16′ |
| Magnitudes | 2.0/9.0 |
| Separation | 18.3″ |
| Type | Double star |
| Other name | Polaris |

William Herschel discovered that Polaris was a binary star in 1780. Although the yellow primary outshines the blue secondary by more than 650 times, a 3-inch telescope at 100× easily will show you Polaris B.

| OBJECT #829 | NGC 956 |
|---|---|
| Constellation | Andromeda |
| Right ascension | 2h32m |
| Declination | 44°36′ |
| Magnitude | 8.9 |
| Size | 6′ |
| Type | Open cluster |

You'll find our next treat 5.7° east-northeast of magnitude 2.2 Almach (Gamma [$\gamma$] Andromedae). This odd cluster has two 9th-magnitude stars at its north and south end, a couple more 10th-magnitude stars, and about a dozen magnitude 12 and fainter stars strewn about.

**Object #830** NGC 957 Anthony Ayiomamitis

| OBJECT #830 | NGC 957 |
|---|---|
| Constellation | Perseus |
| Right ascension | 2h33m |
| Declination | 57°34′ |
| Magnitude | 7.6 |
| Size | 10′ |
| Type | Open cluster |

You'll find our next target about 1.5° east-northeast of the Double Cluster (Object #821). Through an 8-inch scope at 100×, you'll count two dozen stars. The magnitude 8.0 star HIP 11898 sits slightly southeast of the cluster's center. At the cluster's southwest edge sits magnitude 8.5 SAO 23415.

| OBJECT #831 | IC 1805 |
|---|---|
| Constellation | Cassiopeia |
| Right ascension | 2h33m |
| Declination | 61°27′ |
| Magnitude | 6.5 |
| Size | 60′ |
| Type | Emission nebula |
| Other names | The Heart Nebula |

Some nebulae glow because a hot star within them emits enough energy to excite hydrogen atoms. The atoms can't hold the extra energy long, however, and they reemit it as light with a red color.

The Heart Nebula, in contrast, glows because it contains an entire cluster of stars that formed within it. The combined radiation from cluster stars makes such nebulae bigger, brighter, or both. You'll find this object 4.9° east-southeast of magnitude 3.4 Epsilon (ε) Cassiopeiae.

Several of the stars in IC 1805's central cluster are 50 times more massive than the Sun. In addition to powering the nebula's glow, these stars' stellar winds propel hydrogen into space, where gaseous rings overlap other rings. This gives the "heart" its shape.

Because of its size, you can approach observing the Heart Nebula in two ways, with low or high magnification. Each requires at least an 8-inch telescope and a dark sky.

For an overall view, use an eyepiece that provides a 1° field of view. Insert a nebula filter, and identify the heart's brightest regions: the central cluster, the knot of nebulosity to the east, and the crescent of gas to the southwest.

The magnitude 6.7 star cluster NGC 1027 shines 1.2° east of the nebula. This cluster spans 15′. Another bright, condensed nebulous knot, NGC 896, lies 1° southwest of the Heart Nebula. Many observers see this 20′-wide object before they spot IC 1805.

| OBJECT #832 | NGC 972 |
|---|---|
| Constellation | Aries |
| Right ascension | 2h34m |
| Declination | 29°19′ |
| Magnitude | 11.4 |
| Size | 3.4′ by 1.6′ |
| Type | Spiral galaxy |

Through an 8-inch telescope, this galaxy looks like a bright, small oval surrounded by an outer halo, which quickly fades to the black of space. To find NGC 972, head 1.4° east-southeast of the magnitude 5.3 star 12 Trianguli.

| OBJECT #833 | Fornax Dwarf |
|---|---|
| Constellation | Fornax |
| Right ascension | 2h40m |
| Declination | −34°32′ |
| Magnitude | 8.1 |
| Size | 12.0′ by 10.2′ |
| Type | Dwarf elliptical galaxy |

American astronomer Harlow Shapley (1885–1972) discovered the Fornax Dwarf in 1938. At a distance of 438,000 light-years, it is the Milky Way's closest dwarf companion.

Although the Fornax Dwarf's stated magnitude indicates a bright object, its size is such that it covers 17% the area of the Full Moon. The Dwarf's surface brightness, therefore, is low.

I've spotted this object through a 4-inch refractor equipped with a wide-angle eyepiece. Through an 8-inch scope, use an eyepiece that yields a 1° field of view, and slowly sweep the field of view. What you're looking for is a faint haze just brighter than the background sky.

Once you've located the Fornax Dwarf, crank up the magnification and aim for its northern edge. There sits globular cluster NGC 1049. This object shines at magnitude 12.6 and measures roughly 1.2′ across.

| OBJECT #834 | NGC 986 |
|---|---|
| Constellation | Fornax |
| Right ascension | 2h34m |
| Declination | −39°03′ |
| Magnitude | 10.8 |
| Size | 4′ by 3.2′ |
| Type | Spiral galaxy |

Here's a nice treat that lies in southern Fornax right next to that constellation's border with Eridanus. To find NGC 986, first locate magnitude 4.1 Iota (ι) Eridani. Then move 1.6° west-northwest. Small telescopes will show an oval shape elongated northeast to southwest. Through 12-inch and larger scopes, however, you'll spot the two broad but incredibly short spiral arms, one on the north end and the other angling south.

| OBJECT #835 | NGC 1023 |
|---|---|
| Constellation | Perseus |
| Right ascension | 2h40m |
| Declination | 39°04′ |
| Magnitude | 9.3 |
| Size | 8.6′ by 4.2′ |
| Type | Spiral galaxy |
| Other name | The Perseus Lenticular |

The bright lenticular galaxy NGC 1023 measures three times as long as it is wide, stretching in a rough east-west direction. Through small telescopes, the core appears small, almost starlike. Use a 14-inch or larger scope, however, and you'll see that the central region spans half of this galaxy's overall length.

For a bit of a challenge, try to spot the magnitude 13.8 irregular galaxy NGC 1023A, which appears embedded in the eastern end of NGC 1023's halo.

**Object #836** NGC 1027 Anthony Ayiomamitis

| OBJECT #836 | NGC 1027 |
|---|---|
| Constellation | Cassiopeia |
| Right ascension | 2h43m |
| Declination | 61°36′ |
| Magnitude | 6.7 |
| Size | 15′ |
| Type | Open cluster |

You'll find this nice cluster not quite 6° east-southeast of magnitude 3.4 Epsilon (ε) Cassiopeiae. Through an 8-inch telescope at 100×, you'll see the brightest 15 or so members hung in front of a haze of fainter stars. A 12-inch telescope resolves about 20 more stars. The bright star at the heart of NGC 1027 is magnitude 7.0 SAO 12402.

**Object #837** M34 Anthony Ayiomamitis

| OBJECT #837 | M34 (NGC 1039) |
|---|---|
| Constellation | Perseus |
| Right ascension | 2h42m |
| Declination | 42°47′ |
| Magnitude | 5.2 |
| Size | 35′ |
| Type | Open cluster |
| Other name | The Spiral Cluster |

Our next object is a nice small telescope target: open cluster M34 in Perseus. Although this is one of Messier's objects — and a bright one at that — it often gets overlooked by amateur astronomers.

From a dark site you'll find M34 with your naked eyes roughly 5° west-northwest of Algol (Beta [$\beta$] Persei). This magnitude 5.2 cluster contains 10 stars brighter than 9th magnitude spread out over an area 35′ across. That's a bit bigger than the Full Moon.

A 4-inch scope reveals three dozen member stars between 8th and 12th magnitude. The way they loop across the cluster's face led British amateur astronomer Jeff Bondono to give M34 its common name. At medium magnification (say, 100×), look for chains of faint stars crisscrossing the field of view.

| OBJECT #838 | NGC 1052 |
|---|---|
| Constellation | Cetus |
| Right ascension | 2h41m |
| Declination | –8°15′ |
| Magnitude | 10.5 |
| Size | 2.5′ by 2.0′ |
| Type | Spiral galaxy |

At the eastern edge of Cetus, 3.8° west of magnitude 3.9 Azha (Eta [$\eta$] Eridani) lies a nice trio of galaxies highlighted by NGC 1052. Through an 8-inch telescope, NGC 1052 appears bright and oval with an extended central region surrounded by a slight haze.

Only 15′ to the southwest lies the magnitude 11.0 spiral NGC 1042. It's more than 50% bigger than NGC 1052, but you'll need at least a 16-inch scope to bring out any trace of spiral structure.

Even fainter is magnitude 12.2 NGC 1035, 23′ northeast of NGC 1042. This galaxy is three times as long as it is wide, but doesn't show details through most scopes. Some amateurs like to use a magnification of about 100× to catch all three galaxies at once.

| OBJECT #839 | NGC 1055 |
|---|---|
| Constellation | Cetus |
| Right ascension | 2h42m |
| Declination | 0°26′ |
| Magnitude | 10.6 |
| Size | 7.3′ by 3.3′ |
| Type | Spiral galaxy |

Look 39′ east of magnitude 4.1 Delta ($\delta$) Ceti, and you'll find NGC 1055. This galaxy's overall shape looks a lot like the Whale Galaxy (Object #293) to me. Through a 6-inch telescope, the galaxy appears three times as long as wide and aligned nearly east-west. Just to the north of the galaxy you'll find the magnitude 11.2 star GSC 47:1504.

**Object #840** Cetus A (M77) Francois and Shelley Pelletier/Adam Block/NOAO/AURA/NSF

| OBJECT #840 | M77 (NGC 1068) |
|---|---|
| Constellation | Cetus |
| Right ascension | 2h43m |
| Declination | –0°01′ |

| (continued) | |
|---|---|
| Magnitude | 8.9 |
| Size | 8.2′ by 7.3′ |
| Type | Spiral galaxy |
| Other name | Cetus A |

One of the sky's most active galaxies sits in the huge but hard to recognize constellation Cetus the Whale. And it's doubly unfortunate for observers, because much of this galaxy's drama occurs outside the wavelengths of visible light. That fact led observers to begin to call this galaxy by its radio designation as the first strong radio source in the constellation — Cetus A.

Pierre Méchain discovered M77 in 1780. Messier included it in his catalog December 17.

Astronomers classify M77 as a Seyfert galaxy. American astronomer Carl K. Seyfert (1911–1960) pioneered research on nuclear emissions in spiral galaxies. He described galaxies with bright nuclei that emit light with emission-line spectra and exhibit broadened emission lines. These features indicated the galaxies' cores were expelling giant gas clouds at high speeds.

Visually, the galaxy's central area, one-third its total width, commands observers' attention. Although you can spot M77 through any telescope, a 10-inch or larger instrument and a magnification above 300× will reveal the most detail.

Try to ignore the bright core, and search the disk surrounding it for signs of mottled structure. Through even larger scopes, look for the tightly wound spiral arms — the brightest one lies southeast of the core.

| OBJECT #841 | Gamma (γ) Ceti |
|---|---|
| Constellation | Cetus |
| Right ascension | 2h43m |
| Declination | 3°14′ |
| Magnitudes | 3.5/7.3 |
| Separation | 2.8″ |
| Type | Double star |
| Other name | Kaffaljidhmah |

This colorful double is a bit puzzling. The primary appears white or "just a bit" yellow. There's not much dispute about that. The fainter companion, however, should not appear the color observers report. The secondary is a spectral class F star, which is a bit hotter, and therefore should be just a bit whiter than our Sun. Still, most observers see it as some shade of blue.

According to Richard Hinckley Allen, the common name of this star comes from the Arabic Al Kaff al Jidhmah, which represents the whole group that marks the Whale's head.

| OBJECT #842 | NGC 1073 |
|---|---|
| Constellation | Cetus |
| Right ascension | 2h44m |
| Declination | 1°23′ |
| Magnitude | 11.0 |
| Size | 5.0′ by 5.0′ |
| Type | Spiral galaxy |

Our next target lies 1.5° northeast of magnitude 4.1 Delta (δ) Ceti. Telescopes below 10 inches in aperture won't show a lot of details except the bar. Through a 16-inch telescope, the faint spiral arms also pop into view, but the northern arm is definitely brighter than the southern one.

| OBJECT #843 | Musca Borealis the Northern Fly |
|---|---|
| Constellation | Aries |
| Right ascension | 2h46m |
| Declination | 27°36′ |
| Type | Extinct constellation |
| Other name | Apes |

Indulge me on this object. I'm something of a constellation historian, so I thought it would be fun if one of our objects no longer existed in its original form. So, this naked-eye object is the extinct constellation Musca Borealis the Northern Fly. Dutch mapmaker Petrus Plancius (1552–1622) introduced this constellation around 1614 under the name Apes. He formed it from four stars — 33, 35, 39, and 41 Arietis — in the present-day constellation Aries the Ram.

To find these stars, look about 9° east-northeast of Aries' brightest star Hamal (Alpha [α] Arietis). And although I do rate this as a naked-eye object from a dark site, be forewarned: The brightest of these stars, 41 Arietis, shines at only magnitude 3.6, and the faintest, 39 Ari, glows at magnitude 5.3. If you have any trouble seeing the group, use binoculars. The Northern Fly spans only 2.5°.

| OBJECT #844 | NGC 1084 |
|---|---|
| Constellation | Eridanus |
| Right ascension | 2h46m |
| Declination | –7°35′ |
| Magnitude | 10.7 |
| Size | 2.8 by 1.4 |
| Type | Spiral galaxy |

This relatively bright spiral lies 2.9° west-northwest of magnitude 3.9 Azha (Eta [η] Eridani). Through a 4-inch telescope, NGC 1084 has a nearly rectangular shape and is twice as long as it is wide. A 12-inch scope reveals not a lot more, but it does show the core is broad — as much as three-quarters the galaxy's length. The edges of this galaxy are irregular, but I saw no hint of spiral structure.

| OBJECT #845 | NGC 1097 |
|---|---|
| Constellation | Fornax |
| Right ascension | 2h46m |
| Declination | –30°14′ |
| Magnitude | 9.2 |
| Size | 10.5′ by 6.3′ |
| Type | Barred spiral galaxy |
| Other name | Caldwell 67 |

Our next object is a bright barred spiral that sits 2° north of magnitude 4.5 Beta (β) Fornacis. Through an 8-inch telescope, you'll see NGC 1097's core as a bright disk surrounded by an oval haze. Within that oval is the galaxy's faint bar. You won't see much of the faint, thin spiral arms no matter what scope you observe through. Through a 30-inch telescope, I saw what appeared to be the beginning of the northern spiral arm.

Just to the northeast of NGC 1097 sits NGC 1097A. This tiny peculiar elliptical galaxy glows at magnitude 13.2. Some evidence exists for the two galaxies' past interaction.

| OBJECT #846 | IC 1848 |
|---|---|
| Constellation | Cassiopeia |
| Right ascension | 2h51m |
| Declination | 60°26′ |
| Magnitude | 6.5 |

| (continued) | |
|---|---|
| Size | 60′ |
| Type | Emission nebula |
| Common names | The Baby Nebula, the Soul Nebula |

Surroundings mean a lot. Take IC 1848, for example. When viewed or imaged alone, amateur astronomers call it the Baby Nebula because of its shape. Combine it with the Heart Nebula (IC 1805), however, and, collectively, the pair becomes the Heart and Soul Nebula.

The Baby Nebula sits 2.5° east-southeast of the Heart Nebula. Although equally as wide (1°) as IC 1805, the Baby Nebula doesn't cover as much area, so it appears more concentrated.

A nebula-filtered view shows two large regions that form the "baby's" head and body. The head appears denser, while the body surrounds a small star cluster within. To see the stars better, remove the nebula filter. Through a 12-inch or larger telescope, you'll notice brightness differences along the body's edge. Look for two crescent-shaped nebulae, a smaller one to the northeast and a larger one to the west.

| OBJECT #847 | Struve 331 |
|---|---|
| Constellation | Perseus |
| Right ascension | 3h01m |
| Declination | 52°21′ |
| Magnitudes | 5.3/6.7 |
| Separation | 12.1″ |
| Type | Double star |

This nice pair lies near the Pleiades (M45), but the easiest way to find it is to start at magnitude 3.9 Tau ($\tau$) Persei, and head 1.5° northwest. The separation allows splits through even small telescopes. The primary shines lemon yellow, and the secondary is pale blue.

| OBJECT #848 | Beta Persei |
|---|---|
| Constellation | Perseus |
| Right ascension | 3h08m |
| Declination | 40°57′ |
| Magnitude range | 2.1–3.4 |
| Period | 2.867 days |
| Type | Variable star |
| Other name | Algol |

Algol normally shines at a bright magnitude 2.1. Every 2 days, 20 hours, and 49 minutes, however, it dims to magnitude 3.4. So, at maximum, Algol is 3.3 times as bright as when it's at its minimum. This dip in brightness occurs when a faint, unseen star orbiting Algol passes in front of it and blocks some of its light. Each eclipse lasts approximately 10 hours.

The first eclipsing binary star to be discovered (by Italian astronomer Geminiano Montanari in 1667), Algol remains the easiest to observe. Most of the time it appears nearly as bright as magnitude 1.8 Mirfak (Alpha [$\alpha$] Persei), but keep an eye on it and you may catch it when it's fainter than magnitude 3.0 Delta ($\delta$) Persei.

The name Algol comes from the Arabic "Ra's al Ghul," the demon's head. Although some historians have tried to link this name with Algol's changeable brightness, that is incorrect. The name dates from the time of Ptolemy, many centuries before anyone discovered the variability of any star.

| OBJECT #849 | NGC 1201 |
|---|---|
| Constellation | Fornax |
| Right ascension | 3h04m |
| Declination | –26°04′ |

| (continued) | |
|---|---|
| Magnitude | 10.7 |
| Size | 3.6′ by 2.1′ |
| Type | Elliptical galaxy |

Our next target lies 1.3° southeast of magnitude 5.7 Zeta (ζ) Fornacis. This fat, lens-shaped galaxy orients north-south. The extended central region appears featureless, and a thin haze surrounds it.

You'll also note the magnitude 10.7 star GSC 6441:848 lies not quite 4′ to the northeast. This star's magnitude is the same as the galaxy's so compare them to see how point-source magnitudes compare to those of extended sources.

| OBJECT #850 | NGC 1232 |
|---|---|
| Constellation | Eridanus |
| Right ascension | 3h10m |
| Declination | −20°35′ |
| Magnitude | 10.0 |
| Size | 6.8′ by 5.6′ |
| Type | Spiral galaxy |

Although even a 4-inch telescope will reveal NGC 1232, you'll see this classic face-on spiral best through instruments with 12′ of aperture or more. Seeing — the steadiness of the air above your observing site — is the key to discerning the individual spiral arms. Can you see three? Four? Six (or parts thereof)? It all depends on how clear your view is.

Through a large scope, this galaxy's nucleus has a slight east-west elongation. This characteristic puts NGC 1232 in the barred spiral category.

| OBJECT #851 | NGC 1245 |
|---|---|
| Constellation | Perseus |
| Right ascension | 3h15m |
| Declination | 47°15′ |
| Magnitude | 8.4 |
| Size | 10′ |
| Type | Open cluster |

To find NGC 1245, draw a line between magnitude 4.1 Iota (ι) Persei and magnitude 3.8 Kappa (κ) Persei. The cluster lies less than 1° east of the line's center point. Through an 8-inch telescope, you'll see more than 50 stars evenly distributed across its face. The magnitude 8.0 star SAO 38671 gleams at the cluster's southern edge.

| OBJECT #852 | NGC 1252 |
|---|---|
| Constellation | Horologium |
| Right ascension | 3h11m |
| Declination | −57°46′ |
| Size | 6′ |
| Type | Asterism |
| Notes | Grouping of 18–20 stars |

In the far-southern constellation Horologium, you'll find an interesting group of stars 2.5° north-northeast of magnitude 5.1 Mu (μ) Horologii. Eight stars ranging from magnitude 6.0 to magnitude 9.5 form a loose Greek letter Lambda (λ).

But look more closely at magnitude 6.0 GSC 8498:1319, the brightest and most southerly of the Lambda. Some astronomers think the fainter stars near it form an open cluster 2,000 light-years away.

However, proper motion data collected by the Hipparcos satellite indicate that the stars are unassociated. A 4-inch telescope reveals a group of 11th- to 14th-magnitude stars in an area 6′ across.

**Object #853** NGC 1255 Peter and Suzie Erickson/Adam Block/NOAO/AURA/NSF

| OBJECT #853 | NGC 1255 |
|---|---|
| Constellation | Fornax |
| Right ascension | 3h14m |
| Declination | −25°43 |
| Magnitude | 10.7 |
| Size | 4.2′ by 2.7′ |
| Type | Spiral galaxy |

Our next target lies 3.3° north of magnitude 3.9 Alpha (α) Fornacis. Through an 8-inch telescope, you'll see NGC 1255 as an oval with irregular borders. Move up to a 16-inch, and crank the magnification past 350×, and a weak spiral pattern appears. This galaxy's arms curve tightly inward around its core, so they don't display a classic spiral pattern.

| OBJECT #854 | NGC 1261 |
|---|---|
| Constellation | Horologium |
| Right ascension | 3h12m |
| Declination | −55°13′ |
| Magnitude | 8.3 |
| Size | 6.9′ |
| Type | Globular cluster |
| Other name | Caldwell 87 |

This fine globular lies 4.7° north-northeast of magnitude 5.1 Mu (μ) Horologii. You'll first spot this object through a finder scope or even binoculars, and any size telescope shows it well. That said, its halo stars resolve well through a 10-inch or larger scope. You'll also notice the extremely concentrated core. No telescope will allow you to resolve the stars there. The magnitude 9.1 star GSC 8495:1472 lies 3′ northeast of the cluster's center.

**Object #855** Perseus A (NGC 1275) Jeff Cremer/Adam Block/NOAO/AURA/NSF

| OBJECT #855 | NGC 1275 |
|---|---|
| Constellation | Perseus |
| Right ascension | 3h20m |
| Declination | 41°31′ |
| Magnitude | 11.9 |
| Size | 3.2′ by 2.3′ |
| Type | Spiral galaxy |
| Other names | Perseus A, Caldwell 24 |

NGC 1275 is the brightest member of the Perseus Galaxy Cluster. This group is part of the Pisces-Perseus Supercluster, which contains about 1,000 galaxies. The Perseus Galaxy Cluster also goes by the name Abell 426.

That name comes from American astronomer George Abell, who identified and cataloged 2,712 galaxy clusters in 1958. With the inclusion of southern-sky galaxy clusters since then, the catalog has grown to 4,073 galaxy clusters.

To find NGC 1275, look 2° east of Algol (Beta [$\beta$] Persei). Through a telescope, this galaxy appears bright, small, and nearly circular. Don't confuse it with NGC 1272, a similar galaxy just 5′ to the west. NGC 1275 is slightly brighter, at magnitude 11.7. NGC 1272 shines at magnitude 12.

Through a 10-inch telescope, you'll spot a dozen galaxies in a field of view 1° across. Most lie south and west of NGC 1275. Here's a region of sky where increased telescope aperture really pays off. As you look through larger telescopes, you'll see more galaxies, and the ones you've seen already will show more detail.

| OBJECT #856 | Stock 23 |
|---|---|
| Constellation | Camelopardalis |
| Right ascension | 3h16m |
| Declination | 60°02′ |
| Magnitude | 6.5 |
| Size | 14′ |
| Type | Open cluster |
| Other name | Pazmino's Cluster |

This small telescope target lies in the southwestern part of Camelopardalis. It's Pazmino's Cluster, also known as Stock 23. To find it, scan 5.3° northeast of magnitude 3.8 Eta ($\eta$) Persei. The cluster shines at a respectable magnitude 6.5.

Through your finder scope, Stock 23 is an unresolved clump of stars. View it through a 3-inch telescope at a magnification of 50×, however, and you'll spot two dozen stars spread across an area 15′ wide.

Four cluster stars shine brighter than 8th magnitude, including double star ADS 2426, which lies at the center. It's a close double star with a separation of only 7″. If you can't split it at 50×, just double the power, and you'll have no problem.

Observers began calling this cluster Pazmino's Cluster after American amateur astronomer John Pazmino spotted it in 1977. German astronomer Jürgen Stock had cataloged it in the 1950s.

| OBJECT #857 | NGC 1291 |
|---|---|
| Constellation | Eridanus |
| Right ascension | 3h17m |
| Declination | −41°08′ |
| Magnitude | 8.5 |
| Size | 11.0′ by 9.5′ |
| Type | Spiral galaxy |
| Other name | The Snow Collar Galaxy |

You'll find this bright galaxy 3.7° east of magnitude 3.5 Theta ($\theta$) Eridani. It appears slightly oblong, but, apart from a faint outer halo, you'll see no details here through even a medium-sized telescope. With larger apertures, you may see the two faint, broad arcs of light that *Astronomy* magazine Contributing Editor Stephen James O'Meara says look like snowflakes that have fallen on a fur collar.

Note: NGC 1291 is the same galaxy you'll sometimes see labeled as NGC 1269. This is a catalog number error.

| OBJECT #858 | NGC 1300 |
|---|---|
| Constellation | Eridanus |
| Right ascension | 3h20m |
| Declination | −19°25′ |
| Magnitude | 10.4 |
| Size | 5.5′ by 2.9′ |
| Type | Spiral galaxy |

NGC 1300 has a simple shape — that of a squashed letter S — but I bet you'll find yourself returning to view this celestial wonder more than once. Better yet, show it to your friends. It's a classic barred spiral with two arms, both of which originate from the ends of the bar and move out at right angles to the bar.

To find NGC 1300, look 2.3° due north of magnitude 3.7 Tau$^4$ ($\tau^4$) Eridani. Crank up the magnification past 200× and look first for the bright oval nucleus. It's twice as long as it is wide. The next features that will become evident are the beginnings of the spiral arms. They're quite clumpy near the nucleus. Finally, if you're viewing through a 16-inch or larger scope, try to trace the thin spiral arms as they tightly curve past the nucleus on the northern and southern sides.

| OBJECT #859 | NGC 1313 |
|---|---|
| Constellation | Reticulum |
| Right ascension | 3h18m |
| Declination | –66°30′ |
| Magnitude | 8.9 |
| Size | 11.0′ by 7.6′ |
| Type | Spiral galaxy |

Our next object is one of the southern sky's showpiece galaxies, but you're forgiven if you haven't heard of it. NGC 1313 sits in the southwest corner of Reticulum 3.2° southwest of magnitude 3.8 Beta ($\beta$) Reticuli.

Through an 8-inch telescope, the first feature you'll notice is the thick bar that has a slight central bulge that orients north-south. A spiral arm extends eastward from the north end of the bar and westward from the south end. The eastward bar has two distinct sections divided by a dark region. You'll also notice many bright knots along the arms and the bar. Those are star-forming regions — Moreover, a spiral arm divided into two elongated sections extends at a right angle to the east from the north end of the bar. And all this is visible through an 8-inch scope!

If you use a 14-inch or larger scope on this object, look 16′ southeast of NGC 1313 for the magnitude 13.8 edge-on spiral NGC 1313A.

| OBJECT #860 | NGC 1316 |
|---|---|
| Constellation | Fornax |
| Right ascension | 3h23m |
| Declination | –37°12′ |
| Magnitude | 8.9 |
| Size | 11.0′ by 7.6′ |
| Type | Spiral galaxy |
| Other name | Fornax A |

The powerful radio source astronomers call Fornax A is a bright galaxy you can find 1.4° south-southwest of magnitude 6.4 Chi$^1$ ($\chi^1$) Fornacis. This galaxy's spiral arms wrap so tightly around its core that it appears elliptical through most telescopes. NGC 1316 isn't circular, however. It's about half again as long as it is wide, and it orients northeast to southwest. You'll see the broad central region surrounded by a thick halo.

Slightly more than 6′ north of Fornax A lies NGC 1317, a similar spiral that also has tight spiral arms. This galaxy glows at magnitude 11.9.

| OBJECT #861 | NGC 1326 |
|---|---|
| Constellation | Fornax |
| Right ascension | 3h24m |
| Declination | –36°28′ |
| Magnitude | 10.5 |
| Size | 3.9′ by 2.9′ |
| Type | Spiral galaxy |

You'll find our next target 41′ southwest of magnitude 6.4 Chi$^1$ ($\chi^1$) Fornacis. This galaxy appears as a thick oval oriented northeast to southwest. Its broad central region is all most telescopes will show, but it does have a thin halo that high magnification in a large telescope will reveal.

| OBJECT #862 | Alpha Persei Association |
|---|---|
| Constellation | Perseus |
| Right ascension (approx.) | 3h24m |
| Declination (approx.) | 49°52′ |

| (continued) | |
|---|---|
| Magnitude | 1.2 |
| Size (approx.) | 185′ |
| Type | Open cluster |

One naked-eye object you'll want to identify is the Alpha Persei Association. As the name implies, this brilliant stellar group surrounds the star Mirfak (Alpha [$\alpha$] Persei). Another name for this famous cluster is Melotte 20.

This large, scattered group of bright stars appears obvious to the naked eye. I point that out because it lies some 600 light-years away in the rich star fields along our galactic plane.

For the best view of Melotte 20, try binoculars or a rich-field telescope. Keep the magnification under 20×. Even at such low power, you'll see 50 bright stars — most prominent magnitude 1.8 Alpha and magnitude 4.3 Psi ($\Psi$) Persei — mainly to the south and east of Alpha.

All told, more than 100 young stars brighter than magnitude 12 spread across the association's 3° width. The group's total magnitude is an impressive 1.2.

| OBJECT #863 | NGC 1332 |
|---|---|
| Constellation | Eridanus |
| Right ascension | 3h26m |
| Declination | −21°20′ |
| Magnitude | 10.5 |
| Size | 5.0′ by 1.8′ |
| Type | Elliptical galaxy |

If you're moving through this guide from beginning to end, then you've already found Tau$^4$ ($\tau^4$) Eridani. From that magnitude 3.7 star, move 1.6° east-northeast, and you'll land on NGC 1332.

This elongated object appears like a stubby cigar three times as long as it is wide. Through small telescopes the surface brightness remains remarkably constant across NGC 1332's surface. Larger scopes reveal the outer 10% is fainter than the rest, and it fades rapidly with increasing distance from the core.

**Object #864** The Embryo Nebula (NGC 1333) Jay Lavine and Ali Huang/Adam Block/NOAO/AURA/NSF

| OBJECT #864 | NGC 1333 |
|---|---|
| Constellation | Perseus |
| Right ascension | 3h29m |
| Declination | 31°25′ |
| Size | 6′ by 3′ |
| Type | Emission and reflection nebulae |
| Other names | The Embryo Nebula, the Phantom Tiara |

You'll find this nebula in southern Perseus, where its border intersects those of Taurus and Aries. Specifically, look 3.3° west-southwest of magnitude 3.8 Omicron (*o*) Persei.

Through an 8-inch telescope, you'll see a bright haze. It appears brightest at the northeastern end, where the magnitude 10.5 star GSC 2342:624 that illuminates the nebula resides. Along with the nebulae, you'll spot several voids.

This object lets observers try a technique I have used on the Trifid Nebula (M20). Because both objects contain emission and reflection nebulae, a nebula filter will dim the reflection component, increasing the contrast of the emission nebulosity. Remove the filter, and your mind will fool you into thinking the reflection nebulosity has gotten brighter.

*Astronomy* magazine Contributing Editor Stephen James O'Meara gave this deep-sky wonder both of its common names. He noted one magnitude 10.5 star (HIP 16243, which actually lies in Aries, not Perseus) looks like a crown jewel in a phantom diamond tiara. He uses the same star to make the eye of an embryo.

| OBJECT #865 | NGC 1342 |
|---|---|
| Constellation | Perseus |
| Right ascension | 3h32m |
| Declination | 37°22′ |
| Magnitude | 8.9 |
| Size | 11.0′ by 7.6′ |
| Type | Open cluster |

You'll find our next target 5.7° west-southwest of magnitude 3.0 Epsilon (ε) Persei. Through an 8-inch telescope at 150×, you'll spot 50 stars evenly distributed across the face of this cluster. A 12-inch scope shows lines and arcs of the brighter members and brings 50 more stars into view.

| OBJECT #866 | NGC 1350 |
|---|---|
| Constellation | Fornax |
| Right ascension | 3h31m |
| Declination | −31°38′ |
| Magnitude | 10.3 |
| Size | 6.2′ by 3.2′ |
| Type | Spiral galaxy |

Through small to medium-sized telescopes, this galaxy looks like an evenly illuminated white football. Through a 16-inch telescope, you'll spot the long bar and several diaphanous spiral arms that tightly hug the main part of the galaxy. NGC 1350 lies 2.9° southwest of magnitude 5.0 Delta (δ) Fornacis.

| OBJECT #867 | NGC 1360 |
|---|---|
| Constellation | Fornax |
| Right ascension | 3h33m |
| Declination | −25°51′ |

| (continued) | |
|---|---|
| Magnitude | 9.4 |
| Size | 390″ |
| Type | Planetary nebula |

If you expect most planetary nebulae to appear round, you're not alone. NGC 1360, however, is an exception. It appears twice as long as it is wide, extended in a roughly north-south orientation. The northern half glows more brightly than the southern.

Through a 12-inch telescope at a magnification above 200×, both sections have a darker lane crossing them. The northern half's lane enters from the east and is thin. By contrast, the dark region in southern half of NGC 1360 is wide, extending all the way from this object's southern tip to the 11th-magnitude central star. Using an OIII filter will add contrast and help you see the darker and lighter regions better.

To find NGC 1360, look 5.6° northeast of magnitude 4.0 Alpha ($\alpha$) Fornacis.

| OBJECT #868 | NGC 1365 |
|---|---|
| Constellation | Fornax |
| Right ascension | 3h34m |
| Declination | –36°08′ |
| Magnitude | 9.3 |
| Size | 8.9′ by 6.5′ |
| Type | Spiral galaxy |

Barred spiral galaxies form a class of celestial objects that fascinate professional and amateur astronomers alike. Unfortunately for northern observers, the best example — NGC 1365 — languishes in the nearly invisible constellation Fornax the Furnace, more than one-third the way from the celestial equator to the South Celestial Pole.

A barred spiral galaxy contains a lane of stars, gas, and dust slashing across its central region. It has a small central bulge of stellar material. The spiral arms begin at both ends of the bar.

Astronomers using the Hubble Space Telescope revealed NGC 1365 feeds material into its central region, igniting massive star birth and causing its central bulge of stars to grow. The material also fuels a supermassive black hole in the galaxy's core.

Although it's bright, NGC 1365 isn't all that easy to star-hop to. To do so, first find a triangle of three faint stars, magnitude 6.4 Chi$^1$ ($\chi^1$), magnitude 5.7 Chi$^2$ ($\chi^2$), and magnitude 6.5 Chi$^2$ ($\chi^3$) Fornacis. From Chi$^2$, which is the brightest, move 1.3° east-southeast.

Through even a 4-inch telescope at a dark site, you'll see NGC 1365's bar shape and brighter central region. Increase the magnification, and notice how the bar near the core appears dimmer than it does farther out.

An 8-inch scope shows the spiral arms. The brighter one extends northward from the bar's west end. The other arm, only slightly fainter, appears somewhat blotchy, revealing huge star-forming regions within it.

| OBJECT #869 | NGC 1374 |
|---|---|
| Constellation | Fornax |
| Right ascension | 3h35m |
| Declination | –35°14′ |
| Magnitude | 11.0 |
| Size | 2.6′ by 2.4′ |
| Type | Spiral galaxy |

This object lies within the Fornax Galaxy Cluster. It is a round glow that shows no features through amateur telescopes. Observers might be amazed that astronomers classify it as a spiral galaxy, rather than an elliptical one. NGC 1374 falls in Edwin Hubble's class S0, which was the transition between ellipticals and spirals in his famous "tuning fork" diagram.

NGC 1375 lies only 2.5′ to the south. This magnitude 12.2 elliptical glow also shows no details. It measures 2.3′ by 0.9′. To find it, head 1.6° east-northeast from magnitude 5.7 Chi$^2$ ($\chi^2$) Fornacis.

| OBJECT #870 | NGC 1379 |
|---|---|
| Constellation | Fornax |
| Right ascension | 3h36m |
| Declination | −35°27′ |
| Magnitude | 11.0 |
| Size | 2.6′ by 2.5′ |
| Type | Elliptical galaxy |

This galaxy also lies within the Fornax Galaxy Cluster. It appears round, but you'll have to look past the magnitude 7.2 star GSC 7034:577 to see it. You'll find it 1.7° east of magnitude 5.7 Chi$^2$ ($\chi^2$) Fornacis.

| OBJECT #871 | NGC 1380 |
|---|---|
| Constellation | Fornax |
| Right ascension | 3h37m |
| Declination | −34°59′ |
| Magnitude | 10.0 |
| Size | 4.8′ by 2.8′ |
| Type | Spiral galaxy |

From a dark site, you can spot this elongated galaxy through any size telescope. Through a 12-inch scope, NGC 1380 looks like a fuzzy football. The galaxy's brightness remains constant until you're three-quarters of the way to an edge. To find this target, look 1.9° east-northeast of magnitude 5.7 Chi$^2$ ($\chi^2$) Fornacis.

| OBJECT #872 | NGC 1387 |
|---|---|
| Constellation | Fornax |
| Right ascension | 3h37m |
| Declination | −35°31′ |
| Magnitude | 10.8 |
| Size | 3.1′ by 2.8′ |
| Type | Spiral galaxy |

NGC 1387 is yet another mainly featureless elliptical within the Fornax Galaxy Cluster. It's relatively bright, however, so you should have no trouble tracking it down. Look for it 1.9° east of magnitude 5.7 Chi$^2$ ($\chi^2$) Fornacis.

| OBJECT #873 | NGC 1395 |
|---|---|
| Constellation | Eridanus |
| Right ascension | 3h39m |
| Declination | −23°02′ |
| Magnitude | 9.8 |
| Size | 11.0′ by 7.6′ |
| Type | Elliptical galaxy |

NGC 1365 lies 1.8° southeast of magnitude 4.3 Tau$^5$ ($\tau^5$) Eridani. You'll see this galaxy from a dark site through any telescope, but an 8-inch or larger instrument will reveal NGC 1365's bright, non-stellar central region and the halo that surrounds it.

**Object #874** NGC 1398 Sean and Renee Stecker/Adam Block/NOAO/AURA/NSF

| OBJECT #874 | NGC 1398 |
|---|---|
| Constellation | Fornax |
| Right ascension | 3h39m |
| Declination | –26°20′ |
| Magnitude | 9.8 |
| Size | 7.2′ by 5.2′ |
| Type | Barred spiral galaxy |

This spiral galaxy's core is so bright that it masks its delicate spiral arms when viewed through small telescopes. I've only caught glimpses of the arms when I've observed NGC 1398 through a 14-inch telescope at high power. Here's an object that really will reward views through scopes with apertures 20 inches of larger.

This object's bar runs roughly north-to-south and is only slightly fainter than the rest of the wide central region. This galaxy lies 1.6° due north of magnitude 6.0 Tau ($\tau$) Fornacis.

| OBJECT #875 | NGC 1399 |
|---|---|
| Constellation | Fornax |
| Right ascension | 3h39m |
| Declination | –35°27′ |
| Magnitude | 9.9 |
| Size | 8.1′ by 7.6′ |
| Type | Elliptical galaxy |

This really bright galaxy appears ever-so-slightly oblong under high magnifications. To find NGC 1399, draw a line from magnitude 6.4 Chi$^1$ ($\chi^1$) to magnitude 5.7 Chi$^2$ ($\chi^2$) Fornacis, and extend that line 5 times the distance between those two stars.

The large central region takes up three-quarters of this galaxy's diameter. Only a quickly fading fuzz lies outside.

| OBJECT #876 | NGC 1404 |
|---|---|
| Constellation | Fornax |
| Right ascension | 3h39m |
| Declination | −35°35′ |
| Magnitude | 9.7 |
| Size | 4.8′ by 3.9′ |
| Type | Elliptical galaxy |

This galaxy lies just under 10′ south-southeast of NGC 1399. Although both galaxies are similar ellipticals, NGC 1399 is bigger and slightly brighter. The magnitude 8.1 star SAO 194428 lies just 3′ south-southeast of NGC 1404.

| OBJECT #877 | NGC 1407 |
|---|---|
| Constellation | Eridanus |
| Right ascension | 3h40m |
| Declination | −18°35′ |
| Magnitude | 9.8 |
| Size | 4.6′ by 4.3′ |
| Type | Elliptical galaxy |

Our next target forms a pair with another elliptical galaxy, magnitude 10.9 NGC 1400, which lies about 12′ southwest of NGC 1407. A magnification of 100× through an 8-inch telescope will show both objects well. In no size instrument will you see details other than a thin halo around the brighter galaxy. NGC 1407 lies 1.5° southeast of the magnitude 5.2 star 20 Eridani.

| OBJECT #878 | NGC 1421 |
|---|---|
| Constellation | Eridanus |
| Right ascension | 3h43m |
| Declination | −13°29′ |
| Magnitude | 11.4 |
| Size | 3.1′ by 1.0′ |
| Type | Spiral galaxy |

This galaxy lies 1.6° south-southwest of magnitude 4.4 Pi (π) Eridani. It stretches 5 times as long as it is wide and aligns north-south.

Through a small telescope, the core appears stellar, surrounded by a haze. Instruments with apertures above 12′ resolve the thin spiral arms. Crank up the magnification past 200×, and look for bright star-forming regions within the arms. The southern arm stretches twice as far as the northern one, which seems to end abruptly.

| OBJECT #879 | NGC 1433 |
|---|---|
| Constellation | Horologium |
| Right ascension | 3h42m |
| Declination | −47°13′ |
| Magnitude | 10.0 |
| Size | 5.5′ by 3.2′ |
| Type | Barred spiral galaxy |

Our next target makes a worthy observation, but you may hunt for it a while because there's no bright star nearby. To find NGC 1433, look 7.5° southwest of magnitude 3.9 Alpha ($\alpha$) Horologii. The core of this galaxy appears broad and bright. You'll have no trouble spotting the long bar, which orients east-west. Through a 16-inch scope, look for a larger outer halo that begins to show hints of spiral arms.

| OBJECT #880 | NGC 1444 |
|---|---|
| Constellation | Perseus |
| Right ascension | 3h49m |
| Declination | 52°40′ |
| Magnitude | 6.6 |
| Size | 4′ |
| Type | Open cluster |

This cluster is basically the magnitude 6.7 star SAO 24248 and about 10 others. It is bright, however. To find it, look 3.5° northwest of magnitude 4.3 Lambda ($\lambda$) Persei.

| OBJECT #881 | NGC 1448 |
|---|---|
| Constellation | Horologium |
| Right ascension | 3h45m |
| Declination | −44°39′ |
| Magnitude | 10.8 |
| Size | 6.5′ by 1.4′ |
| Type | Spiral galaxy |

Our next target lies 5.8° west-southwest of magnitude 3.9 Alpha ($\alpha$) Horologii. NGC 1448 is one of the "needles" in space — a thin spiral galaxy seen edge-on. Through a 12-inch telescope you'll see a bright spindle four or five times as long as it is wide oriented northeast to southwest. The galaxy brightens gradually from the ends to its broad core. The magnitude 12.9 star GSC 7575:918 shines just southeast of the galaxy's center.

**Object #882** The Pleiades (M45) Tad Denton/Adam Block/NOAO/AURA/NSF

| OBJECT #882 | M45 |
| --- | --- |
| Constellation | Taurus |
| Right ascension | 3h47m |
| Declination | 24°07′ |
| Magnitude | 1.2 |
| Size | 110′ |
| Type | Open cluster |
| Other name | The Pleiades |

One of the finest of naked-eye objects also ranks as the sky's brightest star cluster. It's the Pleiades, also known as the Seven Sisters and M45.

Many ancient astronomers classified the Pleiades as a separate constellation. And why not? This object shines brighter and is more recognizable than many of the 88 constellations that currently fill the sky.

Messier included the Pleiades as the 45th and final entry in the first version of his *Catalog of Nebulae and Clusters of Stars*, which he presented to the Paris Academy of Sciences February 16, 1771. Why he included such a bright, well-known object remains a mystery. Even in Messier's day, most people — and certainly all observers — knew the Pleiades was no comet.

Astronomers call the Pleiades the Seven Sisters, but most people casually glancing at this star cluster see only six stars. Perhaps the hidden sister is too shy? More likely, it's that our observing is too casual.

Amateur astronomers with good vision can spot more than the six, or seven stars. My record is 11 Pleiads naked eye, but I've only tried once. And, on the same night I made that sighting, an observing buddy standing next to me saw 13 stars within the cluster.

Although M45 is a great naked-eye target, it also looks terrific through binoculars. Most amateur astronomers choose binoculars that magnify between 10 and 15 times to observe the Pleiades.

| OBJECT #883 | 32 Eridani |
| --- | --- |
| Constellation | Eridanus |
| Right ascension | 3h54m |
| Declination | −2°57′ |
| Magnitudes | 4.8/6.1 |
| Separation | 6.8″ |
| Type | Double star |

This binary sits in a lonely part of northern Eridanus near the Taurus border. It's worth seeking out, however, because the two stars have a nice color contrast, most often described by observers as yellow and blue.

**Object #884** NGC 1491 Adam Block/Mount Lemmon SkyCenter/University of Arizona

| OBJECT #884 | NGC 1491 |
|---|---|
| Constellation | Perseus |
| Right ascension | 4h03m |
| Declination | 51°19′ |
| Size | 25′ by 25′ |
| Type | Emission nebula |

This nebula lies 1.1° north-northwest of magnitude 4.3 Lambda (λ) Persei. A 10-inch telescope equipped with a nebula filter clearly shows its bright fan shape. Start with a magnification of about 75× to keep the nebula bright, then gradually increase the power.

The nebula appears brightest on the western edge and fades gradually toward the more diffuse eastern side. A 16-inch scope reveals striations that extend away from the nebula's southern tip. Don't expect to see a nebula that even approaches the cataloged size of 25′. Through the eyepiece it scarcely spans 4′.

| OBJECT #885 | NGC 1493 |
|---|---|
| Constellation | Horologium |
| Right ascension | 3h58m |
| Declination | −46°12′ |
| Magnitude | 11.2 |
| Size | 3.5′ by 3.2′ |
| Type | Barred spiral galaxy |

Our next target lies in a region bereft of bright stars 6.4° south-southwest of magnitude 3.9 Alpha (α) Horologii. NGC 1493 is a face-on barred spiral galaxy, but you won't see the bar through a telescope smaller than 30′ in aperture. What you will see is an unusually luminous center. The galaxy appears as a faint, circular smudge that gradually brightens to a central core.

**Object #886** The California Nebula (NGC 1499) Adam Block/NOAO/AURA/NSF

| OBJECT #886 | NGC 1499 |
|---|---|
| Constellation | Perseus |
| Right ascension | 4h01m |
| Declination | 36°37′ |
| Size | 160′ by 40′ |
| Type | Emission nebula |
| Other name | The California Nebula |

Our next celestial wonder is a rarity: a nebula you can spot without a telescope. It's also the only celestial object named for one of America's 50 states.

This unusually shaped nebula resides in one of the Milky Way's outer spiral arms called the Orion arm. The nebula's luminous portion spans 100 light-years. However, this entire region contains vast hydrogen clouds from which many bright, massive stars have formed. Astronomers labeled the family of young stars here the Perseus OB2 association.

The bright star 0.2° south of the nebula is magnitude 4.0 Menkib (Xi [$\xi$] Persei). It belongs to the Perseus OB2 association, and it makes the California nebula glow. This nebula is unusual because it glows strongly not only due to Hydrogen-alpha (H$\alpha$) emission, but also Hydrogen-beta (H$\beta$).

Hydrogen atoms emit light because of the elevation and subsequent drop in energy level of electrons. The electrons gain energy by interacting with energy from Xi Persei. The larger the "fall," or number of energy levels an electron drops, the more energetic is the light released.

In this case, a one-level drop creates H$\alpha$ light with a wavelength of 656.3 nm. An electron dropping two levels creates H$\beta$ light with a wavelength of 486.1 nm. In most nebulae, electrons fall just one level, emitting H$\alpha$ light.

From a dark site, sharp-eyed observers can spot the California Nebula with their naked eyes — almost. Use a nebula filter labeled either "H$\beta$" or "Deep-sky." An OIII filter reduces what you can see.

When you switch to a telescope, select the eyepiece that provides the lowest magnification. Don't forget to attach a filter. If even that view isn't wide enough to take in the whole object, slowly move the telescope back and forth.

**Object #887** The Oyster Nebula (NGC 1501) Adam Block/NOAO/AURA/NSF

| OBJECT #887 | NGC 1501 |
|---|---|
| Constellation | Camelopardalis |
| Right ascension | 4h07m |
| Declination | 60°55′ |
| Magnitude | 11.5 |
| Size | 52″ |
| Type | Planetary nebula |
| Other name | The Oyster Nebula, the Blue Oyster Nebula |

Move 6.9° west of magnitude 4.0 Beta ($\beta$) Camelopardalis to find planetary nebula NGC 1501. Amateur astronomers call this object the Oyster Nebula. *Astronomy* magazine Contributing Editor Stephen James O'Meara adds that, on images, a shell of pale blue gas surrounds NGC 1501's central star. For that reason, he added the "Blue" to the common name.

A 10-inch telescope shows a circular disk. Through a 16-inch scope at magnifications above 350×, however, you'll note that the planetary is ever-so-slightly oval in an east-west orientation.

The magnitude 14 central star is easier to see than its magnitude suggests. It peeks through a slightly darker center that suggests the presence of a thick ring structure. Through the larger instrument, the planetary's face appears patchy, with several small dark areas visible.

| OBJECT #888 | NGC 1511 |
|---|---|
| Constellation | Hydrus |
| Right ascension | 4h00m |
| Declination | −67°38′ |
| Magnitude | 11.1 |
| Size | 3.5′ by 1.3′ |
| Type | Spiral galaxy |

You'll find our next target 3.2° south-southeast of magnitude 3.8 Beta ($\beta$) Reticuli. Two magnitude 14.5 stars, one 1′ east and the other 1′ west, flank NGC 1511. Seen from the edge, the galaxy appears three times as long as it is wide. Its brightness is uniform except at its faint tips.

| OBJECT #889 | Kemble's Cascade |
|---|---|
| Constellation | Camelopardalis |
| Right ascension | 4h00m |
| Declination | 63°00′ |
| Size | 2.5° |
| Type | Asterism |

This object is a chance alignment of stars first described by the late Franciscan amateur astronomer Father Lucian Kemble, who found it while scanning the sky through binoculars. Because Kemble identified it, amateur astronomers now call it Kemble's Cascade.

A magnification of 15× works best for framing the starry chain. The Cascade is 15 stars that stretch 2.5°. Most of the stars range from 7th to 9th magnitude. The exception is the 5th-magnitude sparkler called SAO 12969 that sits in the center.

Want some extra value in your observing? At the southeast end of Kemble's Cascade, and easily visible in the same field of view, sits the tight open cluster NGC 1502 (R.A. = 4h08m; Dec. = 62°20′; Size =7′). You'll need a telescope to see its individual stars, but you won't miss its overall magnitude 5.7 glow. NGC 1502 sometimes goes by the name the Jolly Roger Cluster.

| OBJECT #890 | NGC 1512 |
|---|---|
| Constellation | Horologium |
| Right ascension | 4h04m |
| Declination | –43°21′ |
| Magnitude | 10.2 |
| Size | 8.3′ by 3.6′ |
| Type | Barred spiral galaxy |

Our next object lies 2.1° west-southwest of magnitude 3.9 Alpha ($\alpha$) Horologii. This barred spiral galaxy has a ring around its central region, but you won't see it through amateur telescopes because it lies so close to the core. The core itself appears bright. Look for the nearly stellar nucleus. Through an 8-inch telescope, NGC 1512 looks like an oval twice as long as it is wide. Double the aperture to 16′, and you'll just be able to see the short bar. Notice that the eastern side looks a bit brighter than the western one.

When you're done with NGC 1512, look for magnitude 12.4 NGC 1510 only 5′ to the southwest.

| OBJECT #891 | NGC 1513 |
|---|---|
| Constellation | Perseus |
| Right ascension | 4h10m |
| Declination | 49°31′ |
| Magnitude | 8.4 |
| Size | 12′ |
| Type | Open cluster |

You'll find this nice cluster 1.4° northwest of magnitude 4.1 Mu ($\mu$) Persei, or 1° southeast of magnitude 4.3 Lambda ($\lambda$) Persei. Through a 4-inch telescope, you'll see some three dozen stars spread evenly across the field of view. An 8-inch scope will increase your star count to 50.

**Object #892** The Crystal Ball Nebula (NGC 1514) Adam Block/NOAO/AURA/NSF

| OBJECT #892 | NGC 1514 |
|---|---|
| Constellation | Taurus |
| Right ascension | 4h09m |
| Declination | 30°47′ |
| Magnitude | 10.9 |
| Size | 114″ |
| Type | Planetary nebula |
| Other name | The Crystal Ball Nebula |

Look at our next target through an 8-inch telescope at 200×, and use a nebula filter. You'll see a round haze that amateur astronomers recently began calling the Crystal Ball Nebula. This object is definitely brighter along its rim. Bright knots intermingle with the gas on the northwestern and southeastern sides. The magnitude 9.4 central star is SAO 57020, and, yes, it can be a bit distracting. The filter will help dim it some. If you still have trouble spotting it, try increasing the magnification to 150× or beyond. NGC 1514 sits 3.4° east-southeast from magnitude 2.9 Atik (Zeta [ζ] Persei).

| OBJECT #893 | NGC 1527 |
|---|---|
| Constellation | Horologium |
| Right ascension | 4h08m |
| Declination | −47°53′ |
| Magnitude | 10.7 |
| Size | 4.2′ by 1.8′ |
| Type | Spiral galaxy |

Our next target sits 5.7° south of magnitude 3.9 Alpha (α) Horologii. Through an 8-inch telescope, it appears as a lens-shaped haze with a bright central region. Increasing the scope's aperture won't yield any additional details.

| OBJECT #894 | NGC 1528 |
|---|---|
| Constellation | Perseus |
| Right ascension | 4h15m |
| Declination | 51°12′ |
| Magnitude | 6.4 |
| Size | 18′ |
| Type | Open cluster |

Our next target lies 1.6° east-northeast of magnitude 4.3 Lambda (λ) Persei. Sharp-eyed observers may just detect NGC 1528 as a hazy star from a dark observing site. Through a 4-inch telescope, use 150× to spot 50 member stars. Many stars group into whirls and other patterns. An 8-inch scope will show nearly 100 stars. The brightest star in the cluster, magnitude 8.8 SAO 24496, sits just west of center.

| OBJECT #895 | NGC 1532 |
|---|---|
| Constellation | Eridanus |
| Right ascension | 4h12m |
| Declination | –32°52′ |
| Magnitude | 9.9 |
| Size | 11.2′ by 3.2′ |
| Type | Spiral galaxy |

This object is a double galaxy. It combines the magnificent edge-on spiral NGC 1532 with magnitude 11.7 elliptical NGC 1531, which sits less than 2′ to the northwest. But NGC 1532 is the real treat here. It appears nearly 6 times as long as it is wide.

Through a 16-inch or larger telescope, you'll see the spiral arms extend in a north-northeast to south-southwest orientation. With magnifications of 200× or more, the brilliant core appears surrounded by an oblong haze. And as long as you have a 16-inch scope at your disposal, look for the magnitude 15.2 irregular galaxy PGC 14664. It lies 12.5′ southeast of NGC 1532.

Find this pair 1.5° northwest of magnitude 3.6 Upsilon$^4$ ($\upsilon^4$) Eridani. Star charts also identify this star as 41 Eridani).

**Object #896** Cleopatra's Eye (NGC 1535) Adam Block/NOAO/AURA/NSF

| OBJECT #896 | NGC 1535 |
|---|---|
| Constellation | Eridanus |
| Right ascension | 4h14m |
| Declination | −12°44′ |
| Magnitude | 9.6 |
| Size | 18″ |
| Type | Planetary nebula |
| Other names | Cleopatra's Eye, the Celestial Jellyfish, the Ghost of Neptune Nebula |

This pretty object lies 4° east-northeast of magnitude 3.0 Zaurak (Gamma [γ] Eridani). The planetary is nice and bright, and it takes high powers well.

Through a 6-inch telescope, NGC 1535 has a sharply defined disk surrounded by a faint envelope. Double that aperture to 12′, and you'll begin to pick up this object's color. Now, crank the magnification past 300×, and you'll observe a dark hollow around the central star. At this aperture and power, you'll note that the contrast between the sharp inner disk and the fainter outer shell is at its maximum.

Long known as Cleopatra's Eye, American amateur astronomer Walter Scott Houston called it a "celestial jellyfish." *Astronomy* magazine Contributing Editor Stephen James O'Meara has given this planetary yet another moniker: the Ghost of Neptune. His observations at 72× of NGC 1535 as a pale blue disk led him to label it so.

| OBJECT #897 | NGC 1537 |
|---|---|
| Constellation | Eridanus |
| Right ascension | 4h14m |
| Declination | −31°39′ |
| Magnitude | 10.5 |
| Size | 3.9′ by 2.6′ |
| Type | Elliptical galaxy |

Our next target sits 2.3° north-northwest of magnitude 3.6 Upsilon$^4$ ($\upsilon^4$) Eridani. This galaxy appears oval, about 50% longer than wide. The bright central region takes up three-quarters of the galaxy's length and shows no detail. A thin halo surrounds the core.

| OBJECT #898 | NGC 1543 |
|---|---|
| Constellation | Reticulum |
| Right ascension | 4h13m |
| Declination | −57°44′ |
| Magnitude | 9.7 |
| Size | 7.2′ by 4.9′ |
| Type | Barred spiral galaxy |

This moderately bright galaxy lies 1.6° north-northwest of magnitude 4.4 Epsilon ($\varepsilon$) Reticuli. Through an 8-inch telescope, you'll easily see the galaxy's bright bar. The magnitude 8.7 star SAO 233433 lies 5′ to the south-southwest.

| OBJECT #899 | NGC 1545 |
|---|---|
| Constellation | Perseus |
| Right ascension | 4h21m |
| Declination | 50°15′ |
| Magnitude | 6.2 |
| Size | 12′ |
| Type | Open cluster |

This bright open cluster lies 2.3° east of magnitude 4.3 Lambda ($\lambda$) Persei. Look for it as a faint, fuzzy star with your naked eyes. Through a 4-inch telescope, you'll see about 20 stars. The three brightest form an isosceles triangle near NGC 1545's center. In order of brightness, they are magnitude 7.1 SAO 24556, magnitude 8.1 SAO 24555, and magnitude 9.3 SAO 24549.

Spend some time with this cluster at magnifications above 200×. You'll see many colored stars and also several nice double stars.

| OBJECT #900 | NGC 1549 |
|---|---|
| Constellation | Dorado |
| Right ascension | 4h16m |
| Declination | −55°36′ |
| Magnitude | 9.5 |
| Size | 5.4′ by 4.8′ |
| Type | Spiral galaxy |

This galaxy interacts gravitationally with our next target, NGC 1553. You'll find NGC 1549 sitting 2.6° west-southwest of magnitude 3.3 Alpha ($\alpha$) Doradus. This galaxy appears more elliptical than spiral. It has a large, evenly illuminated central region.

| OBJECT #901 | NGC 1553 |
|---|---|
| Constellation | Dorado |
| Right ascension | 4h16m |
| Declination | –55°47′ |
| Magnitude | 9.1 |
| Size | 6.3′ by 4.4′ |
| Type | Spiral galaxy |

Look 12′ south-southeast of NGC 1549, and you'll spot NGC 1553. This pair is part of the Dorado Group of galaxies. Through any size telescope, NGC 1553 appears oval, bright, and featureless, except for a thin halo that surrounds the bright central region.

| OBJECT #902 | Chi (χ) Tauri |
|---|---|
| Constellation | Taurus |
| Right ascension | 4h23m |
| Declination | 25°38′ |
| Magnitudes | 5.5/7.6 |
| Separation | 19.4″ |
| Type | Double star |

You'll find this star equidistant from the Pleiades (M45) and magnitude 3.5 Epsilon (ε) Tauri. Although most observers see some combination of a yellowish primary and a bluish secondary, others have seen the brighter star as bluish-white and its companion as deep blue. Decide for yourself by cranking up the magnification as much as the sky will allow and moving first the primary, then the secondary, out of the field of view. Seeing the stars alone will give you a better gauge of their true colors.

| OBJECT #903 | NGC 1554/55 |
|---|---|
| Constellation | Taurus |
| Right ascension | 4h22m |
| Declination | 19°32′ |
| Size | 1′ |
| Type | Emission nebulae |
| Other names | Struve's Lost Nebula; Hind's Variable Nebula |

Our next target combines two objects in one. Observers refer to NGC 1554 as Struve's Lost Nebula, and NGC 1555 is Hind's Variable Nebulae. Note the name for NGC 1554, and don't look for it because it's not there. Astronomers now assign the same position to these objects, and it proves a tough catch even through large telescopes.

Both common names refer to the nineteenth-century astronomers who discovered the respective objects. British astronomer John Russell Hind (1823–1895) discovered NGC 1555 October 11, 1852. It remained visible for a few years but then faded from view.

Several noted astronomers, among them Russian astronomer Otto Wilhelm von Struve (1819–1905), subsequently observed Hind's Variable Nebula, but it faded from view by 1868. When examining the region early in 1868, Struve found another small nebula. He gave its position as 4′ to the west-southwest of T Tauri. Subsequent observations showed no object at the position he noted. This object also proved difficult for astronomers to observe consistently until American astronomer Edward Emerson Barnard (1857–1923) looked into the matter in 1890.

Barnard found a position error for the star T Tauri and suggested that other astronomers had been looking in the wrong place for NGC 1554. On March 24 of that year, he glimpsed a faint nebula at the position he calculated along with NGC 1555 through the 36-inch refractor at Lick Observatory. No observer has seen NGC 1554 since. Struve's nebula is indeed lost.

To start your search for Hind's Variable Nebula, head 1.7° west-northwest of magnitude 3.5 Epsilon (ε) Tauri. Near that position, you'll see the magnitude 8.4 star SAO 93887. From there, move 5′ northeast, and you'll encounter the variable star T Tauri, which usually shines at magnitude 9.6. NGC 1554 appears as a faint wisp of nebulosity near T Tauri.

I last observed this object in February 2009. I was using a 30-inch reflector under an ultra-dark sky at the Rancho Hidalgo astronomy community, near Animas, New Mexico. Through that monstrous scope, NGC 1555 appeared as an unevenly lit wedge of light.

| OBJECT #904 | NGC 1559 |
|---|---|
| Constellation | Reticulum |
| Right ascension | 4h18m |
| Declination | −62°47′ |
| Magnitude | 10.4 |
| Size | 4.3′ by 2.2′ |
| Type | Barred spiral galaxy |

For our next target, aim your telescope 0.5° southeast of magnitude 3.3 Alpha (α) Reticuli. Through an 8-inch telescope at a dark site, you'll see a roughly rectangular shape twice as long as it is wide. Only through the largest amateur instruments will you see any mottling.

| OBJECT #905 | Theta (θ) Reticuli |
|---|---|
| Constellation | Reticulum |
| Right ascension | 4h18m |
| Declination | −63°15′ |
| Magnitudes | 6.2/8.2 |
| Separation | 2.9″ |
| Type | Double star |

This far-southern binary combines a blue primary with a white secondary. Theta is a close double, so use a magnification above 150× to assure a clean split. To locate Theta, look 0.9° south-southeast of magnitude 3.3 Alpha (α) Reticuli.

| OBJECT #906 | NGC 1566 |
|---|---|
| Constellation | Dorado |
| Right ascension | 4h20m |
| Declination | −54°56′ |
| Magnitude | 9.4 |
| Size | 7.1′ by 4.8′ |
| Type | Spiral galaxy |

Pan 2° west of Alpha (α) Doradus to find this bright spiral. It belongs to the Dorado Galaxy group, which lies 55 million light-years away. Through a 10-inch telescope, crank the magnification to 200×, and you'll spot the elegant spiral structure — one arms starts at the north end of the central region and curves eastward; the other starts at the south and curves westward. Look for a bright star-forming region that appears like a faint star at the north end of the western spiral arm.

NGC 1566 lies above the center of a right triangle made of three stars: magnitude 8.1 SAO 233486, magnitude 9.9 SAO 233482, and magnitude 10.1 GSC 8505:1410.

# December

| OBJECT #907 | NGC 1582 |
|---|---|
| Constellation | Perseus |
| Right ascension | 4h32m |
| Declination | 43°51′ |
| Magnitude | 7.0 |
| Size | 24′ |
| Type | Open cluster |

Our next target is an attractive cluster that lies 5.4° west of magnitude 3.0 Epsilon (ε) Aurigae. Because of the brightness differences between two groups of stars in this object, NGC 1582 presents, in effect, two "clusters" superimposed. You will need a 10-inch or larger telescope to see what I mean. The first cluster contains 10 stars brighter than 10th-magnitude. The luminary of this group, magnitude 8.6 SAO 39578, sits near the eastern edge. The other cluster contains several dozen fainter stars that fill in the gaps left by the first cluster.

| OBJECT #908 | The Hyades |
|---|---|
| Constellation | Taurus |
| Right ascension | 4h27m |
| Declination | 16°00′ |
| Magnitude | 0.5 |
| Size | 330′ |
| Type | Open cluster |
| Other names | Melotte 25, Caldwell 41 |

The V-shaped Hyades star cluster ($\alpha$, $\theta^1$, $\theta^2$, $\gamma$, $\Delta$, and $\varepsilon$ Tauri) is impossible to miss in Taurus. It's the neighboring cluster of the better-known Pleiades (M45). And although the Hyades is closer, larger,

M.E. Bakich, *1,001 Celestial Wonders to See Before You Die*, Patrick Moore's Practical Astronomy Series,
DOI 10.1007/978-1-4419-1777-5_12, 

and brighter than the Pleiades, it doesn't look as spectacular: The bright stars just spread out too much.

In Greek mythology, the Hyades were the daughters of Atlas and Pleione (or Aethra), and half-sisters of the Pleiades. Regarding their name, Richard Hinckley Allen writes in *Star Names and Their Meanings*, "Ovid called them Sidus Hyantis, after their earthly brother, Hyas, whose name, after all, would seem to be the most natural derivation of the title; and it was their grief at his death which gave additional point to Horace's *tristes Hyadas*, and, in one version of their story, induced Jove to put them in the sky."

The Hyades has a diameter of more than 5°, so it only looks like a cluster to your naked eyes or through binoculars. If you train a telescope toward this object, focus in on the many fine double stars the cluster contains.

Oh, and one further note: Aldebaran (Alpha [α] Tauri), Taurus' brightest star, does not belong to the Hyades. Aldebaran lies only 65 light-years from Earth while the stars of the Hyades cluster are 150 light-years away.

| OBJECT #909 | 1 Camelopardalis |
|---|---|
| Constellation | Camelopardalis |
| Right ascension | 4h32m |
| Declination | 53°55′ |
| Magnitudes | 5.7/6.8 |
| Separation | 10.3″ |
| Type | Double star |
| Notes | Blue and white |

This pretty binary makes a nice target for northern observers. The primary is sky blue and the secondary is white. Because Camelopardalis is such a faint constellation, find 1 Cam by using the relatively bright Lambda (λ) Persei. Look 5.3° northeast of that magnitude 4.3 star.

| OBJECT #910 | NGC 1617 |
|---|---|
| Constellation | Dorado |
| Right ascension | 4h32m |
| Declination | −54°36′ |
| Magnitude | 10.5 |
| Size | 4.3′ by 2.1′ |
| Type | Spiral galaxy |

Our next object sits 0.5° northwest of magnitude 3.3 Alpha (α) Doradus. Through any size telescope, you'll pick out this reasonably bright galaxy. NGC 1617 has an oblong shape twice as long as wide. Its bright central region takes up half the galaxy's diameter. Outside is a thick halo.

| OBJECT #911 | NGC 1624 |
|---|---|
| Constellation | Perseus |
| Right ascension | 4h40m |
| Declination | 50°27′ |
| Size | 5′ by 5′ |
| Type | Open cluster with nebula |

Look 4.6° east-northeast of magnitude 4.1 Mu (μ) Persei, tucked in the far northeastern corner of Perseus, for this intriguing cluster surrounded by a nebula. The cluster holds a dozen faint stars that nevertheless appear distinctly as a cluster because they're tightly packed. A uniform, gauzy glow surrounding the stars suggests dew on the optics.

| OBJECT #912 | NGC 1647 |
|---|---|
| Constellation | Taurus |
| Right ascension | 4h46m |
| Declination | 19°07′ |
| Magnitude | 6.4 |
| Size | 40′ |
| Type | Open cluster |
| Other name | The Pirate Moon Cluster |

From magnitude 0.9 Aldebaran (Alpha [α] Tauri), move 3.5° northeast, and you'll encounter this bright cluster. Sharp-eyed observers can spot it from a dark site with their naked eyes, and, if you can't, binoculars or a finder scope will bring it in.

You'll see 30 stars through a 4-inch telescope, and increasing the aperture increases the number of stars. Increase the magnification past 150×, and look for lots of nice, close pairs. This cluster contains more than its fair share.

*Astronomy* magazine Contributing Editor Stephen James O'Meara bestowed the name Pirate Moon Cluster on NGC 1647. He said that, through 7×50 binoculars, it appeared as a round ghostly glow with an apparent size larger than that of the Full Moon. Furthermore, the cluster's stars congregate into disparate bunches, which, according to O'Meara, mimic the dark and bright features we see on the naked-eye Moon.

| OBJECT #913 | IC 2087 |
|---|---|
| Constellation | Taurus |
| Right ascension | 4h40m |
| Declination | 25°44′ |
| Size | 4′ by 4′ |
| Type | Emission nebula |

Our next treat lies 3.9° east of magnitude 5.5 Chi (χ) Tauri. Through a 12-inch telescope, it appears as a uniform circular haze. More interesting than the nebula, however, is the surrounding starfield, which seems nearly nonexistent. Step down to your lowest-power eyepiece to see this. Pan 1.5° west-southwest, and you'll encounter Barnard 7, one of the hundreds of dark nebulae cataloged by American astronomer Edward Emerson Barnard.

| OBJECT #914 | 55 Eridani |
|---|---|
| Constellation | Eridanus |
| Right ascension | 4h44m |
| Declination | −8°48′ |
| Magnitudes | 6.7/6.8 |
| Separation | 9.2″ |
| Type | Double star |
| Notes | Yellow and white |

This binary features a pair of equally bright stars with distinctively different colors. In fact, if both components were two magnitudes brighter, this would be a showpiece at winter observing events. The primary appears a deep yellow and the secondary is light blue, blue-white, or whitish, depending on your eyes' color-receptor sensitivity. You'll find 55 Eridani 7.7° due west of brilliant Rigel (Beta [β] Orionis). Within your field of view, you'll also see bluish 56 Eridani, which lies 19′ north-northeast of 55 Eridani.

| OBJECT #915 | NGC 1664 |
|---|---|
| Constellation | Auriga |
| Right ascension | 4h51m |

| (continued) | |
|---|---|
| Declination | 43°42′ |
| Magnitude | 7.6 |
| Size | 18′ |
| Type | Open cluster |

You'll find this attractive object 2° west of magnitude 3.0 Epsilon (ε) Aurigae. Through a 4-inch telescope at 100×, you'll see three dozen stars. The background star field in this area is rich, but you'll have no trouble picking out the cluster. The bright star on NGC 1664's southwestern edge is magnitude 7.5 SAO 39807.

| OBJECT #916 | NGC 1672 |
|---|---|
| Constellation | Dorado |
| Right ascension | 4h46m |
| Declination | –59°15′ |
| Magnitude | 9.8 |
| Size | 6.2′ by 3.4′ |
| Type | Spiral galaxy |

Scan 0.5° north-northeast of magnitude 5.3 Kappa (κ) Doradus to locate this four-armed barred spiral galaxy. Through your 12-inch telescope, however, you'll need to crank up the magnification to spot just one arm. It begins on the eastern side and curves toward the north.

**Object #917** IC 342 Ken and Emilie Siarkiewicz/Adam Block/NOAO/AURA/NSF

| OBJECT #917 | IC 342 |
|---|---|
| Constellation | Camelopardalis |
| Right ascension | 4h47m |
| Declination | 68°06′ |
| Magnitude | 8.4 |
| Size | 21.4′ by 20.9′ |
| Type | Spiral galaxy |
| Other name | Caldwell 5 |

Our next deep-sky object is IC 342. Many of you are familiar with the NGC catalog. The Index Catalog (abbreviated IC) actually is an extension of the NGC. English astronomer John Louis Emil Dreyer completed the NGC in 1888. It contains some 7,840 objects.

In 1895 and 1908, he added two appendices to the NGC that he called the *First and Second Index Catalogs of Nebulae and Star Clusters*. Entries are the same deep-sky object types, but with a different catalog designation.

IC 342 lies in the far-northern and ultra-faint constellation Camelopardalis the Giraffe. The galaxy sits 3.2° due south of the magnitude 4.6 star Gamma ($\gamma$) Camelopardalis.

As it stands, IC 342 is a magnitude 8.4 galaxy. But it would be truly spectacular if intervening Milky Way dust and gas didn't dim it by more than 2 magnitudes.

Visually, IC 342 appears 20′ across, but, even at its magnitude, it's not easy to see because its surface brightness is low. Look for a bright central knot about 30″ across in a rich star field. Surrounding this is a fainter halo 2′ across, while an extremely faint, knotty structure extends over a diameter of 20′.

| OBJECT #918 | NGC 1679 |
|---|---|
| Constellation | Caelum |
| Right ascension | 4h50m |
| Declination | −31°59′ |
| Magnitude | 11.5 |
| Size | 3.0′ by 1.5′ |
| Type | Barred spiral galaxy |

Our next target presents a strange face because of several superposed stars. NGC 1679 appears as a thick crescent with a slightly brighter center. A magnitude 12 star lies at the northwest end, and a 13th-magnitude star glimmers in the outer halo to the east. You'll find this galaxy 3.4° east-southeast of magnitude 3.8 Upsilon$^2$ ($\upsilon^2$) Eridani.

| OBJECT #919 | NGC 1714 |
|---|---|
| Constellation | Dorado |
| Right ascension | 4h52m |
| Declination | −66°56′ |
| Size | 1.2′ |
| Type | Emission nebula |

This tiny nebula sits on the western edge of the Large Magellanic Cloud a little more than 6° southwest of magnitude 3.8 Beta ($\beta$) Doradus. Although it measures only 1′ across, its high surface brightness lets you use high magnification for a detailed view. Through a 10-inch telescope, you'll see a circular glow with a bright northern rim. The magnitude 6.3 star GSC 8889:215 lies only 8″ west.

| OBJECT #920 | Lepus |
|---|---|
| Right ascension (approx.) | 5h31m |
| Declination (approx.) | −19° |

| (continued) | |
|---|---|
| Size (approx.) | 290.29 square degrees |
| Type | Constellation |

Another small constellation you should seek out is Lepus the Hare. In the sky, Lepus sits directly below (that is, south of) Orion. It's a mid-sized constellation. Out of the 88 star patterns that cover the sky, Lepus ranks 51st in size. It covers 290 square degrees, or about 0.7% of the sky.

Lepus has two named stars, magnitude 2.6 Arneb (α Leporis) and magnitude 2.9 Nihal (β Leporis). No meteor showers originate from this constellation.

Lepus is completely visible from any latitude south of 63° north, and completely invisible only from latitudes north of 79° north. The best date to see it (when it lies opposite the Sun in the sky as seen from Earth) is December 14. Conversely, don't look for it around June 15, because that's when the Sun is in Lepus' part of the sky.

| OBJECT #921 | NGC 1744 |
|---|---|
| Constellation | Lepus |
| Right ascension | 5h00m |
| Declination | −26°01′ |
| Magnitude | 11.3 |
| Size | 5.1′ by 2.5′ |
| Type | Barred spiral galaxy |

You'll find our next object 3.9° south-southwest of magnitude 3.2 Epsilon (ε) Leporis. NGC 1744 is faint, appearing twice as long as it is wide, oriented north-south. Its core is broad, and the galaxy's edge has an irregular outline.

| OBJECT #922 | NGC 1755 |
|---|---|
| Constellation | Dorado |
| Right ascension | 4h55m |
| Declination | −68°11′ |
| Magnitude | 9.9 |
| Size | 2.6′ |
| Type | Open cluster |

Our next target lies in the deep southern sky. It's an odd open cluster located on the western end of the Large Magellanic Cloud's bar. An 8-inch telescope at 100× reveals 20 magnitude 13 and 14 stars packed into an area 2′ across. There's also a strong background glow, which hints at the existence of many fainter stars. A fainter open cluster, NGC 1749, lies 2′ to the west.

| OBJECT #923 | NGC 1763 |
|---|---|
| Constellation | Dorado |
| Right ascension | 4h57m |
| Declination | −66°24′ |
| Size | 5′ by 3′ |
| Type | Emission nebula |

Let's stay near the LMC for our next object, which is part of a group of four emission nebula in an area less than 0.3° across. NGC 1763 appears as a clumpy haze surrounded by an apparent open cluster. Just 7′ south, you'll find NGC 1760. Move 7′ east-southeast from NGC 1763, and you'll encounter NGC 1769. Finally, NGC 1773 lies 9′ east-northeast of NGC 1763.

| OBJECT #924 | R Leporis |
|---|---|
| Constellation | Lepus |
| Right ascension | 5h00m |
| Declination | –14°48′ |
| Magnitude | 5.5 to 11.7 |
| Period | 432 days |
| Type | Variable star |
| Other name | Hind's Crimson Star |

You'll find our next target in Lepus near that constellation's border with Eridanus. It lies 3.5° west-northwest of magnitude 3.3 Mu ($\mu$) Leporis. Hind's Crimson Star is one of the sky's reddest points of light. To spot it, use low magnification in an 8-inch or larger telescope. When you've centered R Leporis, ever-so-slightly defocus the image to spread out the star's color, making it easier to see.

This star gets its common name from British astronomer John Russell Hind (1823–1895), who discovered it in 1845.

| OBJECT #925 | NGC 1778 |
|---|---|
| Constellation | Auriga |
| Right ascension | 5h08m |
| Declination | 37°01′ |
| Magnitude | 7.7 |
| Size | 8′ |
| Type | Open cluster |

Our next target sits almost 2° east-southeast of magnitude 5.1 Omega ($\omega$) Aurigae. Through a 4-inch telescope, you'll see two dozen stars unevenly spread across this cluster's face. Double the aperture to 8 inches, and you'll raise the star count to 50.

| OBJECT #926 | NGC 1788 |
|---|---|
| Constellation | Orion |
| Right ascension | 5h07m |
| Declination | –3°21′ |
| Size | 5′ by 3′ |
| Type | Reflection nebula |

This reflection nebula sits 2° north of magnitude 2.7 Beta ($\beta$) Eridani. Through a 10-inch telescope, you'll have no trouble spotting NGC 1788. The bright nebula has a diffuse border and features two lobes. The western one surrounds a 10th-magnitude star while the eastern lobe has a small, bright concentration of light at its center. NGC 1788's south end meets dark nebula LDN 1616. Again, you'll have no trouble spotting it because of the lack of background stars over a 5-arcminute-wide region.

| OBJECT #927 | Columba the Dove |
|---|---|
| Right ascension | 5h45m |
| Declination | –35° |
| Size | 270.18 square degrees |
| Type | Constellation |

Columba represents the dove that Noah sent out to test whether the waters from the great biblical flood had receded. It's the only surviving constellation named after an object in the Bible. Columba first appeared in 1592, on a celestial map designed by Dutch astronomer and cartographer Petrus Plancius.

Columba is a constellation most amateur astronomers haven't identified. Well, here's your chance. Find Orion. That's easy enough. Now look south of Orion, and find Lepus. Finally, continue south from Lepus, and you'll end up in Columba.

You'll first notice the constellation's two brightest stars. Phact (Alpha [$\alpha$] Columbae) shines at magnitude 2.6, and Wasn (Beta [$\beta$] Columbae) isn't far behind at magnitude 3.1. From there, you'll find just three other stars brighter than 4th magnitude. Then all you have to do is make a dove out of those stars. Good luck with that!

When you look at Columba, you might want to wave goodbye. This constellation contains the point in the sky away from which our solar system is heading, relative to stars in our neighborhood. Astronomers call this point the solar antapex.

For those of you with large telescopes, there's something at the approximate coordinates of the solar antapex. It's the magnitude 13.2 galaxy IC 2153. Warning: Unless you can set up a large telescope at a dark site, you won't have much luck observing this small faint object.

| OBJECT #928 | NGC 1792 |
|---|---|
| Constellation | Columba |
| Right ascension | 5h05m |
| Declination | −37°59′ |
| Magnitude | 9.9 |
| Size | 5.5′ by 2.5′ |
| Type | Spiral galaxy |

Our next target is a gravitationally disturbed spiral that interacted with nearby NGC 1808 in the recent past. A 4-inch telescope shows only a fat oval devoid of features. Through a 12-inch scope at magnifications above 200×, however, you'll see the galaxy's irregular shape and even illumination. To find this galaxy, look 2.5° south of magnitude 4.6 Gamma$^1$ ($\gamma^1$) Caeli.

| OBJECT #929 | Barnard 29 |
|---|---|
| Constellation | Auriga |
| Right ascension | 5h06m |
| Declination | 31°44′ |
| Magnitude | — |
| Size | 10′ by 10′ |
| Type | Dark nebula |

To find our next target, look 2.4° southeast of magnitude Iota ($\iota$) Aurigae. Here, you're not looking at the stars, rather the dark nebula Barnard 29, a cloud of obscuring material roughly 500 light-years away. It ranks among the most opaque nebulae in the sky, but it's not all that easy to see. Through a 12-inch telescope, B29 appears as a gray, mottled region nearly devoid of stars that blends gradually into its starry surroundings. The darkest area appears 15′ across. To the naked eye, the nebula forms part of a dark lane that runs from magnitude 2.7 Iota ($\iota$) Aurigae to magnitude 1.7 Elnath (Beta [$\beta$] Tauri).

| OBJECT #930 | NGC 1808 |
|---|---|
| Constellation | Columba |
| Right ascension | 5h08m |
| Declination | −37°31′ |
| Magnitude | 9.9 |
| Size | 5.2′ by 2.3′ |
| Type | Spiral (starburst) galaxy |

NGC 1808 is easy to see and accepts high magnifications well because it has a high surface brightness. The galaxy's oval shape — twice as long as it is wide — is apparent, but you'll only see the initial stubs of the faint spiral arms that long-exposure images show stretching around NGC 1808's entire length.

Through a 16-inch or larger telescope, crank up the power, and try to see the dark lanes near the galaxy's outer edge. Astronomers recently discovered this galaxy has a high amount of star-formation occurring within it.

For those of you with the largest amateur scopes, three challenging galaxies lie roughly 10′ southeast of NGC 1808. The brightest, PGC 620467, glows weakly at magnitude 15.6. The other two, PGC 131395 and PGC 16804, are even fainter. Both of these galaxies have magnitudes of 15.9.

| OBJECT #931 | NGC 1817 |
|---|---|
| Constellation | Taurus |
| Right ascension | 5h12m |
| Declination | 16°42′ |
| Magnitude | 7.7 |
| Size | 15′ |
| Type | Open cluster |

Our next object lies 7.5° southwest of magnitude 3.0 Zeta ($\zeta$) Tauri. Don't confuse it with magnitude 7.0 NGC 1807, which measures 12′ across and lies 0.5° to the west-southwest. Through a 4-inch telescope, NGC 1817 displays three dozen stars. A striking chain of stars delineates the cluster's western edge. An 8-inch scope will let you count 100 stars. Although NGC 1807 is brighter than NGC 1817, it doesn't have nearly the appeal of the larger cluster.

| OBJECT #932 | NGC 1832 |
|---|---|
| Constellation | Lepus |
| Right ascension | 5h12m |
| Declination | −15°41′ |
| Magnitude | 11.3 |
| Size | 2.1′ by 1.5′ |
| Type | Spiral galaxy |

Our next target lies 0.5° north-northwest of magnitude 3.3 Mu ($\mu$) Leporis. This fat spiral doesn't show much detail through telescopes smaller than about 16′. Through such a scope at a magnification above 250×, the thin, eastern arm appears better defined and seems detached from the galaxy.

| OBJECT #933 | NGC 1835 |
|---|---|
| Constellation | Dorado |
| Right ascension | 5h05m |
| Declination | −69°24′ |
| Magnitude | 10.1 |
| Size | 1.2′ |
| Type | Globular cluster |

Find the Large Magellanic Cloud, and look within the western part of its bar. NGC 1835 appears round, but high magnification shows faint extensions to the east and west that double its length. Two faint open clusters, magnitude 12.5 NGC 1828 and magnitude 12.6 NGC 1830, lie 6′ to the west.

**Object #934** The Witch Head Nebula (IC 2118) Fred Calvert/Adam Block/NOAO/AURA/NSF

| OBJECT #934 | IC 2118 |
|---|---|
| Constellation | Eridanus |
| Right ascension | 5h07m |
| Declination | –7°13′ |

| (continued) | |
|---|---|
| Size | 180′ by 60′ |
| Type | Emission nebula |
| Other name | The Witch Head Nebula |

After midnight on Halloween, you can spot the Witch Head Nebula, also known as IC 2118, in the constellation Eridanus the River. This reflection nebula is similar to M78 in Orion. The Witch Head Nebula, however, is on a whole new scale: While M78 spans only 8′, IC 2118 stretches 1.5°.

Like M78 and other similar objects, the Witch Head Nebula glows by reflecting starlight. Well, astronomers have identified the star lighting up this cloud. It's Rigel (Beta [$\beta$] Orionis), the upraised left foot of Orion the Hunter.

You'll need to be under a dark sky to spot the Witch Head. Start your search a bit west of a point two-thirds of the way from brilliant Rigel to magnitude 2.8 Beta ($\beta$) Eridani. Through an 8-inch telescope, use an eyepiece that gives low power and a wide field of view. Through such an instrument you should see the brightest parts of the Witch's face.

If you're lucky enough to have a 16-inch or larger telescope, IC 2118 will be quite apparent, but you'll have to move the telescope to scan the Witch's features. The starting point (above) between Rigel and Beta Eridani marks the nebula's northern part.

| OBJECT #935 | NGC 1850 |
|---|---|
| Constellation | Dorado |
| Right ascension | 5h09m |
| Declination | −68°46′ |
| Magnitude | 9.0 |
| Size | 3.4′ |
| Type | Open cluster |
| Other name | Caldwell 18 |

This massive open cluster lies in the northeastern part of the Large Magellanic Cloud's bar. A 6-inch telescope brings out roughly 50 stars glowing at magnitude 13 and 14 in a circular area. The prominent clump of stars on NGC 1850's western edge is a cluster designated NGC 1850A.

| OBJECT #936 | NGC 1851 |
|---|---|
| Constellation | Columba |
| Right ascension | 5h14m |
| Declination | −40°03′ |
| Magnitude | 7.2 |
| Size | 11′ |
| Type | Globular cluster |
| Other name | Caldwell 73 |

Our next object is a nice one through small telescopes. NGC 1851 sits nearly 8° southwest of Phact (Alpha [$\alpha$] Columbae), but you'll see it easily from a dark site through binoculars. This magnitude 7.0 globular is the brightest deep-sky object for more than 20° in any direction.

Through a 4-inch telescope, you'll see a concentrated core that you can't resolve surrounded by many stars you can. NGC 1851's core is difficult to resolve through large telescopes as well because of its distance. It lies 40,000 light-years from the Sun and 55,000 light-years from the Milky Way's center.

| OBJECT #937 | NGC 1857 |
|---|---|
| Constellation | Auriga |
| Right ascension | 5h20m |
| Declination | 39°21′ |
| Magnitude | 7.0 |
| Size | 5′ |
| Type | Open cluster |

Our next target sits 0.8° south-southeast of magnitude 4.7 Lambda ($\lambda$) Aurigae. Through a 4-inch telescope, you'll see several dozen stars. The members of NGC 1857 shine mostly at 13th and 14th magnitude and the southern half of the cluster holds most of the bright stars. The one exception is SAO 57903, a magnitude 7.4 yellow star at the center.

| OBJECT #938 | NGC 1866 |
|---|---|
| Constellation | Dorado |
| Right ascension | 5h14m |
| Declination | −65°28′ |
| Magnitude | 9.7 |
| Size | 4.5′ |
| Type | Open cluster |

From magnitude 3.8 Beta ($\beta$) Doradus, sweep 3.7° south-southwest to reach NGC 1866. This cluster's brightest stars glow at 15th magnitude, so larger apertures can reveal hundreds of stars. At magnifications of 300× and higher, NGC 1866 appears stunning.

| OBJECT #939 | Beta ($\beta$) Orionis |
|---|---|
| Constellation | Orion |
| Right ascension | 5h15m |
| Declination | −8°12′ |
| Magnitudes | 0.1/6.8 |
| Separation | 9″ |
| Type | Double star |
| Other name | Rigel |

A fine small telescope target is Rigel (Beta [$\beta$] Orionis). Rigel marks the brilliant left foot of Orion the Hunter. And it is, indeed, bright. Rigel is the 7th-brightest nighttime star, shining at magnitude 0.12. Its name comes from the Arabic *Rijl Jauzah al Yusra*, the Left Leg of the Jauzah. Regarding the term *Al Jauzah*, in *Star Names and Their Meanings*, Allen writes, "It is often translated Giant, but erroneously, for it, at first, had no personal signification. Originally it was the term used for a black sheep with a white spot on the middle of the body, and thus may have become the designation for the middle figure of the heavens, which from its preeminent brilliancy always has been a centre of attraction."

Point your telescope at Rigel, and insert an eyepiece that provides a magnification of about 100×. Just 9″ to Rigel's south, you'll spot Rigel B, more correctly called Beta ($\beta$) Orionis B. This magnitude 6.7 star is Rigel's companion. Although it's not all that faint, it can be tough to see if you don't use enough magnification. That's because Rigel A shines some 436 times more brightly than Rigel B.

When you do spot Rigel's companion, what color does it appear to you? Magnified through a telescope, Rigel appears white. To me, through my 4-inch refractor, Rigel B has a definite purple cast. Our color receptors vary widely, however, so the color you see could be different.

**Object #940** The Flaming Star Nebula (IC 405) Adam Block/NOAO/AURA/NSF

| OBJECT #940 | IC 405 |
|---|---|
| Constellation | Auriga |
| Right ascension | 5h16m |
| Declination | 34°16′ |
| Size | 30′ by 20′ |
| Type | Emission nebula |
| Other names | The Flaming Star Nebula, Caldwell 31 |

The name Flaming Star Nebula is an exaggeration. This nebula appears as a dim wisp of light glowing in Auriga. To observe this object, first find the star AE Aurigae, which lies 4.2° east-northeast of magnitude 2.7 Iota (*ι*) Aurigae. The star's energy is what causes gas in the nebula to glow.

Most nebulae glow because a bright star or star cluster formed within their midst. Ultraviolet radiation from stars excites hydrogen atoms, causing them to glow. In the Flaming Star Nebula, the star AE Aurigae is the power source. But AE did not form there.

More than 2 million years ago, AE was a hot, young star within the Orion Nebula (M42). Gravitational interactions with nearby stars sent the star on a wild ride into space. By chance, we now see it passing through IC 405. As it does so, it illuminates part of the nebula.

German-born American astronomer John Martin Schaeberle (1853–1924) discovered the Flaming Star Nebula February 6, 1892. He found it on a photograph he took through a 6-inch refractor at Lick Observatory on Mt. Hamilton in California. German astronomer Maximilian Franz Joseph Cornelius Wolf (1863–1932) coined this object's common name in 1903.

Through a 6-inch telescope, the Flaming Star Nebula appears triangular in shape, with one point at the star AE Aur. Increase the telescope's aperture and add a Hydrogen-beta filter to both improve the nebula's contrast and dim AE Aur and other field stars.

| OBJECT #941 | Collinder 464 |
|---|---|
| Constellation | Camelopardalis |
| Right ascension | 5h22m |

| (continued) | |
|---|---|
| Declination | 73°00′ |
| Magnitude | 4.2 |
| Size | 120′ |
| Type | Open cluster |

Collinder 464 (Cr 464) is a fine small telescope target in Camelopardalis. You'll spot it roughly 7° north-northeast of magnitude 4.3 Alpha Camelopardalis.

I hesitate to use that star as a guide because it's actually fainter than the magnitude 4 cluster. Unfortunately, Alpha Cam is the only reasonably bright nearby star. If you're into geometrical figures, Cr 464 makes an equilateral triangle with Alpha Cam and magnitude 4.6 Gamma Cam.

Cr 464 is a large cluster that spans 2°. The stars appear scattered, with the eastern and western sides sharply divided. The west section contains the brightest stars, with five brighter than magnitude 6.5. The five brightest stars in the east half range from magnitudes 6.2 to 7.3. Binoculars reveal the cluster, but I prefer a telescope with a magnification between 25× and 50× to really plumb Cr 464's depths.

| OBJECT #942 | IC 410 |
|---|---|
| Constellation | Auriga |
| Right ascension | 5h23m |
| Declination | 33°31′ |
| Size | 40′ by 30′ |
| Type | Emission nebula |

This wonderful nebula sits 2.4° west-northwest of magnitude 4.7 Chi ($\chi$) Aurigae. IC 410 responds well to a nebula filter, showing richly detailed structure even through small telescopes. The nebulosity glows brightest in a circular area some 5′ in diameter on the northwestern edge. Use a 12-inch scope with an OIII filter, and this object will knock your socks off.

| OBJECT #943 | NGC 1893 |
|---|---|
| Constellation | Auriga |
| Right ascension | 5h23m |
| Declination | 33°24′ |
| Magnitude | 7.5 |
| Size | 12′ |
| Type | Open cluster |

Our last celestial target, IC 410, envelops the open star cluster NGC 1893. Ultraviolet radiation from those stars excite the gas in the nebula, causing it to glow. Through a 10-inch telescope at 100×, you'll see 50 stars irregularly strewn across the nebula's face.

| OBJECT #944 | The Large Magellanic Cloud |
|---|---|
| Constellation | Dorado |
| Right ascension | 5h24m |
| Declination | −69°45′ |
| Magnitude | 0.4 |
| Size | 650′ by 550′ |
| Type | Irregular galaxy |
| Common names | LMC; Nubecula Major |

Southern Hemisphere observers definitely enjoy the sky's better half. Below the celestial equator lie the brightest stars, the center of our galaxy, the best dark nebulae, and the most brilliant celestial wonder — the Large Magellanic Cloud (LMC). I talked about how the Magellanic Clouds got their name in the description for Object #776.

No less than 114 NGC objects lie within the LMC's boundaries. The finest, the Tarantula Nebula (NGC 2070) is itself one of the top 1,001 most spectacular celestial wonders.

If we lived on a planet within the LMC, the Milky Way would dominate the sky. Our galaxy would shine with a magnitude of –2 and would measure 36° long.

Under a dark sky, start observing the LMC without optical aid. Note the galaxy's brightest region, a luminous bar measuring roughly 5° by 1°. Outside the bar, the brightness drops rapidly, but you'll still detect an oval haze measuring 6° by 4°. To extend the LMC's boundary beyond this, use binoculars or a rich-field telescope and pan back and forth.

Through a 6-inch or larger telescope and a magnification around 200×, slowly scan back and forth across the LMC's face. Pause to examine the many star clusters and nebulae. A nebula filter will help you see the nebulae better but will worsen the view of clusters.

**Object #945** M79 Adam Block/NOAO/AURA/NSF

| OBJECT #945 | M79 (NGC 1904) |
|---|---|
| Constellation | Lepus |
| Right ascension | 5h24m |
| Declination | –24°31′ |
| Magnitude | 7.8 |
| Size | 8.7′ |
| Type | Globular cluster |

Globular cluster M79 lies in the small constellation Lepus the Hare. Lepus may be small, but it's easy to find. Just look directly south of Orion.

To spot M79, which, by the way, is the most southerly Messier object in the winter sky, use Alpha ($\alpha$) and Beta ($\beta$) Leporis as pointers. Draw a line from magnitude 2.6 Alpha through magnitude 2.9 Beta and extend that line 3.5°, which is just slightly more than the distance between those two stars.

M79 is among the Milky Way's oldest globular clusters. Situated 60,000 light-years from the galactic center, this object lies 40,000 light-years from Earth. For all of its magnitude 7.8 brightness, M79 is a difficult object to resolve through small telescopes.

A 10-inch instrument shows that the 8.7′-wide globular has a bright, broadly concentrated core nearly devoid of stars. But crank up the magnification beyond 200×, and you'll resolve scores of stars as bright as 13th magnitude at the cluster's edges.

| OBJECT #946 | NGC 1907 |
|---|---|
| Constellation | Auriga |
| Right ascension | 5h28m |
| Declination | 35°19′ |
| Magnitude | 8.2 |
| Size | 6′ |
| Type | Open cluster |

Our next target reveals two dozen stars through a 4-inch telescope at 100×. But use a low-power eyepiece, and you'll sweep up an even-brighter open cluster. NGC 1907 lies 0.5° south-southwest of open cluster M38.

| OBJECT #947 | 118 Tauri |
|---|---|
| Constellation | Taurus |
| Right ascension | 5h29m |
| Declination | 25°09′ |
| Magnitudes | 5.8/6.6 |
| Separation | 4.8″ |
| Type | Double star |

This nice binary is easy to locate. It sits 3.5° south of magnitude 1.7 Elnath (Beta [$\beta$] Tauri). The primary is a bluish-white star, and the secondary glows with a more standard blue color.

**Object #948** M38 Anthony Ayiomamitis

| OBJECT #948 | M38 (NGC 1912) |
|---|---|
| Constellation | Auriga |
| Right ascension | 5h29m |
| Declination | 35°50′ |
| Magnitude | 6.4 |
| Size | 21′ |
| Type | Open cluster |
| Other name | The Starfish Cluster |

Our next target is another of many great open star clusters in Auriga — M38. It's the westernmost of the three Messier clusters in this constellation.

Through a 4-inch telescope, you'll spot some three dozen stars in an area 20′ across. Although this region's background is rich, the cluster stands out well. Crank up the magnification, and you'll identify several nice chains of stars.

Just 0.5° south of M38 lies NGC 1907 (Object #946). Together, these clusters appear like a poor version of the Double Cluster in Perseus. NGC 1907 contains 25 stars, but a 4-inch will show about a dozen within a 4′ span.

This object's common name is the Starfish Cluster, although amateurs rarely use any descriptor other then M38. "Starfish" appears as part of a common name two other times in this book, for NGC 6544 (Object #521) and NGC 6752 (Object #582).

**Object #949** NGC 1931 Al and Andy Ferayorni/Adam Block/NOAO/AURA/NSF

| OBJECT #949 | NGC 1931 |
|---|---|
| Constellation | Auriga |
| Right ascension | 5h31m |
| Declination | 34°15′ |

| (continued) | |
|---|---|
| Size | 4′ by 4′ |
| Type | Emission nebula with associated open cluster |

Our next object sits 0.8° east-southeast of magnitude 5.1 Phi (*Φ*) Aurigae. NGC 1931 surrounds a tiny open cluster that contains only five bright stars. An 8-inch scope at 200× shows the nebula nicely. It orients from the northeast to the southwest and shows non-uniform brightness across its face.

| OBJECT #950 | Delta (δ) Orionis |
|---|---|
| Constellation | Orion |
| Right ascension | 5h32m |
| Declination | –0°18′ |
| Magnitudes | 2.2/6.3 |
| Separation | 53″ |
| Type | Double star |
| Other name | Mintaka |

This is one of the sky's easiest double stars to find. Just locate Orion's belt. Mintaka is the northernmost of the three stars. The brighter component shines with a pure white light. Mintaka B, only 2% as bright as the primary, glows with a deep-blue color. This star's name descends to us from the Arabic *Al Mintakah*, which means "the Belt."

| OBJECT #951 | M1 (NGC 1952) |
|---|---|
| Constellation | Taurus |
| Right ascension | 5h35m |
| Declination | 22°01′ |
| Magnitude | 8.0 |
| Size | 6′ by 4′ |
| Type | Supernova remnant |
| Other name | The Crab Nebula |

One of the sky's most famous objects, and a nice small telescope target, is the Crab Nebula (M1) in Taurus. Although the word "nebula" is part of this object's name, astronomers classify M1 as a supernova remnant.

In 1054, a brilliant new star 4 times brighter than Venus appeared near Taurus the Bull's southern horn. For more than 3 weeks it remained visible during daylight hours, and it took more than a year to fade from view.

William Parsons, Third Earl of Rosse, sketched M1 in 1844 through his 72-inch reflecting telescope. Other astronomers noted the object's crablike appearance and gave M1 its common name.

This supernova remnant has a high surface brightness, so even a 3-inch telescope (and even some binoculars) will reveal it. The object shines at magnitude 8.0. The Crab Nebula has an oval shape. It measures 6′ by 4′ and orients northwest to southeast.

The easiest way to find the Crab Nebula is to start at the 4th-magnitude bluish star Zeta (ζ) Tauri. From there, move 1° to the northwest.

**Object #952** M36 Anthony Ayiomamitis

| OBJECT #952 | M36 (NGC 1960) |
|---|---|
| Constellation | Auriga |
| Right ascension | 5h36m |
| Declination | 34°08′ |
| Magnitude | 6.0 |
| Size | 12′ |
| Type | Open cluster |
| Other name | The Pinwheel Cluster |

M36 is the least spectacular of the Messier trio of open clusters you'll find in Auriga. At magnitude 6.0, however, it still outshines 99% of the sky's star clusters. Through a 4-inch telescope, you'll see several dozen stars strewn across an area 10′ wide.

British amateur astronomer Jeff Bondono gave M36 its common name. Insert an eyepiece that gives a magnification around 100×, and see if you can spot a pinwheel-like pattern in this cluster's stars.

| OBJECT #953 | NGC 1962 |
|---|---|
| Constellation | Dorado |
| Right ascension | 5h26m |
| Declination | –68°50′ |
| Type | Emission nebula |
| Notes | with NGCs 1965, 66, 70 |

Our next targets (four nebulae in a 5′-wide region) lie in the north central region of the Large Magellanic Cloud. Through an 8-inch telescope at low power, NGC 1962 appears circular and featureless. Use a magnification above 200×, and you'll see that the northern rim consists of three enhancements in an arced arrangement.

| OBJECT #954 | NGC 1964 |
|---|---|
| Constellation | Lepus |
| Right ascension | 5h33m |
| Declination | −21°57′ |
| Magnitude | 10.7 |
| Size | 5.0′ by 2.1′ |
| Type | Barred spiral galaxy |

Our next target lies 1.7° southeast of magnitude 2.8 Nihal (Beta [$\beta$] Leporis). Through a 12-inch telescope, you'll see only a haze surrounding a bright core. Less than 2′ northwest of NGC 1964 lies magnitude 10.2 SAO 170546.

| OBJECT #955 | IC 418 |
|---|---|
| Constellation | Lepus |
| Right ascension | 5h28m |
| Declination | −12°42′ |
| Magnitude | 9.3 |
| Size | 12″ |
| Type | Planetary nebula |
| Common names | The Spirograph Nebula, the Raspberry Nebula, the Chameleon Nebula |

This stunning deep-sky object is the Raspberry Nebula (IC 418) in Lepus. Now, it's not stunning because of its size or brightness. It's a small planetary, measuring only 12″ across, and it glows at magnitude 9.3. Much of this brightness comes from the magnitude 10.2 central star that sits within the nebulous disk.

Amateur astronomers who viewed this object through large telescopes dubbed it the Raspberry Nebula because of the pale reddish color they saw. Don't expect to see this color through any telescope with an aperture smaller than about 12 inches. That being said, some observers have noted a faint red when they observed IC 418 at magnifications too low to reveal its disk. So, try decreasing the power, and look for the red.

**Object #956** The Running Man Nebula (NGC 1973/5/7) Peter Spokes/Adam Block/NOAO/AURA/NSF

| OBJECT #956 | NGC 1973/5/7 |
|---|---|
| Constellation | Orion |
| Right ascension | 5h35m |
| Declination | –4°41′ |
| Size | 10′ by 5′ |
| Type | Emission and reflection nebulae |
| Other name | The Running Man Nebula |

In various places, you'll find the Running Man Nebula listed as NGC 1973, NGC 1975, or NGC 1977. In fact, it's all of them. Use a 10-inch or larger telescope, and you'll see an elongated bright nebula with an uneven texture. Two bright stars lie within the nebula: 42 Orionis is the brighter of the pair at magnitude 4.6, and 45 Orionis shines at magnitude 5.2.

Spend some time observing through an eyepiece whose field of view just contains this nebula. Can you see the shape that gives it the name "Running Man"? This feature is much easier to see on images, but under a dark sky you should be able to identify it. Here's a tip: Don't use a nebula filter because it will suppress all the reflection nebulosity that helps give the man his shape.

**Object #957** The Orion Nebula (M42) Adam Block/Mount Lemmon SkyCenter/University of Arizona

| OBJECT #957 | M42 (NGC 1976) |
|---|---|
| Constellation | Orion |
| Right ascension | 5h35m |
| Declination | –5°27′ |

| (continued) | |
|---|---|
| Magnitude | 4 |
| Size | 65′ by 60′ |
| Type | Emission nebula |
| Other name | The Orion Nebula |

This book lists the 1,001 most spectacular celestial wonders by visibility during the year. If we had ranked them according to popularity, however, the Orion Nebula easily would lay claim to the top spot.

Prior to the telescope's invention, celestial cartographers cataloged this region as a star. Bayer, in his *Uranometria*, assigned it the Greek letter Theta ($\theta$). French patron Nicolas-Claude Fabri de Peiresc (1580–1637) first spotted the nebula in 1610.

Next to the Eta Carinae Nebula (Object #164), M42 is the sky's brightest diffuse nebula. Easily seen from a dark site, sharp-eyed observers can detect this cloudy patch even under moderate light pollution. Large telescopes show the full extent of this object, which covers an area 6 times the size of the Full Moon. Although you don't need a nebula filter, using one at a dark location accentuates the contrast between the light and dark regions.

Only 0.1° north-northeast of the Orion Nebula's center lies M43. It is often called De Mairan's Nebula in honor of its discoverer, French mathematician Jean-Jacques D'Ortous de Mairan (1678–1771). Note M43's sharp eastern edge, which is caused by a dark nebula overlapping that area.

| OBJECT #958 | Lambda ($\lambda$) Orionis |
|---|---|
| Constellation | Orion |
| Right ascension | 5h35m |
| Declination | 9°56′ |
| Magnitudes | 3.6/5.5 |
| Separation | 4.4″ |
| Type | Double star |
| Other name | Meissa |

This binary star that marks Orion's head has a blue star as the primary and a pale white companion. Start with a magnification around 100× to split this pair, and increase the power if necessary.

In *Star Names and Their Meanings*, Richard Hinckley Allen states that this star got its name because of an error by the lexicographer al-Firuzabadi (1326–1414), who compiled an extensive Arabic dictionary that served as the basis for subsequent European Arabic dictionaries. According to Allen, the Arabic title *Al Maisan* ("the proudly marching one") originally applied to Gamma ($\gamma$) Geminorum.

| OBJECT #959 | NGC 1981 |
|---|---|
| Constellation | Orion |
| Right ascension | 5h35m |
| Declination | −4°26′ |
| Magnitude | 4.2 |
| Size | 28′ |
| Type | Open cluster |
| Other name | The Coal Car Cluster |

Our next target is open cluster NGC 1981 in Orion. To locate this easy-to-see object, first find the Orion Nebula (M42). After you've soaked up the view from that celestial wonder, look 1° due north. You can think of NGC 1981 as the northernmost "star" in Orion's sword.

NGC 1981 is a bright open cluster. Its magnitude, 4.2, ties it for 11th place among open star clusters. It's large, too. NGC 1981's diameter is nearly that of the Full Moon.

When you observe NGC 1981 through a small telescope, use a magnification near 100×. Be sure to segregate the stars of the cluster from the surrounding star field. Note that the curved line of three magnitude 6.5 stars just to the east do not belong to the cluster.

That said, *Astronomy* magazine Contributing Editor Stephen James O'Meara combines those stars with others in NGC 1981 and draws an antique coal car, from which he extracts his common name for this cluster.

**Object #960** The 13th Pearl Nebula (NGC 1999) Dan and Erica Simpson/Adam Block/NOAO/AURA/NSF

| OBJECT #960 | NGC 1999 |
|---|---|
| Constellation | Orion |
| Right ascension | 5h37m |
| Declination | −6°42′ |
| Size | 2′ by 2′ |
| Type | Reflection nebula |

You'll find NGC 1999 0.8° south-southeast of magnitude 3.0 Iota ($\iota$) Orionis. This object's main feature is a dark obscuration near its center. Because of this void, you might initially think NGC 1999 is a ring-shaped planetary nebula. Crank up the power past 150×, and you'll see the dark, inner nebula's triangular form. Just outside the dark cloud sits V380 Orionis, a variable star that illuminates this nebula.

This nebula has a triangular shape, but there's more. A dark, irregular bar obscures much of the bluish light near NGC 1999's center. This dark cloud is a Bok globule, a region of dust and cold gas — possibly a star-forming region — that obscures the light from objects behind it.

Astronomers named such globules for Dutch-born American astronomer Bart Jan Bok (1906–1983), who pioneered their study. NGC 1999's illumination comes from the star V380 Orionis. It sits just outside and to the east-southeast of the dark central region. The star is so young that the reflection nebula NGC 1999 is material left over from the star's formation.

| OBJECT #961 | Sigma (σ) Orionis |
|---|---|
| Constellation | Orion |
| Right ascension | 5h39m |
| Declination | –2°36′ |
| Magnitudes | 4.0/7.5/6.5 |
| Separation | 2.4″/58″ |
| Type | Double star |

Sigma Orionis isn't just a binary star, it's a nice multiple star system. Most lists mention the three brightest stars, but a fainter one lies nearby. Viewing the wide pair takes only low power. To split the two close bright stars and reveal the fainter fourth component will require you to boost the magnification past 150×. All four stars appear varying shades of white. You'll find Sigma easily with your naked eyes. It lies 0.8° southwest of magnitude 1.7 Alnitak (Zeta [ζ] Orionis).

| OBJECT #962 | NGC 2019 |
|---|---|
| Constellation | Mensa |
| Right ascension | 5h32m |
| Declination | –70°10′ |
| Magnitude | 10.9 |
| Size | 1′ |
| Type | Globular cluster |

Look toward the center of the Large Magellanic Cloud's bar for globular cluster NGC 2019. The reason you'll see this object is because of its small, bright core. Astronomers have found that NGC 2019 has a collapsed core, similar to several other globular clusters in the Magellanic Clouds.

Through an 8-inch telescope, you'll see the lumpy central region easily. Crank the magnification past 200×, and look for the irregular outer boundary. Double the aperture to 16 inches, and individual stars will appear.

| OBJECT #963 | NGC 2022 |
|---|---|
| Constellation | Orion |
| Right ascension | 5h42m |
| Declination | 9°05′ |
| Magnitude | 11.9 |
| Size | 39″ |
| Type | Planetary nebula |

If your telescope lacks a go-to drive, you can still find NGC 2022 easily by using two bright stars. It lies two-thirds of the way from Betelgeuse (Alpha [α] Orionis) to Meissa (Lambda [λ] Orionis). Use at least an 8-inch telescope and high power (more than 250×) on this small planetary. If you can double the aperture to 16 inches, the planetary's outer region will appear slightly brighter, making NGC 2022 look ring-like. Also, through that size scope, you'll have no problem seeing the 15th-magnitude central star.

| OBJECT #964 | NGC 2024 |
|---|---|
| Constellation | Orion |
| Right ascension | 5h42m |
| Declination | –1°51′ |
| Size | 30′ by 30′ |
| Type | Emission nebula |
| Common names | The Flame Nebula, the Tank Tracks, the Ghost of Alnitak |

The Flame Nebula lies not quite 4° north-northeast of the Orion Nebula (M42). If you dial in this distance, however, you'll find yourself staring at magnitude 2.0 Alnitak (Zeta [ζ] Orionis), the southernmost star in Orion's Belt. The Flame Nebula sits only 15′ southeast of this luminary, so increase the magnification past 100× — make that 200× — and move Zeta out of the field of view. You can try using a nebula filter, but even it won't dim dazzling Zeta enough.

When you observe the Flame Nebula, remember that it covers the same area as the Full Moon. So, while its surface brightness is low, there's still lots of detail to be gleaned from this object, especially through telescopes with apertures greater than 10′. Look first for a dark lane that extends north-south, and then spend some time on each side of the lane trying to pull as much detail as you can out of the nebulosity.

In addition to the nebulosity, be sure to spend some time observing Alnitak, a fine triple-star. The two main components, separated by 2.6″, shine at magnitudes 1.9 and 3.4. A 4-inch telescope will separate them easily. The tertiary star, which glows at magnitude 9.5, sits 57″ northeast of the brightest component.

NGC 2024's main common name, the Flame Nebula, is easy to explain when you see it as part of any astronomical image. *Astronomy* magazine Contributing Editor Stephen James O'Meara gave it three other names, but the one I like best is the Ghost of Alnitak. Called that, it brings to mind a similar object, Mirach's Ghost (Object #784).

Famed Canadian astroimager Jack Newton dubbed this object the Tank Tracks Nebula. Through large telescopes, Newton had seen regular spacing in the nebulosity on either side of the dark lane.

| OBJECT #965 | M78 (NGC 2068) |
|---|---|
| Constellation | Orion |
| Right ascension | 5h47m |
| Declination | 0°03′ |
| Size | 8′ by 6′ |
| Type | Reflection nebula |

This small telescope target also is the sky's brightest reflection nebula: M78 in Orion. Through a 4-inch scope at about 120×, you'll spot an 11th-magnitude star. On either side of this star lie the two densest parts of M78.

If your site is dark, you'll see another, much fainter region of nebulosity, NGC 2067, only 4.5′ northwest of M78. A dark lane separates the two. Because so few background stars populate the field of view, astronomers believe M78 sits within a huge, dark cloud that eventually will form stars.

Oh, and don't use any nebular or deep-sky filter when you view M78. Because its light is mainly reflected starlight (composed of all frequencies) a filter will only dim the view.

| OBJECT #966 | NGC 2070 |
|---|---|
| Constellation | Dorado |
| Right ascension | 5h39m |
| Declination | −69°05′ |
| Size | 30′ by 20′ |
| Type | Emission nebula |
| Other names | The Tarantula Nebula, 30 Doradus, the True Lovers' Knot, Caldwell 103 |

Most northern observers, unfortunately, haven't experienced the Tarantula Nebula. Although it lies in the Large Magellanic Cloud, this object looks incredible through medium-sized telescopes. With a true diameter of 1,000 light-years, NGC 2070 would span 20° if it were as close as the Orion Nebula (M42), within the Milky Way.

English astronomer John Flamsteed (1646–1719) cataloged NGC 2070 as the star 30 Doradus. French astronomer Nicolas Louis de Lacaille (1713–1762) first recognized it as a nebula December 5, 1751.

The brightest star cluster in the Tarantula Nebula, and the most remarkable star-forming region anywhere, is R136. Many of its 60 stars are among the most massive, brightest, and hottest stars known.

Even through a 4-inch telescope, NGC 2070 shows loops and filaments. A dense central bar runs north to south. R136 is easy to spot as a 1′-wide region of several dozen bright stars. The longest filament begins near the cluster's center and extends 7′ to the south. It then extends eastward and loops an equal distance to the north.

Two well-defined dark bays, one slightly darker than the other, lie east of R136. The appearance of nebulous "ropes" encircling darker regions led English astronomer William Henry Smyth to describe this nebula as the True Lover's Knot. According to some accounts, sixteenth-century Dutch sailors tied similar knots to remind them of lovers they'd left behind.

| OBJECT #967 | Zeta (ζ) Orionis |
|---|---|
| Constellation | Orion |
| Right ascension | 5h41m |
| Declination | −1°57′ |
| Magnitudes | 1.9/4.0/9.9 |
| Separation | 2.4″, 58″ |
| Type | Double star |
| Other name | Alnitak |

Like Mintaka (Delta [δ] Orionis), you'll find Alnitak in Orion's belt. Alnitak is the easternmost and southernmost of the three bright stars. Zeta Orionis is a triple star system, but if you think it's a double, I understand. The bright light-blue A component has a nearby 4th-magnitude companion, which also is blue. A 3-inch telescope will split this pair if your magnification exceeds 150×. Larger scopes will show the 10th-magnitude tertiary star — also blue — nearly 1′ away.

The name Alnitak comes from the Arabic *Al Nitak*. It means "the girdle."

**Object #968** The Horsehead Nebula (B33) Adam Block/Mount Lemmon SkyCenter/University of Arizona

| OBJECT #968 | Barnard 33 |
|---|---|
| Constellation | Orion |
| Right ascension | 5h41m |
| Declination | −2°28′ |
| Size | 6′ by 4′ |
| Type | Dark nebula |
| Other name | The Horsehead Nebula |

If you can see the Flame Nebula (Object #964) easily, try for the wonder many observers consider the ultimate challenge object — the Horsehead Nebula. This dark protrusion sits in front of emission nebula IC 434. Only 15′ to its northeast lies NGC 2023, a 10′-wide dot of bright nebulosity.

Although some observers have viewed the dark protrusion through telescopes as small as 5′ in aperture, a 12-inch or larger scope will bring out the horsehead shape. A standard nebula or deep-sky filter will help, but to make this wonder really pop, use a Hydrogen-beta filter.

| OBJECT #969 | NGC 2090 |
|---|---|
| Constellation | Columba |
| Right ascension | 5h47m |
| Declination | −34°15′ |
| Magnitude | 11.0 |
| Size | 4.5′ by 2.3′ |
| Type | Spiral galaxy |

From magnitude 2.7 Phact (Alpha [α] Columbae), pan 1.5° east to our next target. Through any telescope, NGC 2090 appears lens shaped and oriented north-south. The core is broad and evenly illuminated. You'll need a 12-inch scope to see the thin, diffuse halo.

**Object #970** The Salt-and-Pepper Cluster (M37) Anthony Ayiomamitis

| OBJECT #970 | M37 (NGC 2099) |
|---|---|
| Constellation | Auriga |
| Right ascension | 5h52m |
| Declination | 32°33′ |
| Magnitude | 5.6 |
| Size | 20′ |
| Type | Open cluster |
| Other name | The Salt and Pepper Cluster |

Want to make a model of a celestial object at home? Lay a piece of black paper on a table, and sprinkle a quarter teaspoon of salt on it. Then shine a bright light on the crystals. Voila! M37. Well, not exactly, but you get the idea.

Hodierna discovered this star cluster prior to 1654. Messier independently discovered it in 1764. Admiral Smyth probably described it best when he wrote, "A magnificent object; the whole field being strewd, as it were, with sparkling gold dust and the group is resolvable into about 500 stars from 10–14 mag. besides outliers."

Through any size telescope, M37 displays an even distribution of stars not found in many other clusters. Although it sits squarely within Milky Way, M37's edge is easy to discern.

A 3-inch scope reveals 50 stars. The brightest member — a magnitude 9 orange star — sits near the cluster's center.

With many open clusters, increasing the size of the telescope worsens the view; as the cluster gets magnified, the stars thin. Also, larger scopes reveal more background stars, making it harder to see the cluster's outline.

Not with M37. With this wonder, larger scopes only reveal more stars. Through a 10-inch scope, you'll count 200 stars, and a 16-inch will reveal 500.

| OBJECT #971 | NGC 2100 |
|---|---|
| Constellation | Dorado |
| Right ascension | 5h42m |
| Declination | –69°14′ |
| Magnitude | 9.6 |
| Size | 2.8′ |
| Type | Open cluster |

Our next target lies 0.3° east-southeast of the Tarantula Nebula (NGC 2070). Through an 8-inch telescope, you'll see two dozen stars around the outskirts of this cluster. It has a compact core that requires high magnification to resolve.

| OBJECT #972 | Sh 2–276 |
|---|---|
| Constellation | Orion |
| Right ascension | 5h48m |
| Declination | 1°00′ |
| Size | 600′ by 30′ |
| Type | Emission nebula |
| Other name | Barnard's Loop |

Because Barnard's Loop is so huge — it spans some 20° and covers much of Orion — you'll need a telescope/eyepiece combination that yields a wide field. The nebula's brightest part stretches 6° and tapers at the northwestern end. Head to a dark site, and take your time when you search for Barnard's Loop. You'll be moving your scope around a lot, but try to keep the movements slow ones. Even with a nebula filter in place, you'll see just a slight diffuse brightening in that part of the sky.

| OBJECT #973 | NGC 2112 |
|---|---|
| Constellation | Orion |
| Right ascension | 5h54m |
| Declination | 0°24′ |
| Magnitude | 8.4 |
| Size | 11′ |
| Type | Open cluster |

This small telescope target is open cluster NGC 2112 in Orion. This object, which lies 2,800 light-years away, glows at magnitude 9.0.

To find NGC 2112, move 4° northeast from Zeta (ζ) Orionis, the bottom star in Orion's belt. A 4-inch telescope at 100× reveals two dozen faint stars strewn across an area 8′ in diameter. Although the cluster appears loose, a strong background glow hints at the presence of dozens of unresolved stars. For those, however, you'll need a larger scope.

| OBJECT #974 | NGC 2126 |
|---|---|
| Constellation | Auriga |
| Right ascension | 6h03m |
| Declination | 49°54′ |
| Magnitude | 10.2 |
| Size | 6′ |
| Type | Open cluster |

Our next target lies midway between magnitude 1.9 Menkalinan (Beta [β] Aurigae) and magnitude 3.7 Delta (δ) Aurigae. Through a 6-inch telescope, you'll see about 20 stars. The magnitude 6.0 star SAO 40801 lies 3′northeast.

| OBJECT #975 | NGC 2129 |
|---|---|
| Constellation | Gemini |
| Right ascension | 6h02m |
| Declination | 23°19′ |
| Magnitude | 6.7 |
| Size | 6′ |
| Type | Open cluster |

Our next object lies less than 0.5° from the intersection of the borders of Taurus, Orion, and Gemini. To find it, look 3.3° west-northwest of magnitude 3.3 Propus (Eta [η] Geminorum).

A 4-inch telescope will show you two dozen stars. The brightest, SAO 77842, shines at magnitude 7.4.

| OBJECT #976 | Epsilon (ε) Monocerotis |
|---|---|
| Constellation | Monoceros |
| Right ascension | 6h24m |
| Declination | 4°36′ |
| Magnitudes | 4.5/6.5 |
| Separation | 13.4″ |
| Type | Double star |

Our next target is a nice bright binary that's an easy split through any telescope. To find it, star-hop from Betelgeuse (Alpha [α] Orionis). Just look 7.6° east-southeast of that brilliant star.

| OBJECT #977 | NGC 2141 |
|---|---|
| Constellation | Orion |
| Right ascension | 6h03m |
| Declination | 10°26′ |
| Magnitude | 9.4 |
| Size | 10′ |
| Type | Open cluster |

This open cluster lies 0.8° north of magnitude 4.1 Mu ($\mu$) Orionis. Although this star cluster contains many stars, they're all faint and closely packed, so you'll need an 8-inch telescope to resolve them. At 200×, the stars appear evenly distributed across NGC 2141's face. A nice background of fainter stars hovers near the limit of vision. Want to see them? Move up to a 12-inch or larger scope.

**Object #978** The Dusty Hand Galaxy (NGC 2146) Adam Block/NOAO/AURA/NSF

| OBJECT #978 | NGC 2146 |
|---|---|
| Constellation | Camelopardalis |
| Right ascension | 6h19m |
| Declination | 78°21′ |
| Magnitude | 10.6 |
| Size | 5.4′ by 4.5′ |
| Type | Barred spiral galaxy |
| Other name | The Dusty Hand Galaxy |

Our next target is a spiral galaxy undergoing a major burst of star formation. Finding it without a go-to drive isn't easy because of the lack of bright stars in the area. Look 11.6° northeast of magnitude 4.6 Gamma ($\gamma$) Camelopardalis. NGC 2146 displays a broad central region. Telescopes with apertures larger than 10 inches show some mottling close to the core and a dark, subtle lane near the southwestern edge.

The Dusty Hand Galaxy got its common name from images that showed a system of three dust lanes, which may be spiral arms. Astronomers have found clear indications of starburst activity, such as a strong galactic wind. In most cases where other galaxies have displayed strong star-forming activity, a companion is responsible, usually seen merging with the main galaxy. Although NGC 2146 shows no obvious signs of a merger, that scenario seems to be the most plausible explanation. Probably the encounter happened long ago that no traces are left.

| OBJECT #979 | Theta ($\theta$) Aurigae |
|---|---|
| Constellation | Auriga |
| Right ascension | 6h00m |
| Declination | 37°13′ |
| Magnitude range | 2.6/7.1 |
| Separation | 3.6″ |
| Type | Double star |

The components of this relatively close binary exhibit a marked difference in brightness. The primary outshines the secondary by 63 times. The brighter star appears white with perhaps a tinge of blue. The secondary appears light-orange.

**Object #980** M35 Anthony Ayiomamitis; NGC 2158 Adam Block/Mount Lemmon SkyCenter/University of Arizona

| OBJECT #980 | M35 (NGC 2168) |
|---|---|
| Constellation | Gemini |
| Right ascension | 6h09m |
| Declination | 24°20′ |
| Magnitude | 5.1 |
| Size | 28′ |
| Type | Open cluster |

It's hard to resist a two-for-one sale, and that's exactly what our next wonder offers. M35 lies 2.3° northwest of magnitude 3.3 Eta ($\eta$) Geminorum. From a dark site, you'll spot the cluster easily without optical aid. Point a telescope at M35, however, and not only will it explode into stars, you'll see a second open cluster, magnitude 8.6 NGC 2158.

Philippe Loys de Chéseaux discovered this cluster in late 1745 or early 1746. English astronomer John Bevis also found it before 1750, the year he featured it in his star atlas, *Uranographia Britannica.* Messier, who added it to his catalog August 30, 1764, credited Bevis for the discovery.

M35 contains two dozen stars brighter than 9th magnitude, most of which reside near the cluster's center. In this region, look for a string of stars some 10′ long shaped like a saxophone. The stars align in many other patterns; some of the arcs extend nearly 1°. Fainter members push the total number of visible stars past 200.

If you use a low-power eyepiece to engulf M35, NGC 2158 will look like a small, fuzzy ball. This cluster rewards high magnification. Through a 14-inch telescope at 500×, you'll count 30 stars. Increase the aperture to 20 inches, however, and you'll double that number.

| OBJECT #981 | NGC 2169 |
|---|---|
| Constellation | Orion |
| Right ascension | 6h08m |
| Declination | 13°58′ |
| Magnitude | 5.9 |
| Size | 6′ |
| Type | Open cluster |
| Other name | "37" Cluster |

To find our next target, first locate magnitude 4.5 Xi ($\xi$) and magnitude 4.4 Nu ($\nu$) Orionis. NGC 2169 forms the top of an isosceles triangle roughly 0.8° from each star. Well, sort of. Actually, the cluster lies to the south of the stellar pair. From a dark site, sharp-eyed observers will spot NGC 2169 with their naked eyes.

The cluster isn't huge — it spans only 6′ — but it contains seven stars brighter than 9th magnitude. Use an eyepiece that gives a 1° field of view and enjoy the wide view of NGC 2169 against the Milky Way's myriad background stars.

Oh, I almost forgot. The common name the "37" Cluster comes from imaginative observers viewing this cluster at low power and seeing the number 37 in it. (Some see the letters "LE.") To do this, you need a view that puts north up and east to the left. Also, a telescope/eyepiece combination that yields a field of view of about one-quarter degree seems to work best.

**Object #982** NGC 2170 Doc. G. and Dick Goddard/Adam Block/NOAO/AURA/NSF

| OBJECT #982 | NGC 2170 |
|---|---|
| Constellation | Monoceros |
| Right ascension | 6h08m |
| Declination | −6°24′ |
| Size | 2′ by 2′ |
| Type | Reflection nebula |

Our next object lies 1.8° west of magnitude 4.0 Gamma ($\gamma$) Monocerotis and sits within a group of reflection nebulae. Through an 8-inch telescope, you'll see a bright circular haze that surrounds a magnitude 9.5 star. Only 0.5° east lies reflection nebula NGC 2182. It appears fainter than NGC 2170, measures 1′ across, and surrounds a magnitude 9.3 star.

| OBJECT #983 | NGC 2175 |
|---|---|
| Constellation | Orion |
| Right ascension | 6h10m |
| Declination | 20°30′ |
| Size | 40′ by 30′ |
| Type | Emission nebula |
| Other name | The Monkey Face Nebula |

Our next celestial target combines an emission nebula with an open star cluster. Astronomers divide the two into NGC 2174 and NGC 2175, but you'll most often see it referred to as NGC 2175. You'll spot it easily 2.2° southwest of magnitude 3.3 (Eta [$\eta$] Geminorum).

This nebula is a roughly circular object with an indentation on its western side, and it lies in a rich star field. A 4-inch telescope will reveal NGC 2175, but you'll need a 16-inch or larger telescope to see the full extent of the nebulosity. That size instrument will begin to show you the shape that gives NGC

2175 its common name. The 7th-magnitude star you'll spot at the nebula's center is a chance alignment. It lies much closer than NGC 2174. Use a nebula filter such as an OIII to reduce the star's glare and improve the nebula's contrast.

**Object #984** NGC 2182 Adam Block/NOAO/AURA/NSF

| OBJECT #984 | NGC 2182 |
|---|---|
| Constellation | Monoceros |
| Right ascension | 6h10m |
| Declination | −6°20′ |
| Size | 2.5′ by 2.5′ |
| Type | Reflection nebula |

You'll find our next target 1.3° west of magnitude 4.0 Gamma (γ) Monocerotis. The magnitude 9.3 star GSC 4795:1776 appears to be involved with the nebulosity. Look 20′ to the east-northeast for another reflection nebula, NGC 2183.

| OBJECT #985 | NGC 2186 |
|---|---|
| Constellation | Orion |
| Right ascension | 6h12m |
| Declination | 5°28′ |
| Magnitude | 8.7 |
| Size | 5′ |
| Type | Open cluster |

This nice cluster displays some two dozen stars through an 8-inch telescope at 100×. A pair of stars, magnitudes 9.3 and 9.8, sit at NGC 2186's center. You'll find this object 4.6° east-southeast of magnitude 0.5 Betelgeuse (Alpha [α] Orionis).

| OBJECT #986 | NGC 2188 |
|---|---|
| Constellation | Columba |
| Right ascension | 6h10m |
| Declination | −34°06′ |
| Magnitude | 11.6 |
| Size | 5.5′ by 0.8′ |
| Type | Barred spiral galaxy |

Our next object has a disk that inclines only 3° from edge-on. It appears moderately bright through an 8-inch telescope, five times as long as it is wide, and roughly oriented north-south. NGC 2188 shows an even brightness distribution. A 16-inch scope will let you see the truncated southern edge.

**Object #987** NGC 2194 Doug Matthews/Adam Block/NOAO/AURA/NSF

| OBJECT #987 | NGC 2194 |
|---|---|
| Constellation | Orion |
| Right ascension | 6h14m |
| Declination | 12°48′ |
| Magnitude | 8.5 |
| Size | 8′ |
| Type | Open cluster |

Our next object lies near Orion's uplifted arm 1.5° south-southeast of magnitude 4.5 Xi (ξ) Orionis. This reasonably bright open cluster is an appealing small-telescope target. At 150×, you'll spot several dozen stars between 10th and 13th magnitude spread over an area 6′ across. Look for a concentration of 11th-magnitude stars at the cluster's heart.

| OBJECT #988 | NGC 2204 |
|---|---|
| Constellation | Canis Major |
| Right ascension | 6h16m |
| Declination | −18°40′ |
| Magnitude | 8.6 |
| Size | 10′ |
| Type | Open cluster |

Our next target sits 1.8° west-southwest of magnitude 2.0 Mirzam (Beta [$\beta$] Canis Majoris). Through an 8-inch telescope, you'll see three dozen stars, the brightest of which sits on the northwestern edge. That's the magnitude 8.8 star SAO 151278. Just 12′ to the north-northwest sits magnitude 6.0 SAO 151274.

| OBJECT #989 | NGC 2207 |
|---|---|
| Constellation | Canis Major |
| Right ascension | 6h16m |
| Declination | −21°22′ |
| Magnitude | 10.8 |
| Size | 4.8′ by 2.3′ |
| Type | Spiral galaxy |

From magnitude 2.0 Mirzam (Beta [$\beta$] Canis Majoris), pan 4° south-southwest to NGC 2207, a spiral galaxy interacting with its neighbor, the magnitude 11.7 spiral IC 2163. Through a 10-inch telescope, NGC 2207 shows a bright core surrounded by a thin halo. The galaxy orients east-west. IC 2163 lies 1′ east and shows a broadly concentrated glow 2′ across.

| OBJECT #990 | NGC 2214 |
|---|---|
| Constellation | Dorado |
| Right ascension | 6h13m |
| Declination | −68°16′ |
| Magnitude | 10.9 |
| Size | 3.6′ |
| Type | Open cluster |

This small cluster lies 4.5° east-northeast of the Large Magellanic Cloud. For a better marker, look 0.7° north-northeast of magnitude 5.1 Nu ($\nu$) Doradus. Through a 4-inch telescope, NGC 2214 appears as a bright haze. Even through a 12-inch scope, the cluster's stars are hard to resolve except at the edge.

| OBJECT #991 | 9–12 Geminorum Cluster |
|---|---|
| Constellation | Gemini |
| Right ascension | 6h18m |
| Declination | 23°38′ |
| Type | Asterism |

A nice small telescope target is the unusually named 9–12 (pronounced "9 through 12") Geminorum Cluster. The name has nothing to do with any date. Rather, it refers to numbers given to four stars by the first British Astronomer Royal, John Flamsteed (1646–1719). During his surveys of the constellations, Flamsteed assigned numbers to the stars in each constellation by increasing right ascension. His entire list contains 2,554 stars.

Although most astronomers no longer refer to the majority of stars by their Flamsteed numbers, a few famous exceptions exist: 61 Cygni was one of the first stars to have its proper motion determined;

51 Pegasi, in 1995, became the first Sun-like star found to have a planet orbiting it; and 47 Tucanae, also known as NGC 104, is the sky's second-brightest globular cluster.

So, the four stars in the 9–12 Geminorum Cluster are 9, 10, 11, and 12 Geminorum. Their magnitudes are 6.3, 6.6, 6.9, and 7.0, respectively. This small cluster, which has a total listed magnitude of 5.7, also carries the designation Collinder 89. It lies about 1.5° northwest of magnitude 2.9 Mu ($\mu$) Geminorum. The 9–12 Geminorum Cluster measures 1° across and makes a nice binocular target.

| OBJECT #992 | NGC 2215 |
|---|---|
| Constellation | Monoceros |
| Right ascension | 6h21m |
| Declination | −7°17′ |
| Magnitude | 8.4 |
| Size | 10′ |
| Type | Open cluster |

Our next target lies 2° west of magnitude 4.7 Beta ($\beta$) Monocerotis. Through a 6-inch telescope, you'll see 25 stars. The even distribution and similar magnitudes leads the eye to pattern forming. Do you see a curved letter M?

| OBJECT #993 | NGC 2217 |
|---|---|
| Constellation | Canis Major |
| Right ascension | 6h22m |
| Declination | −27°14′ |
| Magnitude | 10.2 |
| Size | 5.0′ by 4.5′ |
| Type | Spiral galaxy |

This object lies 3° north of magnitude 3.0 Zeta ($\zeta$) Canis Majoris. A 12-inch telescope at 250× shows a short bar elongated east-west. The galaxy has an even brightness distribution across its central region. To the north and south you may see the faint halo.

| OBJECT #994 | NGC 2232 |
|---|---|
| Constellation | Monoceros |
| Right ascension | 6h27m |
| Declination | −4°45′ |
| Magnitude | 3.9 |
| Size | 29′ |
| Type | Open cluster |

You'll see this cluster with your naked eyes from any dark site. The blue-white central star is magnitude 10 Monocerotis, which shines at magnitude 5.1. A 6-inch telescope at 100× shows the cluster as a group of a dozen stars elongated north-south. Through a 12-inch scope, dozens more faint stars pop into view.

| OBJECT #995 | Beta ($\beta$) Monocerotis |
|---|---|
| Constellation | Monoceros |
| Right ascension | 6h29m |
| Declination | −7°02′ |
| Magnitudes | 4.7/5.2 |
| Separation | 7.3″ |
| Type | Double star |

Beta Monocerotis is a star you won't tire of observing. It's a close triple star whose magnitudes are 4.7, 5.2, and 6.1. Astronomers refer to them as the A, B, and C components, respectively. The separations are A-B = 7″; B-C = 3″; A-C = 10″. All three stars appear white.

**Object #996** NGC 2236 Mark and Patricia Wessels/Adam Block/NOAO/AURA/NSF

| OBJECT #996 | NGC 2236 |
|---|---|
| Constellation | Monoceros |
| Right ascension | 6h30m |
| Declination | 6°50′ |
| Magnitude | 8.5 |
| Size | 6′ |
| Type | Open cluster |

Our next target lies 2.7° north-northeast of magnitude 4.5 Epsilon (ε) Monocerotis. Through a 6-inch telescope, you'll see an unevenly distributed group of two dozen stars. Those faint points surround a 9th-magnitude luminary that lies at NGC 2236's heart.

**Object #997** The Rosette Nebula (NGC 2237–9) Adam Block/NOAO/AURA/NSF

| OBJECT #997 | NGC 2237–9 |
|---|---|
| Constellation | Monoceros |
| Right ascension | 6h32m |
| Declination | 5°03′ |
| Size | 80′ by 60′ |
| Type | Emission nebula |
| Other names | The Rosette Nebula, Caldwell 49 |

Most observers recognize the magnificent Rosette Nebula as a single deep-sky wonder. But this object wasn't found all at once.

Sir William Herschel discovered open cluster NGC 2244 in 1784. This naked-eye star cluster lies at the heart of the Rosette Nebula. In 1864, German astronomer Albert Marth (1828–1897) discovered NGC 2238. To complete the picture, American astronomer Lewis Swift (1820–1913) discovered NGC 2237 in 1883 and NGC 2246 in 1886.

From a dark site, you'll first spot NGC 2244. A 4-inch telescope will reveal two dozen stars in an oval region elongated northwest to southeast; a half dozen of these surpass 8th magnitude. Larger scopes reveal countless fainter background stars.

To best observe the Rosette Nebula, use a magnification of 50× and insert a nebula filter to dim NGC 2244's stars. The nebula's western side appears brighter. Here, the ring's inner wall appears straight with a tiny bit of scalloping; the outer wall looks thin with an extension to the northwest.

The Rosette's eastern side is much wider. A nebulous wall with a well-defined border forms its northern edge. Although the Rosette is an emission nebula, your eye will also notice the many small dark nebulae superimposed on the bright background.

| OBJECT #998 | NGC 2243 |
|---|---|
| Constellation | Canis Major |
| Right ascension | 6h30m |
| Declination | –31°17′ |
| Magnitude | 9.4 |
| Size | 13′ |
| Type | Open cluster |

This faint cluster lies 2.3° east-southeast of magnitude 3.0 Furud (Zeta [ζ] Canis Majoris). An 8-inch telescope reveals a dozen or so faint stars against a hazy glow. A 16-inch scope reveals 30 stars, but resolving them is still a problem.

| OBJECT #999 | NGC 2244 |
|---|---|
| Constellation | Monoceros |
| Right ascension | 6h32m |
| Declination | 4°52′ |
| Magnitude | 4.8 |
| Size | 23′ |
| Type | Open cluster |
| Other name | Caldwell 50 |

Our next object is the open cluster associated with the Rosette Nebula (NGC 2237–9). You'll spot it easily with your naked eye from a dark observing site. Through a 6-inch telescope, you'll see two dozen bright stars and some 100 fainter ones. A half dozen of the bright stars shine at magnitude 8 or brighter. The cluster's main portion appears oval and oriented northwest to southeast.

| OBJECT #1000 | NGC 2251 |
|---|---|
| Constellation | Monoceros |
| Right ascension | 6h35m |
| Declination | 8°22′ |
| Magnitude | 7.3 |
| Size | 10′ |
| Type | Open cluster |

Our 1,000th object lies 9.8° east of magnitude 0.5 Betelgeuse (Alpha [α] Orionis). This bright cluster looks nice even through a 4-inch telescope. At 100×, you'll see two dozen stars randomly distributed and of unequal brightnesses. Move up to an 8-inch scope, and the star count will rise to 50.

**Object #1001** Hubble's Variable Nebula (NGC 2261) Carole Westphal/Adam Block/NOAO/AURA/NSF

| OBJECT #1001 | NGC 2261 |
|---|---|
| Constellation | Monoceros |
| Right ascension | 6h39m |
| Declination | 8°44′ |
| Size | 3.5′ by 1.5′ |
| Type | Emission and reflection nebulae |
| Other names | Hubble's Variable Nebula, Caldwell 46 |

All too soon we've come to the end of our list. It's nice to end with a bang, and although you can spot NGC 2261 through a 3-inch telescope, you'll need at least a 12-inch scope at 250× to give this object some pop.

Hubble's Variable Nebula is a fascinating reflection nebula associated with the variable star R Monocerotis. NGC 2261 appears triangular, almost comet-like, with the comet's "head" pointing southward. The nebula's brightness appears even across its face and, except for the northern side, all edges look sharp.

Why is the nebula variable? A recent theory states that dense knots of opaque dust pass close to R Monocerotis. As they do, they cast slowly moving shadows that fall on the dust in the nebula that reflects most of the light.

William Herschel discovered Hubble's Variable Nebula in 1783. The nebula's common name comes from American astronomer Edwin Hubble (1889–1953), who studied it extensively starting in 1916 at Yerkes Observatory in Wisconsin.

And here's an additional bit of trivia for you: According to the California Institute of Technology, Hubble's Variable Nebula was the first object photographed by the 200-inch Hale Telescope at Palomar Observatory. Hubble sat in the prime focus cage of that instrument and recorded an image of it January 26, 1949.

# Index



M.E. Bakich, *1,001 Celestial Wonders to See Before You Die*, Patrick Moore's Practical Astronomy Series, DOI 10.1007/978-1-4419-1777-5, © Springer Science+Business Media, LLC 2010

Printed in the United States of America